Rajiv J. Kapadia

Digital Filters

Related Titles

Taylor, F. J., Mellott, J.

Digital Filters

Principles and Applications
with MATLAB

2011
ISBN: 978-0-470-77039-9

Leis, J.

**Digital Signal Processing Using
MATLAB for Students and
Researchers**

2011
ISBN: 978-0-470-88091-3

Adali, T., Haykin, S.

Adaptive Signal Processing

Next Generation Solutions

2010
ISBN: 978-0-470-19517-8

Siarry, P. (ed.)

**Optimisation in Signal and
Image Processing**

E-Book 2010
ISBN: 978-0-470-39394-9

Grewal, M. S., Andrews, A. P.

Kalman Filtering

Theory and Practice Using MATLAB

2008
ISBN: 978-0-470-17366-4

Sayed, A. H.

**Fundamentals of Adaptive
Filtering**

E-Book 2008
ISBN: 978-0-470-35751-4

Najim, M. (ed.)

**Digital Filters Design for Signal
and Image Processing**

2006
ISBN: 978-1-905209-45-3

Shenoi, B. A.

**Introduction to Digital Signal
Processing and Filter Design**

2005
ISBN: 978-0-471-46482-2

Rajiv J. Kapadia

Digital Filters

Theory, Application and Design of Modern Filters

WILEY-VCH

WILEY-VCH Verlag GmbH & Co. KGaA

The Author

Prof. Rajiv J. Kapadia
120 Mapleridge Drive
Mankato, MN 56001
USA

For a collection of problems and projects,
please go to the book's catalogue entry on
www.wiley-vch.de/publish/dt/books/by
SubjectEE00/ISBN3-527-41148-8.

Library of Congress Card No.: applied for

British Library Cataloguing-in-Publication Data
A catalogue record for this book is available from the
British Library.

**Bibliographic information published by
the Deutsche Nationalbibliothek**
The Deutsche Nationalbibliothek lists this publication in the Deutsche Nationalbibliografie; detailed
bibliographic data are available on the Internet at
http://dnb.d-nb.de.

© 2012 Wiley-VCH Verlag & Co. KGaA,
Boschstr. 12, 69469 Weinheim, Germany

Print ISBN: 978-3-527-41148-1

Composition Thomson Digital, Noida, India
Printing and Binding Markono Print Media Pte Ltd, Singapore
Cover Design Adam-Design, Weinheim, Germany

Printed in Singapore
Printed on acid-free paper

Contents

1
Background and Introduction

1.1
Introduction

Ever since the beginning of time engineers have been trying to get information from the source of the information to the sink of the information. To get from the source to the sink, the information travels through a *channel*. The source produces the information, the sink uses this information in a way that it is useful, and the channel is supposed to deliver the information without corrupting the information. In this classical problem, as the information travels across the channel, the noise that is present in the channel gets added to the information. This noise may come from the channel itself or from the sensor that produces the information or from the electronic devices that are present between the source and the sink. Independent of where the noise comes from and contaminates the information, it is the job of the engineer to separate the signal from the noise. Noise in a signal comes in different forms. Noise may be just an unwanted signal that occupies a different frequency band, or noise may be additive, and just rides on the signal.

In this book, we will examine various methods that an engineer uses to separate the signal from the noise. We will do this most of the time by the use of a suitable filter, so most of this book is devoted to designing or understanding the design of filters. The filters that we will be working with are discrete filters, linear filters that can be either deterministic filters or stochastic filters. A filter is a linear filter when the output from the filter is some linear combinations of the input to the filter.

In the first half of the book, we will be working with deterministic signals so the results will be definite and the filters that we design and use will be deterministic filters. In the second half of the book, we will introduce stochastic signals and systems. The filters that we use with these signals are stochastic filters. These filters operate on the statistical properties of the signals; hence we use the statistical properties such as the mean and the autocorrelation of the signal to design the filter. Thus, designing a filter is both an art and a science.

The modern approach to the design and use of filters covers advanced theories such optimization and prediction. Optimization is introduced in Chapter 9 on Weiner filter that uses the mean square value to optimize the filter. The filter coefficients are optimized so that the output from the filter has minimum error from the expected output.

Digital Filters: Theory, Application and Design of Modern Filters, First Edition. Rajiv J. Kapadia.
© 2012 Wiley-VCH Verlag GmbH & Co. KGaA. Published 2012 by Wiley-VCH Verlag GmbH & Co. KGaA.

The optimization is done in the *mean square sense(MSS)*. The MSS is used as it produces a bowl-shaped surface over which we want to determine a minimum. This surface has a unique minimum so the result is guaranteed. In studying the Weiner filter, we will assume that the stochastic properties of the signal are stationary in the wide sense (WSS) at least. If the input signal is not WSS, then the Kalman filter theory needs to be used.

Toward the end of the book, we study the predictor filters. These filters try to predict what the next signal value will be based on the statistical properties of the signals received in the recent past. When we are designing predictors, we develop a modular structure that consists of a cascade connection of a number of similar stages known as the *lattice predictor*. The lattice predictor is uniquely characterized by the reflection coefficients. There is one reflection coefficient per stage. The lattice predictor characterized by reflection coefficients is unique in the sense that it provides both the forward and the backward predictor; the lattice stage can be increased or decreased without affecting the specifications of the rest of the predictor stages. When the lattice predictor is used, we get an important benefit, the backward prediction errors represent uncorrelated samples of the input signal, while the input signal is itself a set of correlated samples. This transformation of correlated samples to uncorrelated samples can be used to provide an efficient method for the estimate of the desired response.

1.2
How is Digital Processing Done?

Since the availability of digital computers in the 1950s, the processing of signals has migrated from the analog domain to the digital domain. Initially, the digital processing was limited to processing analog signals using analog domain algorithms modified to work on the digital information that was being processed. This showed the engineers the power of using the digital processing instead of analog processing. So quickly engineers designed and discovered digital algorithms that would perform tasks that were similar to the analog algorithms. One such example is of designing FIR filters. Even though FIR filters were known, their designs with optimization such as equiripple design were not practical. Digital algorithms made this possible.

To process the analog signals in the new digital domain, we have to have a system that will resemble like the one shown in Figure 1.1. In the figure, we first have to sample the analog signal at regular periodic interval of time. This is done by the sample and hold device. This device will sample the analog signal at regular periodic interval of time and present this value to the analog to digital (A/D) converter. The expected waveforms from each processing block are shown in Figure 1.2. The analog signal to be processed is shown in Figure 1.2a.

The A/D converter will take the sampled signal and quantize it and convert it to a digital representation. The analog signal and the sampled signal are shown in Figure 1.2b.

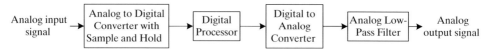

Figure 1.1   Scheme for processing an analog signal in the digital domain.

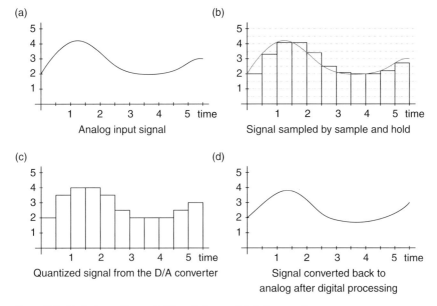

Figure 1.2 Typical waveforms at different stages of digital processing.

Note that when the sample is taken it is the same value as the input signal. This value is held till the next sample is taken. The sampled signal is then quantized. Quantizing values are shown in the figure also. The process of quantization takes the signal that is present and estimates its value to the closest value that it can represent. Notice that there are many sample values that are not the same as the quantization levels, so this process introduces some error, but the error can be controlled up to some minimum point by increasing the quantization levels that is also the same as increasing the number of bits used to represent the sampled value. Once the analog signal is converted to digital, it is processed by the digital processor.

After the signal is processed in the digital domain, it is ready to be converted to the analog domain to be used in the application of interest. This is done by first converting the digital value to the analog value by the D/A converter. The output from the D/A converter is a quantized value of the expected output, the signal from the D/A represents a staircase waveform. This waveform is passed through a low-pass filter to convert the staircase wave-form into a smooth waveform that is expected. The smooth value of the output is shown in Figure 1.2d. This is the signal we were after, this is the result of processing the signal in the digital domain.

1.3
What is Filtering?

Look at the signal path from the source to the sink as shown in Figure 1.3. The message at the source goes through the transmitter where it is converted from the native format to a format

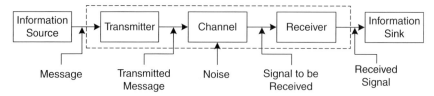

Figure 1.3 Signal path from source to the sink.

that is ideal for transmission through the channel that is to be used. As the information travels through the channel, the information is corrupted. Of many sources of corruption, there are two main sources of corruption. These two sources are intersymbol interference and noise.

1.3.1
Intersymbol Interference

Ideally, we want the channel to deliver all the information that is entered as it is entered. For the channel to be able to do this, the channel should have an impulse response that is of the form given in Eq. (1.1)

$$h(t) = A\delta(t-\tau) \tag{1.1}$$

In Eq. (1.1), we see that an ideal channel has two very important properties, first it will at most delay the information before it is delivered to the destination and second it may alter the amplitude of the signal. Other than these changes, an ideal channel will not alter the transmitted signal. In Eq. (1.1), τ represents the delay that the channel requires to deliver the information and A represents the amplitude of the delivered signal, and finally $\delta(t)$ represents the impulse function. Converting Eq. (1.1) to the frequency domain, we see why this ideal channel cannot be found. Equation (1.2) says that all frequencies will undergo a linear phase shift. It does not matter if the frequency is a low frequency, a frequency in the midband, or a very high frequency. The channel has to provide a phase shift equal to $\omega\tau$, an amount that is proportional to the frequency.

$$H(j\omega) = A\,e^{(-j\omega\tau)} \tag{1.2}$$

There is no channel that can do this and as a result the symbols cannot remain as impulses any more. Since the symbols cannot remain as impulses any more and they begin to spread, they will start to interfere with their neighbors. This is intersymbol interference. If the smearing of the successive symbols is severe enough, then the result will be that the symbols will be indistinguishable.

1.3.2
Noise

We can view noise as unwanted change in the energy spectrum of the signal. Some form of change in energy is present in every communication channel. The energy spectrum in the

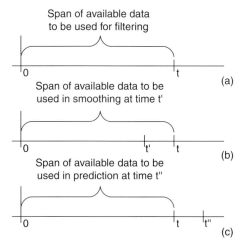

Figure 1.4 Three different forms of filtering: (a) smoothing, (b) filtering, and (c) prediction.

signal at the output of the channel is not the same spectrum that is present at the input of the channel. This noise can be internal to the system as thermal noise generated by the electronics getting hot or some other internal noise or the noise can be external to the channel and can come from the environment or from other signals that are present in the same channel from other sources. Whatever the source of the noise, it is unwanted and the filter is used to remove or reduce the noise.

The effect of the two impairments is that the signal received is generally not the same as the signal transmitted, it is distorted and it is noisy. It is the job of the filter to take the received signal and present as its output an estimate of the transmitted signal at the output. In estimating the received signal, we see that there are several types of operations depending on which information is available and which information is to be estimated. Look at Figure 1.4 to see the various types of estimations.

Filtering is an operation that takes the past inputs from beginning to some time t to extract the required information at time t. We can consider this as real-time operation as we use all the information available till now to decide on what is the best estimate of the present value.

Smoothing is an operation that takes the past inputs from beginning to some time t to estimate the information at some earlier time t' that occurred prior to time t. So during smoothing, we are, in fact, using the received sample and a few samples after the sample of interest to refine our estimate (make it better) of the sample of interest, and since we are using samples that arrive after the sample of interest, there is some delay involved in arriving at a smooth or a better estimate of the received sample. This operation is not considered as real-time operation as we are trying to predict the value that occurred in the past.

Prediction is an operation that takes the past inputs from beginning to some time t to estimate the information at some earlier time t'' that will occur at some future time after time t. This is forecasting since we are truly trying to predict what the received signal will be at some time in the future. We can consider this as real-time operation as we use all the information available till now to decide on a future value.

1.4
Linear Filters

The filter we use as low-pass filter in the digital processing of signals shown in Figure 1.1 or the filter used for estimation, for smoothing, or for prediction can be organized in various ways. The choice of the filter structure has a profound effect on the operation of the algorithm employed. Engineers have classified filters in various ways. In this book, we will be dealing with only linear filters, and we will not deal with nonlinear filters. We say a filter is a linear filter if the filtered, smoothed, or the predicted value that is the output from the filter obeys the principle of superposition, it is a linear combination of the applied input to the filter. If this is not true, then the filter is a nonlinear filter. Another important classification of filters is based on the memory requirement of the filter. Filters that require a finite amount of memory are known as *finite impulse response(FIR)* filters. When the size of the impulse response is infinite, we get a filter that is known as *infinite impulse response(IIR)*. The IIR filters require infinite but diminishing memory.

1.4.1
The FIR Filter

The operation of a digital filter involves three basic tasks. These are multiplication of the filter input with the filter weight, addition of all the multiplications together, and finally shifting or delaying the inputs so the next iteration can be performed. All these operations for an FIR filter are shown in the transverse structure of Figure 1.5.

The figure shows a transverse FIR filter of order 3. In the filter, the blocks labeled z^{-1} represent the delay elements. So when we shift the inputs for the next iteration, the input to each of the z^{-1} block is forwarded to the next z^{-1} block. Thus, it is the input to the z^{-1} blocks that is remembered. Since this is an order 3 filter, we need three memories or we need to remember the past three inputs. The coefficients $h(n)$ represent the filter coefficients. The filter coefficients are multiplied by the input and presented to the summing blocks.

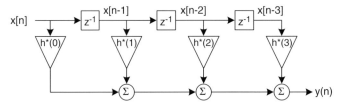

Figure 1.5 A transverse FIR filter.

The summing blocks add the individual products and present the sum to the output. The output of the transverse filter can be written as shown in Eq. (1.3)

$$y[n] = \sum_{i=0}^{N-1} h^*(i)x[n-i]$$

$$y[n] = \sum_{i=0}^{N-1} h^*(i)x[n]z^{-i}$$

(1.3)

Typically, the coefficients of the filter can be complex as much as the inputs can be complex. To emphasize this, the filter coefficients are often written as complex conjugates as we have done in Figure 1.5. Eq. (1.3) is known as the convolution sum as it convolves the input with the filter coefficients.

Notice carefully that the FIR filter is made up of only feed forward branches; this filter has no feedback loops present. As a result of the feedforward structure, the filter is made up of only zeros. The all-zero filter is inherently stable and the impulse response for any input persists for only N sample periods after the input, where N is the order of the filter.

1.4.2
The IIR Filter

Just like the FIR filter, the IIR filter also requires three basic operations that are the delay, the multiplication, and the summation. This time, however, there are two sets of coefficients. One set of coefficients represents the feedback loop, while the other set of coefficients represents the feedforward path of the signal. In the figure, we have shown that the filter has the same number of feedforward and the feedback paths, but this is not a requirement. Usually, there are more feedback paths than there are feedforward paths. The feedback paths are created because of the poles that are present in the filter, while the feedforward paths are created because of the zeros that are present in the filter. It is the presence of the feedback paths that makes the impulse response infinite.

The presence of the feedback paths introduces poles in the filter and we know that when the poles lie outside the unit circle, the system is an unstable system. So an IIR filter has the potential to become unstable (it may break into oscillations); therefore, in the design of these filters extra care has to be exercised so that the filter is always stable. The equation for the output of this filter can be written as shown in Eq. (1.4). The structure shown in Figure 1.6 is just one of the many possible filter structures that can be used for IIR filters.

$$y[n] = \frac{\sum_{k=0}^{N-1} b_k z^{-k}}{1 + \sum_{i=0}^{N-1} a_i z^{-i}} x[n]$$

(1.4)

Another popular structure that is used divides all the poles and zeros in second-order poles and zeros. Then, each second-order pole–zero combination is implemented using a structure similar to the one shown in Figure 1.6. There may be an odd number of poles and/or zeros in one section, but all the other sections have an even number of poles and zeros.

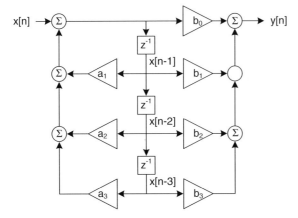

Figure 1.6 An IIR filter.

1.4.3
How and Where Filters Operate

In the previous section, we saw two different types of filters. Both these filters were defined using time as the variable. The signal that was received and the processed signal output were both time dependent. So we say that the filter operates on time. The same filter can be used when the signal is a spatial signal. Some examples of spatial signals are a row or a column from the image sensor of your digital camera, or a scanline from a scanner. So most of the filters that we come across in this book are applicable for both temporal and spatial signals.

1.5
Multirate Filters

The filters that we have seen so far in this book have all be single-rate filters. The single-rate filters operate on a single sampling rate. The sampling rate at the input is the same as the sampling rate at the output. No samples are added that increases the sampling rate and no samples are discarded that reduces the sampling rate. The signal processing operations performed are shown in Figures 1.5 and 1.6, where the signal can be multiplied, added, or delayed (or stored). Sometimes, it is required that we change the sampling rate of a signal because we want to use the signal that was sampled at a particular sampling rate in a device that is operating at a different sampling rate. One example of this comes up very frequently in the many different audio devices that we have and use everyday. Each audio device operates on its own sampling rate; for example, the audio CD operates at a sampling rate of 44.1 kHz sampling rate, while the digital audio tape (DAT) operates at 48 kHz sampling rate. Conversion of sampling rate among these devices is absolutely necessary if we want to experience the same audio quality when a file from one device is transferred to the other device. If we do not change the sampling rate, then the receiving device will alter the sampling rate and hence the frequency of the output signal will be altered.

Sometimes, it is advantageous to alter the sampling rate of the signal, we might want to increase the sampling rate of the signal because it makes the design of the subsequent filter easier. The filter design is easier since the transition band between the baseband spectra and the image will be wider and hence the filter does not need a steep transition band. We might want to decrease the sampling rate of the signal because the signal was sampled at a much higher rate than necessary and the extra information is not required. If the sampling rate was not decreased, then all the computations would have to be performed at a higher sampling rate, which requires a more powerful processor. These and other reasons make multirate sampling not only advantageous but also a requirement for efficient processing of signals.

The multirate filters use all the signal processing operations performed in the single-rate filters and are shown in Figures 1.5 and 1.6 such as multiplication, addition, and the delaying of signals. In addition to these devices, the multirate signal processing filters use two more signal processing operations that are interpolation that increases the sampling rate and decimation that reduces the sampling rate.

The interpolators, which are also known as upsamplers, increase the sampling rate by inserting zero-value samples between the input samples; an upsampler is shown in Figure 1.7. To upsample a signal by L we insert $L-1$ zero-value samples between two consecutive samples of the original signal. Inserting zero-value samples reduces the time delay between two consecutive samples by the rate of increase in the sampling rate as shown in Eq. (1.5)

$$x_u[n] = \begin{cases} x[n/L] & n = 0, \pm1, \pm2, \dots \\ 0 & \text{otherwise} \end{cases} \tag{1.5}$$

The sampling interval of the signal before upsampling in Figure 1.7 is T; this signal is shown on the left-hand side of the figure. When the signal goes through the upsampler, we insert L-1 zero-valued samples between two consecutive samples, and after upsampling the upsampled signal has a sampling rate $T/2$; the upsampled signal is shown on the right-hand side of the figure. Notice that the sampling interval between any two samples of the upsampled signal is half that of the original signal since the upsampling is by a factor of 2. Due to the insertion of zero-valued samples in the signal, the sampling rate of the signal is increased. As we will see in Chapter 6 later, this increase in the sampling rate by inserting zero-valued samples introduces image spectra in the frequency domain. So every time we use an upsampler we will need to follow it up with an image elimination filter. The purpose of this filter is to remove the image spectra.

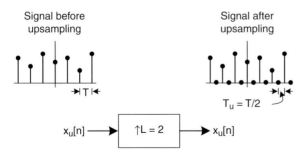

Signal before upsampling

Signal after upsampling

T

$T_u = T/2$

$x_u[n] \longrightarrow \boxed{\uparrow L = 2} \longrightarrow x_u[n]$

Figure 1.7 An upsampler: upsampling by $L = 2$.

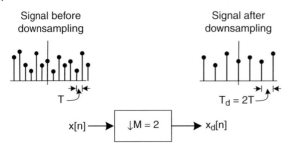

Figure 1.8 A downsampler: downsampling by $M = 2$.

The decimators that are also known as downsamplers decrease the sampling rate by removing and discarding M-1 consecutive samples and retaining every Mth sample; a figure of a downsampler is shown in Figure 1.8. We remove M-1 samples, so we in effect increase the time delay between two samples. The increase in the sampling time equals the rate of downsampling. The decrease in the sampling rate is shown in Eq. (1.6)

$$x_d[n] = x[Mn] \tag{1.6}$$

The sampling interval of the signal before downsampling in Figure 1.8 is T; this signal is shown on the left-hand side of the figure. When the signal goes through the downsampler, every other sample is thrown away and the downsampled signal has a sampling rate $2T$; the downsampled signal is shown on the right-hand side of the figure. Notice that the sampling interval between any two samples of the downsampled signal is twice that of the original signal since the downsampling is by a factor of 2. Due to the removal of the M-1 samples from the signal, the sampling rate of the signal is decreased. As we will see in Chapter 6 later, this decrease in the sampling rate by removing some samples could introduce aliasing in the frequency domain if the sampling rate drops down below the Nyquist sampling rate. So every time we use a downsampler, we will need to lead it with an image aliasing filter. The purpose of this filter is to band limit the signal so that there is no aliasing in the spectra after the signal is downsampled.

1.5.1
Altering the Sampling Rate by a Fraction

Using the upsampler or the downsampler individually will alter the sampling rate by an integer multiple. This is useful by itself, but the real power of multirate sampling is revealed when we have to alter the sampling rate of a signal by a fraction. For example, when we want to convert a signal from a CD to be stored in a DAT, we need to convert the sampling rate from 44 100 to 48 000. This requires that we increase the sampling rate by 160/147. This cannot be achieved by any device. In Chapter 6, we will see in detail how this is done, but for now we will look at one possible way this can be done. Look at Figure 1.9 to see how this sampling rate increase can be achieved.

We take the signal sampled at 44.1 kHz and upsample it by a factor of 160. This is done by inserting 159 zero-valued samples between any two consecutive samples. This in effect increases the sampling rate from 44.1 kHz to 7056 kHz. This also introduces many image

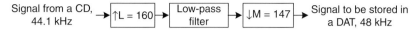

Figure 1.9 A fractional sampling rate converter.

spectra. So the upsampled signal is passed through a low-pass filter. If we design this filter properly, the filter can serve two purposes, both as an image elimination filter and as an antialiasing filter. So the filter will not only eliminate the image spectra but will also band limit the signal. Since the signal is band limited, we can downsample the signal so that the resulting sampling rate of the signal will be 48 kHz. Now this signal can be stored in a DAT and it can be used so that it has all the qualities that were present in the signal when the signal was on the CD.

The low-pass filter in Figure 1.9 serves several purposes. We have seen how the filter is used both as an antiimaging filter and as an antialiasing filter. This same filter serves another purpose also. We know that the upsampler has inserted 159 zero-value samples between any two consecutive samples. If any of these zero-value samples are still present in the signal after the downsampling, then there will be unwanted noise present and the signal will be corrupted. The low-pass filter prevents this from happening. The low-pass filter serves as an interpolator and "fills" in the values that the upsampler introduced. Thus, when the downsampler sees the signal it does not see all the zero-value samples. They have been "filled" in by the "interpolator," which is the low-pass filter.

1.5.2
Subband Decomposition

In practice you often encounter signals that have most of their energy within a particular frequency region. One such dramatic example is shown in Figure 1.10. In Figure 1.10, we see that most of the signal energy is contained in the region $0 \leq \omega \leq \pi/2$. So in this case, we can simply decimate the signal by half and achieve signal compression by 50%. By doing this, we would lose some information and if this loss is not critical, then we have achieved signal compression. If the signal in the region $\pi/2 \leq \omega \leq \pi$ cannot be ignored, then we cannot decimate the signal by half without causing aliasing. Such a small amount of energy is preventing us from achieving 50% compression.

Consider the following way to get around the problem. We first split the signal into two frequency bands by passing the signal through the two filters shown in Figure 1.10b. Now the signal $X_0(e^{j\omega})$ has almost no energy in the high-frequency region $\pi/2 \leq \omega \leq \pi$, while the signal $X_1(e^{j\omega})$ has almost no energy in the low-frequency region $0 \leq \omega \leq \pi/2$. Now both the signals can be decimated by a factor of 2 and there will be no aliasing. Next, we can code the signal with most of the energy in the region $0 \leq \omega \leq \pi/2$ using 16 bits per sample, while the low-energy signal in the region $\pi/2 \leq \omega \leq \pi$ can be coded with fewer bits, say 4 bits per sample. This way we will use fewer bits to code and transmit the signal achieving signal compression without losing fidelity. There are many such examples, such as the design of transmultiplexers that convert frequency division multiplexed signals into time division multiplexed signals, the analog voice privacy systems where multirate signal processing is a very elegant answer.

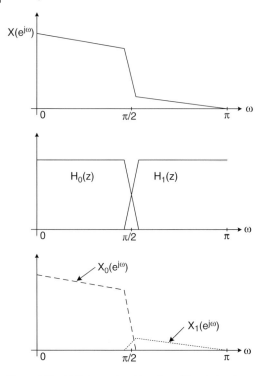

Figure 1.10 Splitting a signal into its subband signals.

1.6
A Classical Filtering Model

In this section, we synthesize an optimum linear filter. This is a common synthesis problem that arises not only in electrical engineering but also in many disciplines such as financial systems, life cycle systems, drug delivery, and so on. Our approach taken here is general and it is directly applicable to other signal processing applications. We will develop the detailed theory toward the end of the book in Chapters 7–9. At that time, we will use several *tricks of the trade* to reach the results that we need. In this section, we will define a classical signal processing problem and develop the result later in the book.

In Figure 1.11, the signal $s[n]$ is the transmitted signal, and as this signal travels toward the destination through a channel as shown in Figure 1.3, it gets corrupted by noise. Different types of noise that corrupt the signal were discussed in Section 1.3. So the received signal $x[n]$ is corrupted by noise and can be written as shown in Eq. (1.7)

$$x[n] = s[n] + n[n] \qquad n \geq 0 \tag{1.7}$$

Our goal is to receive the signal that was originally transmitted, so the job of the time invariant linear filter in Figure 1.11 is to determine as best as it can what the original transmitted signal is. Equation (1.7) shows that the noise is an additive process and it gets

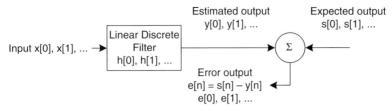

Figure 1.11 Optimum filtering problem.

added to the transmitted signal. This noise process is assumed to be a *white noise process*. The white noise process is discussed in detail in Section 7.5. White noise process is assumed to have zero mean and it has a constant variance for zero lag. In addition to this, we will assume that each sample of the noise process is independent of all the other samples. So we will use the observation $x[n]$ to estimate the transmitted quantity $s[n]$. As the signal passes through the filter, we can define the output of the filter as shown in Eq. (1.8)

$$
\begin{aligned}
y[n] &= \sum_{k=0}^{\infty} h^*[k]x[n-k] \\
y[n] &= \sum_{k=0}^{\infty} h^*[k]\{s[n-k] + w[n-k]\} \\
y[n] &= \sum_{k=0}^{\infty} h^*[k]s[n-k] + \sum_{k=0}^{\infty} h^*[k]w[n-k] = y_s[n] + y_w[n]
\end{aligned}
\tag{1.8}
$$

The filter response in Eq. (1.8) is written as a summation of response from two different inputs, one is a response from the noise input that has corrupted the signal transmitted, while the other is the response that we want. This ability to separate the two responses arises from the fact that we are first using a linear filter and that the noise corrupting the signal is additive in nature. Our goal in using this filter is to somehow make the response $y_s[n]$, which is the response to the signal input, so that it resembles the signal $s[n]$ itself, while the response $y_w[n]$, which is the response to the noise input, so that it is as small as possible. We aim to do this by selecting the filter coefficients according to some plan.

Due to the random nature of the noise signal, these objectives are not immediately achievable, but we can do a very creditable job in the long run. So we are looking for the solution for a large value of samples that have gone through the filter. This is necessary since noise added is known only in the statistical sense. Since we are looking for a solution in the statistical sense, it will be useful to examine the mean value and the variance of the output from the filter.

$$
\begin{aligned}
\mathcal{E}\{y[n]\} &= \mathcal{E}\{y_s[n]\} + \mathcal{E}\{y_w[n]\} \\
\mathcal{E}\{y[n]\} &= \mathcal{E}\{y_s[n]\} + \mathcal{E}\left\{\sum_{k=0}^{\infty} h^*[k]w[n-k]\right\} \\
\mathcal{E}\{y[n]\} &= \mathcal{E}\{y_s[n]\} + \sum_{k=0}^{\infty} h^*[k]\ \underbrace{\mathcal{E}\{w[n-k]\}}_{\substack{\text{Expected value of}\\\text{noise which is zero}}}
\end{aligned}
\tag{1.9}
$$

Using the fact that the noise has zero mean, we can write the expectation of the output for a large value of n as shown in Eq. (1.9). So the expected value of the output from the filter is the response of the filter for the required signal. Next, we look at the variance of the output from the linear filter. To do this, we will use a very special result from linear filter theory that relates the input *power spectrum density(PSD)* to the output PSD. The result is developed later in Section 7.6, and this equation is also given in Eq. (7.47) and in Eq. (1.10)

$$\underbrace{\Gamma_y(\omega)}_{\text{Output PSD}} = \left|H\left(e^{j\omega}\right)\right|^2 \underbrace{\Gamma_x(\omega)}_{\text{Input PSD}} \tag{1.10}$$

In Eq. (1.10), the quantity $\Gamma_y(\omega)$ represents the PSD of the output and $\Gamma_x(\omega)$ represents the PSD of the input and this input PSD is multiplied by magnitude squared value of the filter transfer function. So if the filter is going to achieve its goal of removing the noise from the input signal, then the filter that we design will be selected to satisfy the following two conditions simultaneously:

1) The mean value of the output should be equal to the mean value of the transmitted signal.
2) The PSD of the output must be the same as the PSD of the transmitted signal everywhere except for zero lag, where it is equal to the sum of the signal PSD and the noise PSD.

1.7
An Optimum Solution to the Classical Problem

In the previous section, we used only the statistical properties of the signal to design the filter that will separate the signal from the additive noise. The filter designed in this manner could be a suboptimum filter; this filter does a good job on average, it may miss by a lot some time, and then at other times it will be right on the money. Such a filter is not an optimum filter. To design an optimum filter, we will consider an unconstrained causal linear time invariant filter like the one shown in Eq. (1.8). In this filter, we want to measure how good is our estimate $y_s[n]$ as compared to the transmitted signal $s[n]$. To measure this, we will define an error signal as the difference between the desired output (which is the same as the transmitted signal) and the estimate of the desired signal (which is the same as the output signal from the filter) as shown in Eq. (1.11) and in Figure 1.11

$$\underbrace{e[n]}_{\substack{\text{error}\\\text{signal}}} = \underbrace{s[n]}_{\substack{\text{desired}\\\text{signal}}} - \underbrace{y[n]}_{\substack{\text{received}\\\text{signal}}} \tag{1.11}$$

Our objective now is to select the filter impulse response in such way as to make the error signal minimum according to some optimizing measure. To define the measure for optimum response from the filter, we will first define a cost function. Then when we minimize the cost function according to our chosen measure, we will have an optimum filter. The measure that we will choose is the *mean square value of the error(MSE)*. We choose the

MSE because this will give us a quadratic function that resembles a bowl-shaped surface. A bowl-shaped surface has a unique minimum so the solution that we will get will be an optimum result.

Minimizing the MSE value will lead to a set of equations that we call as the *normal equations*. These equations are discussed in detail in Eq. (8.11). The normal equations are a set of N simultaneou equations that relate the autocorrelation of the input signal for varying amount of lag to the cross correlation between the input signal and the desired signal. The solution to the normal equations will lead us to a filter that is known as the Weiner filter.

1.7.1
Improving the Optimum Filter

The Weiner Filter requires the knowledge of the statistical properties of the signal. These statistical properties of the signal may not be known always, or over a long transmission the statistical properties of the signal may drift. Even a small change in the statistical properties will render the Weiner filter unusable. When the statistical properties of the signal change, the filter designed using the original normal equations is no longer an optimum filter. So when the properties of the signal are not known originally (or they drift over time), then we have to somehow obtain the information that is not available. One way to do this is to use sufficient number of the initial input samples to estimate the statistical properties of the filter. Once we have obtained the estimate of the statistics of the signal, we can use this knowledge to solve the normal equations and design the Weiner filter.

This two-stage approach, which uses the "plug and play" method to design the optimum filter, requires some time. This introduces delay in the processing of signals. If the filter is to be used in real time, then the plug and play two-stage approach cannot be used. It would be very desirable if we could somehow, with each received input, make a better estimate of the statistics of the signal. Then, having made a better estimate we use this latest estimate to alter the filter so that it comes closer to the optimum filter. We alter the filter by adjusting the impulse response of the filter with each input that is received. This way the filter gradually *adapts* itself to the optimum filter.

A system that uses a filter that adapts itself we have a system that is a self-tuning system. The self-tuning is done by a recursive algorithm that looks at the old value of the impulse response and computes the adjustment before the next input arrives. Then after the next input, the filter does this all over again. This will go on till there is no change in the impulse response of the filter when the filter will have reached the optimum value. The recursive algorithm makes it possible for the filter to operate in an environment where there is no complete knowledge of the statistics of the signal or it is simply not available.

The recursive algorithm generally begins with some arbitrary initial condition if nothing is known about the signal or if there is some prior knowledge, then that can be used to improve the initial conditions. Then, recursively as we gather knowledge about the signal we adjust the impulse response of the filter. This way we approach the optimum filter. In a stationary environment, the optimum solution reached will be identical to the Weiner filter even though we began with no knowledge of the statistics of the signal.

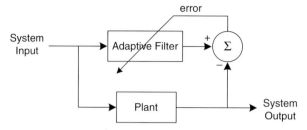

Figure 1.12 Adaptive filter used for system identification.

1.8
Classes of Applications of Adaptive Filters

Now that we have seen that the adaptive filters can operate in an environment where the statistics of the input signal are not known, it would be interesting to see what kinds of situations can take advantage of this type of a filter. The fact that the filter can converge to the optimum filter makes it a very powerful signal processing tool in many diverse fields such as sonar, control systems, biomedical engineering, and so on. Each of these fields has applications that vary in nature, but all the applications have a known desired response that is compared to the output of the adaptive filter to generate the error signal; this error signal is then used to determine the adjustment in the weights that are then adjusted before the next error is computed.

The similarities in the use of the adaptive filters lead to four different ways the adaptive filter is used in any of these fields: to identify the unknown system as shown in Figure 1.12, to inverse model the given system as shown in Figure 1.13, to predict the next random input to the system as shown in Figure 1.14, or to cancel the interference as shown in Figure 1.15.

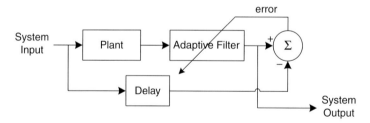

Figure 1.13 Adaptive filter used for inverse modeling.

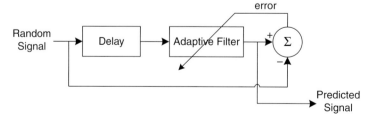

Figure 1.14 Adaptive filter used for predicting signal.

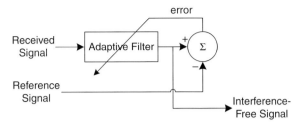

Figure 1.15   Adaptive filter used for canceling interference.

1.8.1
System Identification

In this application of the adaptive filter, we are interested in determining the transfer function of the plant. To do this, the input to the plant is allowed to go through both the plant and the adaptive filter. The two outputs are compared and the error configured. The error signal is used to update the filter coefficients. When the adaptive filter converges to a stable value, the transfer function of the adaptive filter is the "same" as the transfer function of the plant and the plant has been identified.

1.8.2
Inverse Modeling

In this application, we have a noisy plant and we are interested in canceling the effects of the noisy plant. When we are able to do this, we will have an ideal transmission medium. This form of the adaptive filter has huge application in satellite communication. As the signal travels from earth to the satellite and back, it undergoes corruption due to environment conditions. So at the receiving station, the inverse plant can be used to cancel the corruption due to environment conditions and thus the received signal will appear as if it has just gone through an ideal channel.

1.8.3
Prediction Modeling

In this application, we are interested in predicting what the next value of the random signal will be. The prediction works as follows: we estimate the value that we expect to receive, then when we receive the next random value we compare it with the predicted value, this gives us the required error signal that we use to adjust the coefficients of the adaptive filter. Prediction like this is used in varied fields such as weather forecasting, seismology, and many others.

1.8.4
Interference Canceling

The adaptive filter can be used to cancel the interference that may be present in the signal as the signal travels between the source of the information and the sink of the information.

In this the reference signal is present at the sink of the information and it is used to tune the adaptive filter. One example of this is the communication channel. In the channel the source will first transmit a known sequence of message and the receiver will compare the received signal with the known signal to tune the adaptive filter so that the adaptive filter will cancel the interference in the communication channel.

1.9
Chapter Summary

In this chapter, we introduced you to what you will come across and what to expect in the rest of the book. First, we discuss why and how we use digital processing. Digital processing is convenient because we can easily transmit and receive the information without any error or corruption. In addition to this, we can use software algorithms to build filters only if we have digital processing.

Next, we discuss filters in general; the filters that we will examine will be all linear, time-invariant filters that are mostly finite in their impulse response. Finite impulse response filters are always stable as they do not have any feedback paths, and because there are only feedforward paths, the filters only consist of zeros, there are no poles.

Then, we look at efficient ways of implementing these filters. This leads us to multirate filters. These filters change the sampling rate of the signal so that different techniques can be used that either permit us to use the signal in different devices or permit us to compress the signal efficiently.

Toward the end of the book, we develop optimum filters. These filters are known as Weiner filters and they operate on stochastic signals.

2
Discrete Time Signals and Systems

2.1
Introduction

In this chapter, we introduce you to a special class of systems. These systems are discrete systems that are linear and shift invariant. Throughout this book, we will study the linear shift-invariant (LTI) systems in various domains. The motivation to study the LTI systems arises because a large majority of systems that we will come across are LTI systems, and also there are many time domain tools that are very well known and useful in analyzing the LTI systems. With the availability of computers everywhere and digital signal processing being the way we design and implement these systems, it is even more important to study the LTI systems. In this chapter, we examine these systems in the time domain. Discrete systems are described by difference equations and the solution of the difference equation will give us the response from the difference system. The input–output relations of a discrete system are also described by the convolution sum. So we will examine the properties of the convolution operation and see how the convolution sum is evaluated.

In the study of any system we like to see how the system will perform under different situations, what is the effect on the output when a particular variable is changed, and how much control does this other variable have on the system? Answers to these and other similar questions can be obtained from experience, by building the system, or by using software to simulate the system. In this book, we will take the simulation route and use the MATLAB program to perform the necessary simulations. To introduce you to this software tool throughout this chapter and the rest of the book, I have placed MATLAB scripts that will demonstrate the concept that is being covered in the section of the text. In the appendix at the end of the book, there is a concise tutorial of the MATLAB program.

2.1.1
Discrete Time Signals

Discrete time signals are obtained most of the time from analog signals by the sampling process. The sampling of an analog signal is done usually by multiplying an analog signal by

Digital Filters: Theory, Application and Design of Modern Filters, First Edition. Rajiv J. Kapadia.
© 2012 Wiley-VCH Verlag GmbH & Co. KGaA. Published 2012 by Wiley-VCH Verlag GmbH & Co. KGaA.

an impulse train. The impulse train is a signal that is made up of impulses that are of unit strength and equally spaced in the time domain. The sampling process is shown in Eq. (2.1).

A continuous time signal $x(t)$, when sampled by a periodic impulse train with interval $1/T$, becomes a discrete time signal.

$$x[n] = x(t)_{t=nT} = x[nT] = x[n]|$$
(2.1)

In the sampling process shown in Eq. (2.1), if the analog signal to be sampled is a complex signal, then the sampled signal will also be a complex signal, so a sampled signal in general can be real or imaginary or complex; therefore, a general signal $x[n]$ will be written as shown in Eq. (2.2).

$$x[n] = \{x_{re}[n] + x_{im}[n]\}$$
(2.2)

The discrete time signal we have just created by sampling has values only at the sampling instants; in between the sampling instants, the signal is undefined. It is not zero.

In Section 2.1, we talked about an impulse train. Here, we define what an impulse signal is and how we get a train of impulses that we can use to sample an analog signal. An impulse signal in the discrete domain is a signal that has a defined value at only one sampling instant. So by definition, we can write an impulse signal as shown in Eq. (2.3).

$$\delta[n] = \begin{cases} 1 & n = 0 \\ 0 & \text{otherwise} \end{cases}$$
(2.3)

The impulse function in Eq. (2.3) is activated when the argument of the function is zero. The argument of the function in this case is the variable n. A unit impulse signal is shown in Figure 2.1a.

Often, we would like to start or activate the impulse function at some time other than at the origin. Say, we are interested in starting the impulse at sampling instant 4 as shown in Figure 2.1e. To get the impulse shifted, we need to modify the argument of the function so that the argument will be zero when $n = 4$. This is done as shown in Eq. (2.4).

$$\delta[n-4] = \begin{cases} 1 & n = 4 \\ 0 & \text{otherwise} \end{cases}$$
(2.4)

The next signal we talk about is a unit step signal. This signal can be thought of as a sequence of impulses that are shifted in time. A step signal in the discrete domain is very mach like a step signal in the analog domain except it is defined only at the sampling instants. So by definition, we can write a step signal as shown in Eq. (2.5)

$$\mu[n] = \begin{cases} 1 & n \geq 0 \\ 0 & \text{otherwise} \end{cases}$$
(2.5)

The step function in Eq. (2.4) is activated when the argument of the function is zero. The argument of the function in this case is the variable n. A unit step signal is shown in Figure 2.1b.

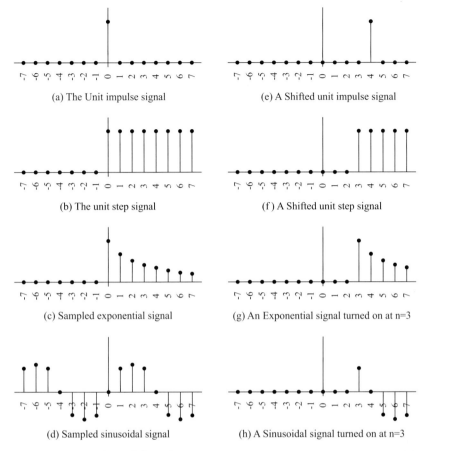

(a) The Unit impulse signal

(e) A Shifted unit impulse signal

(b) The unit step signal

(f) A Shifted unit step signal

(c) Sampled exponential signal

(g) An Exponential signal turned on at n=3

(d) Sampled sinusoidal signal

(h) A Sinusoidal signal turned on at n=3

Figure 2.1 Standard sampled signals.

Often, we would like to start or activate the step function at some time other than at the origin. Say, we are interested in starting the step function at sampling instant 4 as shown in Figure 2.1f. To get the step function shifted, we need to modify the argument of the function so that the argument will be zero when $n = 3$. This is done as shown in Eq. (2.6).

$$\mu[n-3] = \begin{cases} 1 & n \geq 3 \\ 0 & \text{otherwise} \end{cases} \tag{2.6}$$

The exponential signal, like the step signal, is defined only at the sampling instants. We define an exponential signal as shown in Eq. (2.7).

$$x_{\exp}[n] = a^{-n} \tag{2.7}$$

The exponential function in Eq. (2.7) is a decaying exponential when $n > 0$ and an increasing exponential when $n < 0$. Just like we were able to turn on the step function at $n = 0$, sometimes we want to turn on an exponential at $n = 0$. We can do this by multiplying the exponential with a step function. This is shown in Eq. (2.8). An exponential signal that is turned on at $n=0$ is shown in Figure 2.1c.

$$x_{\exp}[n] = \alpha^{-n}\mu[n] \qquad (2.8)$$

Often, we would like to start or activate the exponential function at some time other than at the origin. Say, we are interested in starting the exponential function at sampling instant 3 as shown in Figure 2.1g. To get the exponential function shifted, we need to modify the argument of the switch function so that the argument of the switch function will be zero when $n=3$. The switch function in this case is the step function, the exponential function is active all the time, while the switch function turns on at $n=3$ in this example. This is done as shown in Eq. (2.9).

$$x_{\exp}[n-3] = \alpha^{-n}\mu(n-3) \qquad (2.9)$$

The sinusoidal signal, like the step and the exponential, is also defined at the sampling instants. We define the sinusoidal signal as shown in Eq. (2.10). This function is active for all this when n is either positive or negative.

$$x_{\sin}[n] = \sin(2\pi\omega n) \qquad (2.10)$$

Often, we would like to start or activate the sinusoidal function at some time other than at the origin. Say, we are interested in starting the sinusoidal function at sampling instant 3 as shown in Figure 2.1h. To get the sinusoidal function started at some specific time, we need to add a switch function so that the argument of the switch function will be zero when $n=3$. The switch function in this case is the step function, the sinusoidal function is active all the time, while the switch function turns on at $n=3$ in this example. This is done as shown in Eq. (2.11).

$$x_{\sin}[n-3] = \sin(2\pi\omega n)\mu(n-3) \qquad (2.11)$$

Notice that when we decide to turn on a function at some specific time, we use a switch function that is turned on or off at a specific time, while the function stays the same. In the case of the impulse or the step function, we wanted to shift the function itself, so we modified the argument of the function and did not need a switch. Another reason we did not need a switch is because the step and the impulse function by themselves are switch functions, so we can turn them on at any point we want to turn them on.

2.2
Operations on Signals

In this section, we came across some simple signals. From these simple signals, we can get different signals by carrying out some simple operations on the signals. The operations on the signals are performed term by term and are examined in the following section. The basic operations on the signals are shown in Figure 2.2. The operations are performed term by term and are discussed below.

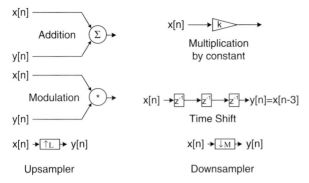

Figure 2.2 Basic operations on a sequence.

Addition or subtraction is one operation that is performed on two sequences. The value of sequence $x[n]$ is added to the value of sequence $y[n]$ to get a result sequence. Generally, we align the two sequences but this is not necessary. There can be an offset between the two sequences also. An example of the addition of two sequences is shown in Eq. (2.12). In Eq. (2.12), the sequence $x[n]$ is added to the sequence $y[n]$ to give a result sequence $z[n]$.

$$z[n] = x[n] + y[n] \tag{2.12}$$

A simple MATLAB script to perform this operation is given in MATLAB 2.1. In the MATLAB script we have first defined the sequences x and y to be length 5. The sequences are added together to give the sequence z that is also of length 5 and each element of the sequence z corresponds to the sum of the elements of sequences x and y. This is shown in the result sequence.

MATLAB 2.1

```
X = [ 1 4 2 6 -1] ;
Y = [ 2 5 1 3 2] ;
Z = x+y
z =
    3 9 3 9 1
```

Multiplication by a constant. The next operation on a sequence is a multiplication of every element in a sequence by the same constant. In this operation, we multiply each individual element of the sequence by a fixed constant to get the resulting sequence. An example of multiplication by a constant is shown in Eq. (2.13). In Eq. (2.13), the sequence $x[n]$ is multiplied by a constant k to get the result sequence $z[n]$.

$$z[n] = k * x[n] \tag{2.13}$$

A simple MATLAB script to perform this operation is given in MATLAB 2.2. In the MATLAB script, we have first defined the sequence $x[n]$ and the constant k and then performed the required operation to get the sequence z. The result is also shown and the result sequence can be seen to be k times the individual values of the sequence x.

MATLAB 2.2

```
x = [ 1 4 2 6 -1] ;
k = 3;
z = k* x
z =
       3 12 6 18 -3
```

Modulation. In the modulation operation, we multiply the two sequences with each other element by element. This operation is also known as the inner product, sometimes. An example of modulation is shown in Eq. (2.14). In Eq. (2.14), each element of sequence $x[n]$ is multiplied by a corresponding element of the sequence $y[n]$ to get the result sequence $z[n]$.

$$z[n] = x[n] * y[n] \tag{2.14}$$

A simple MATLAB script to perform this operation is given in MATLAB 2.3. In the MATLAB script, we have first defined the two sequences $x[n]$ and $y[n]$. Next, we perform the modulation operation to get the sequence z. Notice that the operator used to perform the term-by-term multiplication is a modified multiplication operator. This modification needs to be used whenever we need to perform a term-by-term operation in MATLAB. The result is also shown; notice that each term in the result sequence is obtained by multiplying the corresponding terms in the two sequences x and y.

MATLAB 2.3

```
x = [ 1 4 2 6 -1] ;
y = [ 2 5 1 3 2]
z = x.* y
z =
       2 20 2 18 -2
```

Time shift. The time shift operation simply moves the sequence from some time domain starting point to a different starting point. The sequence is always delayed and the delay operator is written as z^{-1}. Every occurrence of the delay operator will delay the sequence by one sampling instant; so, if we want to delay the sequence by four sampling time instants, we will need to use four delay operations. The delayed sequence is written as shown in Eq. (2.15).

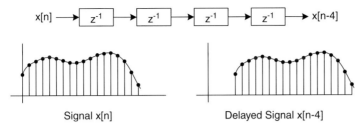

Figure 2.3 Signal $x[n]$ and its delayed version $x[n-4]$.

In Eq. (2.15), the sequence $z[n]$ is a delayed version of the sequence $x[n]$ and the delay period is four sampling units (Figure 2.3).

$$z[n] = x[n-4] \tag{2.15}$$

Upsampler. The upsampling process introduces zero-valued samples in the sequence between two consecutive samples of the sequence. So, if the upsampler is of order L, then $(L-1)$ zero-valued samples are introduced between two consecutive samples of the original sequence. The upsampling process on the sequences is a special process since this process changes the length of the sequence. All the other operations that we have seen so far keep the number of samples in the result sequence the same as the number of samples that enter in the operation. This is not so with the upsampler. So the upsampling process actually changes the sampling rate of the sequence, the resulting sequence operates at a higher sampling rate. The upsampling process is shown in Figure 2.4a. The upsampling process is shown in Eq. (2.16).

$$z[n] = \begin{cases} x[n/L] & n = 0, \pm L, \pm 2L \ldots \\ 0 & \text{otherwise} \end{cases} \tag{2.16}$$

A simple MATLAB script to perform upsampling operation is given in MATLAB 2.4. In the MATLAB script, we have first defined the sequence $x[n]$ and the upsampling index L. Next,

(a)

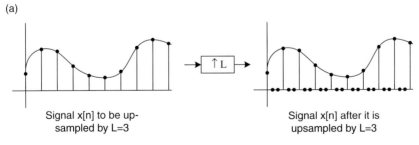

Signal x[n] to be up-
sampled by L=3

Signal x[n] after it is
upsampled by L=3

Figure 2.4 (a) The upsampling process. (b) The downsampling process.

(b)

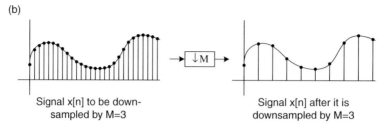

Signal x[n] to be down-
sampled by M=3

Signal x[n] after it is
downsampled by M=3

Figure 2.4 *(Continued)*

we define a sequence Z that is all zeros and has length equal to L times the length of sequence x. Finally, we insert the values of sequence x at every Lth location in the sequence z. The result is as shown and it shows that there are zeros between two successive samples of the sequence x. The sampling operation is indicated by an uparrow with a number next to it. This is also shown in Figure 2.4a.

MATLAB 2.4

```
x = [ 1 4 2 6 -1] ;
L = 2;
z = zeros (1, L* length (x));
z (1:L:L* length (x)) = x,

z =
    1 0 4 0 2 0 6 0 -1 0
```

Downsampler. The downsampling process removes samples from the sequence and leaves behind a smaller sequence. So, if the downsampler is of order M, then $(M-1)$ samples are removed and the Mth sample of the original sequence is left in. The downsampling process on the sequences is a special process since this process changes the length of the sequence. Most of the other operations that we have seen so far keep the number of samples in the result sequence the same as the number of samples that enter in the operation. This is not so with the downsampler. So, the downsampling process actually changes the sampling rate of the sequence and makes it a slower sampling rate. The downsampling process is shown in Figure 2.4b and in Eq. (2.17).

$$z[n] = x[Mn] \qquad\qquad (2.17)$$

A simple MATLAB script to perform downsampling operation is given in MATLAB 2.5. In the MATLAB script, we have first defined the sequence $x[n]$ and the downsampling index M. Next, we insert every Mth values of sequence x into the sequence z. The result is also shown; note that there are fewer samples in z than there are in x. The sampling operation is indicated by a downarrow with a number next to it. This is also shown in Figure 2.4b.

MATLAB 2.5

```
x = [ 1 4 2 6 -1 -2] ;
M = 2;
z = x (1:M:length(x)),
z =
    1 2 -1
```

2.3
Symmetry in Signals

Discrete time signals are classified in various different ways. Here, we examine what type of symmetry exists in these discrete signals. When we examine the symmetry classification of signals, we first define the midpoint of the signal. Generally, this midpoint of the signal is chosen to be the origin or $n = 0$.

2.3.1
Odd and Even Sequences

Discrete time signals can be classified as either odd sequences or as even sequences. The odd and the even sequences have property as shown in Eq. (2.18).

$$x[n] = x[-n] \quad \text{even sequence}$$
$$x[n] = -x[-n] \quad \text{odd sequence}$$

(2.18)

In general, a discrete time sequence is neither even nor odd, but any general discrete time sequence can be separated into its even part and its odd part sequence. Separating a general sequence into its even part and its odd part is shown in Eq. (2.19).

$$x[n] = x_{\text{ev}}[n] + x_{\text{od}}[n]$$

(2.19)

To separate a general signal into its even part and its odd part, we follow the operation shown in Eq. (2.20).

$$x_{\text{ev}}[n] = 0.5\{x[n] + x[-n]\}$$
$$x_{\text{od}}[n] = 0.5\{x[n] - x[-n]\}$$

(2.20)

When the sequence $x[n]$ is a complex sequence, the center of symmetry of the even part of the sequence, which occurs at $n = 0$, is identical to the original sequence. The center of symmetry of the odd part of the sequence, which occurs at $n = 0$, is identically equal to zero. In Eq. (2.20), $x[-n]$ is a time-reversed signal $x[n]$. MATLAB can be used to easily separate the even part and the odd part of a signal. MATLAB 2.6 gives us a simple script to separate the signal $x[n]$ into its even part and its odd part. In MATLAB 2.6, we first define a random sequence of nine entries that is a complex sequence. Next, we obtain another sequence that is a time-reversed signal. When we combine the signals as required in Eq. (2.20), we get the even part and the odd part of the original signal.

MATLAB 2.6

```
x = rand(1,9) + i*rand(1,9);
y = x(9:-1:1);
z_even = 0.5*(x+y),
z_odd = 0.5*(x-y),
```

2.3.2
Conjugate Symmetric and Conjugate Antisymmetric Sequences

A discrete time sequence can also be classified as a conjugate symmetric and conjugate antisymmetric sequence. A conjugate symmetric and the conjugate antisymmetric sequences have the property as shown in Eq. (2.21).

$$x[n] = x^*[-n] \qquad \text{conjugate symmetric}$$
$$x[n] = -x^*[-n] \quad \text{conjugate antisymmetric}$$

(2.21)

In general, a discrete time sequence is neither conjugate symmetric nor conjugate antisymmetric. A general sequence can be separated into its conjugate symmetric and its conjugate antisymmetric sequence as shown in Eq. (2.22).

$$x[n] = x_{CS}[n] + x_{CA}[n]$$

(2.22)

To separate the conjugate symmetric and the conjugate antisymmetric sequences from a complex, sequence is done as shown in Eq. (2.23).

$$x_{CS}[n] = 0.5\{x[n] + x^*[-n]\}$$
$$x_{CA}[n] = 0.5\{x[n] - x^*[-n]\}$$

(2.23)

When the sequence $x[n]$ is a real sequence, the conjugate symmetric and the even sequence are identical, and the conjugate antisymmetric sequence and the odd sequence are identical.

In a general sequence, the conjugate symmetric sequence displays the real part of the sequence as an even sequence, while the imaginary part of the sequence is an odd sequence. The point of symmetry, which occurs at $n = 0$ of a conjugate symmetric sequence, is a purely real number.

In a general sequence, the conjugate antisymmetric sequence displays the real part of the sequence as an odd sequence, while the imaginary part of the sequence is an even sequence.

The point of symmetry, which occurs at $n = 0$ of a conjugate antisymmetric sequence, is a purely imaginary number.

MATLAB can be used to easily separate the even part and the odd part of a signal. MATLAB 2.7 gives us a simple script to separate the signal $x[n]$ into its conjugate symmetric part and its conjugate antisymmetric part. In MATLAB 2.7, we first define a random sequence of nine entries that is a complex sequence. Next, we obtain another sequence that is a time-reversed signal. When we combine the signals as required in Eq. (2.23), we get the conjugate symmetric part and the conjugate antisymmetric part of the original signal. The command conj in the MATLAB script 2.7 gives us the conjugate of the complex number.

MATLAB 2.7

```
x = rand(1,9) + i* rand(1,9);
y = x(9:-1:1);
z_CS = 0.5* (x+conj(y)),
z_CA = 0.5* (x- conj(y)),
```

2.4
Energy and Power Signals

Signals can be classified as either energy signals or as power signals. A signal is an energy signal if the energy in the signal is finite. The energy in a signal is defined as the sum of the absolute value squared of all the terms in the signal as shown in Eq. (2.24).

$$\varepsilon_x = \sum_{n=-\infty}^{\infty} |x[n]|^2 \qquad (2.24)$$

An infinite length sequence with finite values may or may not be an energy signal. The signal is defined as an energy signal if the summation in Eq. (2.24) is finite. If the summation is not finite, the signal is not an energy signal. It is interesting to note that we define energy in a limited portion of the signal, say from $n = -k$ to $n = k$ in the same way as we defined the energy in the entire sequence as described in Eq. (2.24), this is shown in Eq. (2.25).

$$\varepsilon_{x,k} = \sum_{n=-k}^{k} |x[n]|^2 \qquad (2.25)$$

In a sequence the power is defined as average energy per sample in the sequence as the sequence tends to infinite length; this is shown in Eq. (2.26).

$$\mathcal{P}_x = \underset{k \to \infty}{\text{Limit}} \frac{1}{2k+1} \sum_{n=-k}^{k} |x[n]|^2 \qquad (2.26)$$

We know that power is defined as energy per unit time, the power in a signal is defined as the average energy per sample of the sequence as the sequence tends to be infinite in length, and so we can use the energy definition in Eq. (2.25) to define power as shown in Eq. (2.27).

$$P_x = \underset{k \to \infty}{\text{Limit}} \frac{1}{2k+1} \varepsilon_{x,k} \tag{2.27}$$

The average power in an infinite signal may be finite or infinite. We can modify the definition of average power in a periodic signal so that we do not have to sum an infinite number of terms. In a periodic signal since all the periods are the same, we can determine the average power in one period. This then will be the average power of the entire sequence. This is shown in Eq. (2.28); note that the period of the sequence in Eq. (2.28) is M samples.

$$P_x = \frac{1}{M} \sum_{n=0}^{M-1} |x[n]|^2 \tag{2.28}$$

2.5
The Concept of Frequency in Discrete Time Systems

The concept of frequency and periodicity is very simple to understand and familiar to everyone. In discrete time systems, there are many issues that lead to several different view points that should be made clear right away. While talking about frequency and periodicity, we will limit our discussion to sinusoidal signals only as there is most confusion with these signals.

When we consider analog signals, we see that for any value of the frequency f the analog sinusoidal is a periodic signal. If we increase the frequency f in an analog signal, the number of cycles completed in a fixed amount of time also increases. Similarly, decreasing the frequency will decrease the number of cycles completed in a unit of time. Also for every unique frequency value, the analog signal will have a unique wave shape that can be clearly distinguished. All these facts are shown in Figure 2.5.

The analog sinusoidal signal is represented as shown in Eq. (2.29). In Eq. (2.29), the parameter A is the amplitude of the sinusoidal signal, F is the frequency in Hertz, and Ω is the frequency in radians per second. The two frequencies are related to each other by a simple relation $\Omega = 2\pi F$, the last parameter θ is the offset in the sinusoid.

$$x_a(t) = A\cos(2\pi F t + \theta) = A\cos(\Omega t + \theta) \tag{2.29}$$

Sometimes, we will find it very convenient to express the sinusoid as a complex exponential. To represent the sinusoid as a complex exponential, we use Euler's identity. This is repeated for you in Eq. (2.30).

$$x_a(t) = Ae^{j(\Omega t + \theta)} = A\cos(\Omega t + \theta) + jA\sin(\Omega t + \theta) \tag{2.30}$$

Another concept of frequency is interesting and is frequently very convenient to use. Frequency is inherently a positive quantity. However, we will find it convenient to sometime represent frequency as a negative quantity. For example, if we wanted to represent the analog

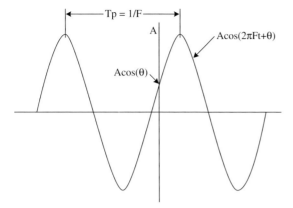

Figure 2.5 An analog sinusoidal signal.

signal $A \cos(\Omega t + \theta)$ using complex exponentials, then we would represent this signal as shown in Eq. (2.31).

$$A \cos(\Omega t + \theta) = \frac{1}{2} \left\{ A e^{j(\Omega t + \theta)} + A e^{-j(\Omega t + \theta)} \right\} \tag{2.31}$$

In Eq. (2.31), the second term represents negative frequency. We accept this if we visualize frequency as a rotating vector. So, when the vector is rotating in the counterclockwise direction, we represent the rotation as positive and the rate of rotation as a positive value. This is the frequency of this vector. With this interpretation, a vector rotating in the clockwise direction is rotating in the negative direction and the rate of rotation is again the frequency of this vector, is a negative value, and hence a negative frequency. So the frequency of a rotating vector can be either positive or negative, and the range of frequency is from $-\infty$ to $+\infty$.

A discrete time sinusoidal can be expressed as $x[n] = A \cos(\omega n + \theta)$. In this definition, n is always an integer, it is called the sample number, ω is the frequency in radians per sample, A is still the amplitude, and θ is the phase offset in radians. Just as we did in the analog case, we can define the sequence in terms of frequency of cycles per sample. This is shown in Eq. (2.32). In Eq. (2.32), f is cycles per sample.

$$A \cos(\omega n + \theta) = A \cos(2\pi f n + \theta) \tag{2.32}$$

In Figure 2.6, we see a sampled version of the analog signal in Figure 2.5. Often, there is a need to know if there exist a relation between the analog frequency Ω and the discrete frequency ω. Here, we first try to determine this relation. To do this, we will define F_s as the sampling frequency. Then, if we sample the analog signal with the sampling frequency F_s, the discrete signal that we get will be as shown in Eq. (2.33).

$$A \cos(2\pi F t + \theta) = A \cos(2\pi F n T_s + \theta) = A \cos\left(2\pi \left(\frac{F}{F_s} \right) n + \theta \right) \tag{2.33}$$

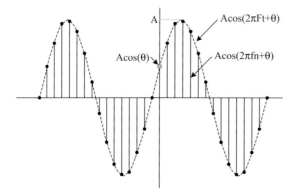

Figure 2.6 A discrete sinusoidal signal.

We see from Eq. (2.33) that the frequency of the discrete sinusoid is the ratio of the analog frequency and the sampling frequency. This is very interesting. Consider two different analog signals, one at frequency F_1 and another at frequency $F_2 = 2^* F_1$. We know that both these analog signals will be different and the second signal will have twice as many cycles as the first signal in the same time interval. Now, if we were to sample both these signals so that signal with frequency F_1 is sampled at frequency of F_s and the signal with frequency $F_2 = 2^* F_1$ is sampled at frequency of $2F_s$, then the two sampled signals will have a frequency of F_1/F_s and we will not be able to tell the two sampled signals apart! Another question that you might have is as follows: Even though the analog signal was a periodic signal and we sampled this signal with a periodic sampling function, is the resulting sampled signal a periodic signal? To answer these questions consider the following.

We know that a discrete time signal $x[n]$ is periodic with period N if

$$x[n+N] = x[n] \qquad (2.34)$$

for all n. Then, the smallest value of N that satisfies the relation in Eq. (2.34) is the period of the discrete time signal. Consider a sinusoidal signal $x[n] = A \cos(\omega n + \theta)$ and to see if this signal is periodic we substitute this in Eq. (2.34) to get

$$A\cos(\omega(n+N)+\theta) = A\cos(\omega n+\theta)\cos(\omega N) + A\sin(\omega n+\theta)\sin(\omega N)$$
$$A\cos(\omega(n+N)+\theta) = A\cos(\omega n+\theta) \qquad (2.35)$$

Equation (2.35) holds only under the condition that $\omega N = 2\pi f N = 2\pi r$. When we have $\omega N = 2\pi r$, $\cos(\omega N) = 1$ and $\sin(\omega N) = 0$, so Eq. (2.35) holds. From Eq. (2.35), we get the required relation as given in Eq. (2.36).

$$f = \frac{r}{N} \qquad (2.36)$$

Since both r and N are integers, we must have f a rational fraction for the discrete time signal to be periodic. Now that we know that when a discrete time sinusoid is periodic, $A \cos(\omega(n + N) + \theta) = A \cos(\omega n + \theta)$. So discrete time sinusoids whose frequencies are separated by 2π are indistinguishable from each other; this is also obvious from

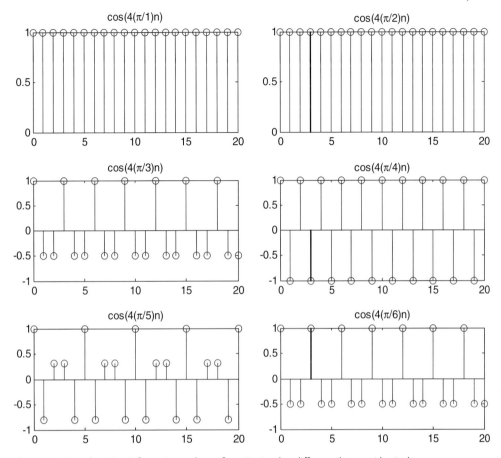

Figure 2.7 Signal $\cos(\omega_0 n)$ for various values of ω_0. Notice that different plots are identical.

Eq. (2.35). Figure 2.7 shows several sinusoids, and even though we begin with the same analog frequency for the sinusoid and sample it with different sampling frequencies, we get results that are some times identical and sometimes different.

Consider the plots with titles $\cos(4(\pi/3)n)$ and the plot $\cos(4(\pi/6)n)$. We see that the two plots are identical. To show that the two plots are indeed identical, we need to show that the two frequencies are 2π radians per sample apart. This can be shown as in Eq. (2.37).

$$\cos((4/3)\pi n) = \cos((2-2/3)\pi n)$$

$$\cos((4/3)\pi n) = \cos(2\pi n)\cos((2/3)\pi n) + \sin(2\pi n)\sin((2/3)\pi n) \qquad (2.37)$$

$$\cos((4/3)\pi n) = \cos((2/3)\pi n)$$

Similarly, you can show that the plot $\cos(4(\pi/1)n)$ and the plot $\cos(4(\pi/2)n)$ are identical. Can you determine at least one other frequency for which you will get a plot that is identical to the plot $\cos(4(\pi/5)n)$?

From the above discussion, we can make some statements about sampled signals.

1) A discrete sinusoidal is periodic only if its frequency f is a rational fraction.
2) Discrete sinusoids whose frequency is separated by 2π are identical.
3) There are only a limited number of sampled periodic sinusoids possible for any sampling interval.

Just as we represented the analog sinusoid as complex exponentials, we can represent the discrete signals also as complex exponentials. Let us choose the frequency of the discrete time signal $f = 1/N$, where N is the period of the discrete signal. Then, a set of harmonically related exponentials can be written as shown in Eq. (2.38).

$$x[n] = Ae^{j2\pi k f_0 n} \qquad k = 0 \pm 1, \pm 2, \ldots \tag{2.38}$$

In Eq. (2.38) as k takes on various values, we see that for a while we keep on getting different sinusoids. But when $k > N$, the sinusoids start to repeat themselves as shown in Eq. (2.39).

$$x_{(N+k)}[n] = Ae^{j2\pi(N+k)f_0 n} = A \underbrace{e^{j2\pi(N)f_0 n}}_{\text{This equals 1}} e^{j2\pi(K)f_0 n} = Ae^{j2\pi(K)f_0 n} = x_k[n] \tag{2.39}$$

This clearly shows us that there are only N unique sinusoids when we have a discrete signal with period N.

2.6
Discrete Time Systems

When a signal $x[n]$ goes through a system and comes out as a different signal, we represent the operation as a transformation as shown in Eq. (2.40).

$$y[n] = T\{x[n]\} \tag{2.40}$$

In Eq. (2.40), the operator $T(\cdot)$ transforms the input sequence into an output sequence, so the operator T is the system. A discrete system may be a hardware system such as a microprocessor or a shift register or a software system such as an algorithm. Either of these systems comes as various different types and they all modify the input in a desirable manner to get the required output. Here, we will examine the properties of discrete systems in general. Some examples of systems are given as follows:

1) $y[n] = 3x[n]$
2) $y[n] = 0.85x[n] + 0.25x[n-1]$
3) $y[n] = nx[n]$
4) $y[n] = \max[x[n-1]; x[n]; x[n+1]]$
5) $y[n] = 4\{x[n]\}^2$

Each of the above systems will transform the input in its prescribed manner to give us an output. In what follows we will classify the various ways in which the inputs are modified to give an output.

2.6.1
Static versus Dynamic

Systems can be classified as either static or memoryless and dynamic or having memory. A static system is like an instantaneous system, the output of the system depends only on the current input. If the output of the system does depend on either the past or the future inputs, then the system is a dynamic system. A dynamic system requires memory. If the output depends only on a few past or future inputs, then the system is said to have limited memory. If the output depends on all of the past or future inputs, then it is said to have infinite memory. In the five systems defined above, system 1, 3, and 5 are static systems. These systems give you the output as soon as the present input is available. They do not need to remember either the past inputs or the past outputs. Systems 2 and 4 are dynamic systems; to determine the output at any time system 2 needs to know the past input and system 4 needs to know the past input and the future input.

2.6.2
Time Variant versus Time Invariant

These systems are sometimes also known as shift invariant systems. A system is shift invariant if and only if it obeys the following: a relaxed system is shift/time invariant if and only if $y[n] = T\{x[n]\}$ implies that $y[n - k] = T\{x[n - k]\}$ for all inputs $x[n]$ and all shifts k. To determine shift invariance, we first determine the output when the input is shifted; next, we shift the output and determine the required input for this shifted output. If both the inputs are the same for all values of the shift parameter, then the system is shift invariant. Typical input–output relations in a shift invariant system is shown in Figure 2.8.

Consider system 2 in Section 2.6. We first shift the input by k units of time; this gives us a system output as shown in Eq. (2.41).

$$y[n, k] = 0.85x[n-k] + 0.25x[n-k-1] \tag{2.41}$$

Next, we shift the output by k units of time and determine the required input for this output; this is shown in Eq. (2.42).

$$y[n-k] = 0.85x[n-k] + 0.25x[n-k-1] \tag{2.42}$$

From Eqs. (2.41) and (2.42), we see that the result is the same irrespective of the shift, hence the system is shift invariant. On the other hand, when we consider system 3 we get an output when we shift the input k units of time as shown in Eq. (2.43).

$$y[n, k] = nx[n-k] \tag{2.43}$$

Next, we shift the output by k units of time and determine the required input for this output; this is shown in Eq. (2.44).

$$y[n-k] = [n-k]x[n-k] \tag{2.44}$$

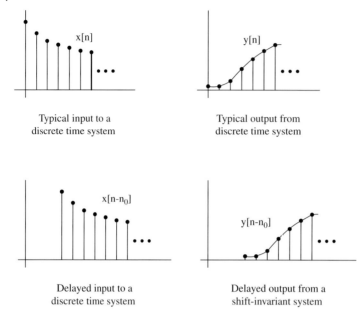

Figure 2.8   Input–output relations in a shift-invariant system.

From Eqs. (2.43) and (2.44), we see that the result is not the same irrespective of the shift, hence the system is shift variant.

2.6.3
Linear versus Nonlinear

To test a system for linearity, we need to test if the system obeys the law of superposition. This is easily done by using two different input sequences and following the operations depicted in Figure 2.9. The superposition theorem says that if the sum of two different outputs is the same as the output from the sum of the two corresponding inputs, then the system is linear. This is shown in Figure 2.9.

In Figure 2.9, we begin with two different inputs $x_1[n]$ and $x_2[n]$. When these two sequences go through the system, we get two different outputs $y_1[n]$ and $y_2[n]$. Adding the two outputs together, we get the output $y_a[n]$. Next, we begin again and this time we first add the two inputs together. After the added inputs go through the system, we obtain an output $y_b[n]$. If the output $y_a[n]$ is equal to the output $y_b[n]$, then the system is a linear system and if the two outputs are not equal, then the system is a nonlinear system. Of the five systems listed in Section 2.4, systems 1, 2, and 3 are linear and the other two systems are nonlinear.

To test for linearity of a system, we follow the method indicated in Figure 2.9. To test for linearity, we will test system 2 above. We first calculate output $y_1[n]$ when the input is $x_1[n]$, so $y_1[n] = 0.85x_1[n] + 0.25x_1[n-1]$. Next, we compute $y_2[n]$ when the input is $x_2[n]$

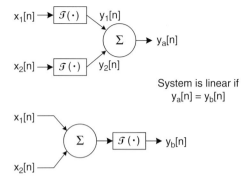

Figure 2.9 Determining if a system is linear or not.

so $y_2[n] = 0.85x_2[n] + 0.25x_2[n-1]$. When we add the two individual outputs, we get the output $y_a[n]$ as shown in Eq. (2.45).

$$y_a[n] = y_1[n] + y_2[n]$$
$$y_a[n] = 0.85(x_1[n] + x_2[n]) + 0.25(x_1[n] + x_2[n-1]) \tag{2.45}$$

To calculate $y_b[n]$, we first add the two inputs together so the input to the system will be $(x_1[n] + x_2[n])$. With this as the input, we calculate the output as shown in Eq. (2.46).

$$y_b[n] = 0.85(x_1[n] + x_2[n]) + 0.25(x_1[n] + x_2[n]) \tag{2.46}$$

Since $y_a[n]$ is equal to $y_b[n]$, we can say that the system is a linear system.

To verify that system 3 is a linear system, we first calculate $y_a[n]$. To do this again we will assume that we have two inputs $x_1[n]$ and $x_2[n]$. This gives us $y_a[n] = nx_1[n] + nx_2[n]$. Next, we determine $y_b[n]$ and this is $y_b[n] = n(x_1[n] + x_2[n])$. Since $y_a[n]$ and $y_b[n]$ are the same, the system is a linear system. Spend some time to think why system 4 is a nonlinear system.

2.6.4
Causal versus Noncausal

A system is considered a causal system if the present output of the system depends only on past and the present inputs. The present output does not depend on any future inputs. In terms of trying to determine if a discrete system is causal or not, we examine all the input terms that are required to compute the required output. If the index of any of the input terms points to a sample that occurs after the required output (in the future), then the system is a noncausal system. If the index of all the input terms refer to the samples that have already arrived in the past, then the system is a causal system.

To determine if a given system is causal or not consider the following: you are given two different input sequences that are identical from $n=0$ to $n=n_0$: $x_1[n] = x_2[n]$ for $n \le n_0$. For both these inputs, we determine the output from a system that we want to test for causality. The output for the two inputs are given in Eqs. (2.47) and (2.48).

$$y_1[n_0] = \sum_{k=-\infty}^{n_0} h[k]x_1[n_0-k]$$

$$y_1[n_0] = \sum_{k=-\infty}^{-1} h[k]x_1[n_0-k] + \sum_{k=0}^{n_0} h[k]x_1[n_0-k]$$

(2.47)

$$y_2[n_0] = \sum_{k=-\infty}^{n_0} h[k]x_2[n_0-k]$$

$$y_2[n_0] = \sum_{k=-\infty}^{1} h[k]x_2[n_0-k] + \sum_{k=0}^{n_0} h[k]x_2[n_0-k]$$

(2.48)

The two outputs $y_1[n_0]$ and $y_2[n_0]$ must be equal to each other since the two inputs are the same till $n=n_0$ and the system is the same. When we examine Eqs. (2.47) and (2.48), we have separated the convolution summation from $-\infty$ to -1 and the other one from 0 to n_0. The summations from $k=0$ to $k=n_0$ are identical to each other since the two inputs in this summation are the same. So, for the two outputs to be the same, the first summation that goes from $-\infty$ to -1 must also be equal. In these two summations, we cannot depend on the inputs $x_1[n]$ to be equal to $x_2[n]$. So, the only way these two summations are equal is when the system function $h[k]=0$ for $k<0$. This then is the condition for causality as shown in Eq. (2.49). Often, however, we can determine if a system is causal or not by simple examination. If the output does not depend on any future inputs, then the system is a causal system.

For a causal system we must have $h[n]=0$ for $n<0$ (2.49)

It should be apparent to every one that in real-time signal processing, we cannot observe inputs from the future so all systems that work on real time must be causal systems. If we do not require the system to operate in real time, if we can store the inputs and then process after we have received all the inputs, then we can build systems that are noncausal since we already know what the future inputs will be. We often have such a situation in some special applications such as image processing or in geophysical signal processing.

Among the five systems presented in Section 2.4, system 4 is a noncausal system, while all the other systems are causal systems.

2.6.5
Stable versus Nonstable

Stable systems are the systems that we want. Stability in a system is an absolutely essential quality. Systems that are unstable exhibit erratic and extreme behavior. Over the years, there have been many different definitions of stability of a system. Here, we will work with the most popular definition of stability. This definition is very easy to understand and to verify. We refer to the bounded input–bounded output (BIBO) definition of stability. The BIBO stability can be defined as follows: For any bounded input if the output from the system is also bounded then the system is said to be BIBO stable.

For a discrete time system, the input–output relations are written as shown in Eq. (2.50) and also in Figure 2.10. In Eq. (2.50), the input is $x[n]$ and the system impulse response is

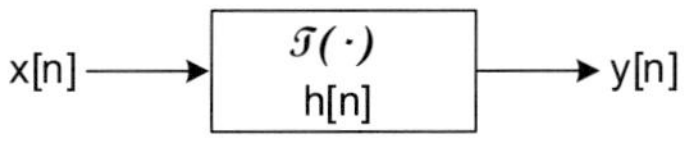

$$y[n] = \sum_{k=0}^{N-1} x[k]h[n-k]$$

Figure 2.10 Input–output relations of a typical discrete system.

$h[n]$. Using the input and the impulse response of the system, we compute the output from the system as shown in Figure 2.10. In equation we see that the absolute value of the summation will always be greater than the output; so, if the absolute value of the summation is finite, then the output of the system will be finite and the system will be stable.

$$y[n] \leq \left| \sum_{k=0}^{N-1} x[k]h[n-k] \right| \leq M \left| \sum_{k=0}^{N-1} h[n-k] \right| \tag{2.50}$$

Continuing with this reasoning, we make the input $x[n]$ the largest value it can ever be, and this is denoted as M in Eq. (2.50). If the input M is always a constant, then we can take it out of the summation sign and we get the last inequality. This last inequality shows that if the impulse response of the system is a finite value, then the output will also be a finite value and the system will be stable. So, we get a very simple test to determine if the system is BIBO stable; this test is shown in Eq. (2.51).

$$\left| \sum_{n=0}^{N} h[n] \right| < \infty \tag{2.51}$$

The test is that if the system impulse response is finite then the system is BIBO stable.

2.7
Analysis of Shift-Invariant Linear System

When we have an arbitrary input sequence, how do we determine the response of a system? To answer this question, we will follow a traditional path where we determine the response to a system for an input that is an impulse and then we will define an arbitrary sequence that is to be an input to the system as a sum of weighted shifted impulses. Finally, we will use the superposition theorem for linear systems to determine the response of our system to an arbitrary input sequence.

2.7.1
Input Sequence as a Sum of Shifted Weighted Impulses

In Section 2.1.1, we saw how we can represent an impulse and a shifted impulse. A shifted impulse $\delta[n-k]$ has value of 1 at $n=k$ and it is zero everywhere else. If we multiply this impulse by the sequence $x[n]$ as shown on the left-hand side of Eq. (2.52) then we can write this product as shown on the right-hand side of Eq. (2.52).

$$x[n]\delta[n-k] = x[k]\delta[n-k] \tag{2.52}$$

This equation holds because the impulse function is zero everywhere except at $n = k$, so the impulse picks out the one value from the sequence $x[n]$ where the impulse function has a nonzero value. If we were to shift the impulse at, say, some other value m, so the impulse is $\delta[n - m]$, then the product on the left-hand side of Eq. (2.52) would give us $x[m]\,\delta[n - m]$ on the right-hand side of Eq. (2.52). So the multiplication of the sequence $x[n]$ by the impulse picks out the one value from the sequence where the impulse has a nonzero value.

If we were to have an infinite many impulses each shifted by a different amount, then we can write the sequence $x[n]$ as a weighted shifted sum as shown in Eq. (2.53).

$$x[n] - \sum_{k=-\infty}^{\infty} x[k]\delta[n-k] \tag{2.53}$$

Equation (2.53) tells us that the right-hand side is a sum of infinite many sequences that are arranged in a very special manner. Each sequence has only one nonzero value, and these nonzero values occur in different locations. When we add all these sequences together, we get the original sequence $x[n]$. We will use this concept of $x[n]$ being made up of several impulses in the next section, where we determine the response to the system using superposition.

2.7.2
Response to a Linear Shift Invariant System

Now that we have the input represented as a sum of shifted impulse sequences, let us determine the response to a single impulse, we will call this the impulse response of the system. By definition then Eq. (2.54) shows us the impulse response of the system when the impulse is active at $n = k$.

$$y[n, k] = h[n, k] = T[\delta[n-k]] \tag{2.54}$$

Scaling this impulse by one of the samples from the input sequence will give us the response of that particular sample from the sequence. So the response to a scaled impulse can be written as shown in Eq. (2.55).

$$y[n, k] = x[k]h[n, k] = T[x[k]\delta[n-k]] \tag{2.55}$$

Equation (2.55) shows the response to an arbitrary weighted impulse occurring at $n = k$. What we want is the response to the entire sequence $x[n]$. In Eq. (2.53), we wrote the entire input sequence as a sum of weighted shifted sequences. Since the system that we are examining is a linear system, we can use superposition and hence the sum of individual responses is equal to the response of the sum of inputs. So, we can write the response of the system as shown in Eq. (2.56).

$$y[n, k] = x[n]h[n, k] = \sum_{k=-\infty}^{\infty} x[k]h[n, k] \tag{2.56}$$

In Eq. (2.56), the term $h[n,k]$ is known as the impulse response of the system. If the system that we are examining is a time-invariant system, then we can write $h[n,k]$ as $h[n-k]$, else this would not be true. If the system is not shift invariant, then we would have infinite impulse responses each for a unique value of k. So, we can write the response of a linear shift invariant system as shown in Eq. (2.57).

$$y[n] = T[x[k]\delta[n-k]] = \sum_{k=-\infty}^{\infty} x[k]h[n-k] \qquad (2.57)$$

The last summation is a very special summation and it is known as the convolution sum. It is a very interesting sum, and to complete the sum we first reverse the impulse response of the system, next we shift the reversed impulse response by k units, then we multiply the reversed shifted impulse response by the corresponding input samples from the input sequence, and finally we add all the multiplication results to get the response of the system at that instant in time. When we examine Eq. (2.57), we see that the relaxed linear time-invariant system is completely defined by $h[n]$, the response to $\delta[n]$ that is its impulse response. Another important property of the impulse response we saw in Section 2.4.5, where we used the impulse response to determine if the system is stable or not. When the absolute sum of the impulse response is finite, a relaxed shift-invariant linear system is a BIBO stable system.

2.8
The Convolution Sum

It is interesting how the convolution sum is computed. To understand all the steps involved in the computation, look at Figure 2.11; we begin the summation by first time reversing the impulse response of the system. This will give us $h[-n]$. Next, we shift $h[-n]$ one unit at a time. After each shift, we multiply $x[k]$ with $h[k-n]$ where they overlap. Finally, we add the result of the individual multiplications before we shift $h[-n]$ one more time.

In Figure 2.11, we have created a table to demonstrate how to complete the convolution sum. In step 1, the sequences $x[k]$ and $h[-n]$ are listed. Since there are only zero terms overlapping, the result of convolution is zero. This is $y[0]$. In step 2, we have shifted $h[-n]$ by one unit of time; this time there is one overlap, so we complete the product and the result of convolution is obtained by adding all the individual products. This is $y[1]$. This way we continue to shift $h[-n]$ by one unit every time and forming the product and then adding the individual products. With each shift, we get the next output from the linear shift-invariant system. This continues till all the $h[-n]$ are shifted so far that there is no overlap any more. The entire result is shown in the result column of Figure 2.11.

2.8.1
Properties of Convolution

Understanding how convolution operates will also give us an idea of how the various discrete time systems can be interconnected.

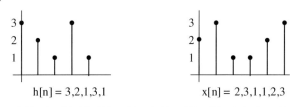

$$h[n] = 3,2,1,3,1 \qquad x[n] = 2,3,1,1,2,3$$

		2	3	1	1	2	3	Result	
Step 1	1,3,1,2,3	0	0	0	0	0	0	0	y[0]
Step 2	1,3,1,2	3	0	0	0	0	0	6	y[1]
Step 3	1,3,1	2	3	0	0	0	0	13	y[2]
Step 4	1,3	1	2	3	0	0	0	11	y[0]
Step 5	1	3	1	2	3	0	0	14	y[3]
Step 6		1	3	1	2	3	0	20	y[4]
Step 7		0	1	3	1	2	3	20	y[5]
Step 8		0	0	1	3	1	2	12	y[6]
Step 9		0	0	0	1	3	1	10	y[7]
Step 10		0	0	0	0	1	3	11	y[8]
Step 11		0	0	0	0	0	1	3	y[9]

Figure 2.11 Performing the convolution sum.

2.8.1.1 Commutative Law
In the convolution sum, we first flipped and shifted the impulse response of the system. If we interchange the meaning of the two sequences, $x[n]$ and $h[n]$, then we can write as shown in Eq. (2.58) and in Figure 2.12.

$$x[n] * h[n] = h[n] * x[n]$$

$$\sum_{k=-\infty}^{\infty} x[k]h[n-k] \qquad \text{let } k = n-k \qquad (2.58)$$

$$\sum_{k=-\infty}^{\infty} x[n-k]h[n-(n-k)] = \sum_{k=-\infty}^{\infty} h[k]x[n-k]$$

Equation (2.58) and Figure 2.12 show us that it does not matter which function we keep stationary and which function is first flipped and then shifted. Prove this to yourself by interchanging the places for the two functions $x[n]$ and $h[n]$ in Figure 2.11.

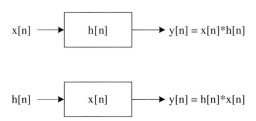

Figure 2.12 Convolution is commutative.

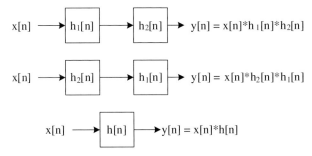

Figure 2.13 Convolution is associative.

2.8.1.2 Associative Law

What is the system response when $x[n]$ is the input to one system and the output of this first system is the input to the next system as shown in Figure 2.13. In this case, what is the relation of the output to the input? Is it possible to replace the two interconnected systems with just one composite system? What is the impulse response of the composite system? All these questions are answered by the associative law. Equation (2.59) also gives us details about the associative law.

$$[x[n] * h_1[n]] * h_2[n] = x[n] * [h_1[n] * h_2[n]] = x[n] * h[n] \tag{2.59}$$

From Eq. (2.59), we see that the composite system is $h[n] = h_1[n] * h_2[n]$. As we have seen in the associative law, either we can have $h_1[n]$ first or $h_2[n]$ first or we can just combine the two systems into one system.

2.8.1.3 Distributive Law

What is the system response when $x[n]$ is the input to two systems and the output of both the individual systems is then added together as shown in Figure 2.14. In this case, what is the relation of the output to the input? Is it possible to replace the two parallel systems with one composite system? These questions are answered by the distributive law. Equation (2.60) gives us the details of the distributive law. From Eq. (2.60), we see that the composite system is $h[n] = h_1[n] + h_2[n]$.

$$y[n] = x[n] * h_1[n] + x[n] * h_2[n] = x[n] * [h_1[n] + h_2[n]] = x[n] * h[n] \tag{2.60}$$

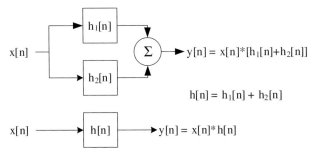

Figure 2.14 Convolution is distributive.

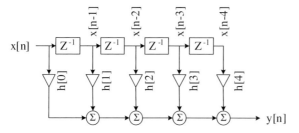

Figure 2.15 Implementing the convolution sum.

2.9
Systems Described by Difference Equations

In the previous section, we saw systems that are described by the convolution sum. We can describe any system, big or small, by the convolution sum by increasing or decreasing the limits of the summation. The convolution sum also suggests how the system can be implemented. To implement the system described by the convolution sum, we first form the product of the impulse response term and the input, then adding the various products gives us the output at that particular instant. To form the various products, we need to memorize the inputs and the impulse response terms. This is shown in Eq. (2.61) and in Figure 2.15.

$$y[n] = \sum_{k=0}^{4} h[k]x[n-k] \tag{2.61}$$

Figure 2.15 shows us how we would implement the convolution sum. In Figure 2.15, the boxes that represent delay are labeled as z^{-1}. The product is formed at the triangular boxes and the sum is performed in the circular summing blocks. The delay blocks also represent the memory that is required because the previous inputs have to be remembered.

The system shown in Eq. (2.61) is a finite impulse response system. The number of delay elements and hence the memory elements are finite and they equal the size of the convolution sum. As the size of the summation increases, the amount of memory required also increases.

Another important characteristic that you should notice is that in the system shown in Figure 2.15 the signal is always traveling in the forward direction. This system does not have any feedback in the block diagram. Systems like this are known as finite impulse response (FIR) systems. An important property of FIR systems is that we can determine any output at any time. If you need to know what is the output when $n = 257$, we can directly go and determine output 257 directly, we do not need to go through all the previous outputs from $n = 0$ to $n = 256$ to get $n = 257$.

There are other systems that have feedback built in the system representation. Such a system is shown in Figure 2.16. If we tried to implement the system shown in Figure 2.16 by using the convolution sum shown in Eq. (2.61), then we would need infinite numbers of delays and infinite amount of memory. Infinite delays and infinite memory is not feasible for practical systems that have to operate in real time. This means that we have to determine an

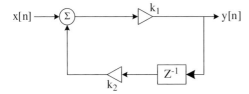

Figure 2.16 An infinite impulse response system.

alternative way to represent systems that have feedback. These systems are known as infinite impulse response (IIR) systems.

We represent IIR systems, like the one shown in Figure 2.16, with an equation like the one shown in Eq. (2.62).

$$y[n] = k_1 x[n] + k_1 k_2 y[n-1] \qquad (2.62)$$

Now all the information about all the previous inputs is contained in the term $y[n-1]$, so we do not have to memorize all the previous inputs. In the system described by Eq. (2.62), we must begin to evaluate the system from its beginning; so, if we need to determine, say, an output from the system at $n = 257$, then we have to begin with the output at $n = 0$, then 1, then 2, and so on till we arrive at the output at $n = 257$. This is because the system is a recursive system and the present output depends on the past output. We will also have to have an initial value of the past output at $n = 0$. This is known as the initial condition of the system. The equation that describes the system is known as a difference equation. When you were studying analog or continuous time systems, you described the system by a differential equation. When the system that you have is a discrete time systems, you describe it by a difference equation. Just as solving the differential equation gave us the response to the continuous time system, solving the differential equation will give us the response to the discrete time system.

2.9.1
Systems Described by Constant Coefficient Difference Equation

Here, we describe linear time-invariant systems that are described by a difference equation with constant coefficients. First, we will see what the response looks like and that will give us an idea of how to develop the solution to such systems in general. To begin the method of solution and show the important points of such systems, we begin with a simple second-order system described by a constant coefficient difference equation shown in Eq. (2.63) and Figure 2.17.

$$y[n] = x[n] + a_1 y[n-1] + a_2 y[n-2] \qquad (2.63)$$

In Figure 2.17 and Eq. (2.63), the multipliers a_1 and a_2 are constants. If we apply an input $x[n]$ for $n \geq 0$ to the system and we assume the existence of initial conditions $y[-1]$ and $y[-2]$, then we can calculate the output from the system. Suppose, we assume that we know the

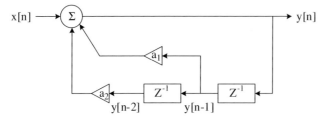

Figure 2.17 Block diagram of a system represented by a constant coefficient difference equation.

constants a_1 and a_2, then we can develop the response of the system shown as follows.

$$y[0] = x[0] + a_1 y[-1] + a_2 y[-2]$$

$$
\begin{aligned}
y[1] &= x[1] + a_1 y[0] + a_2 y[-1]\\
&= x[1] + a_1\{x[0] + a_1 y[-1] + a_2 y[-2]\} + a_2 y[-1]\\
&= x[1] + a_1 x[0] + \left(a_1^2 + a_2\right) y[-1] + a_1 a_2 y[-2]
\end{aligned}
$$

$$
\begin{aligned}
y[2] &= x[2] + a_1 y[1] + a_2 y[0]\\
&= x[2] + a_1\{x[1] + a_1 x[0] + \left(a_1^2 + a_2\right) y[-1] + a_1 a_2 y[-2]\}\\
&\quad + a_2\{x[0] + a_1 y[-1] + a_2 y[-2]\}\\
&= x[2] + a_1 x[1] + \left(a_1^2 + a_2\right) x[0] + \left(a_1^3 + 2a_1 a_2\right) y[-1] + \left(a_1^2 a_2 + a_2^2\right) y[-2]
\end{aligned}
$$

$$
\begin{aligned}
y[3] &= x[3] + a_1\{x[2] + a_1 x[1] + \left(a_1^2 + a_2\right) x[0] + \left(a_1^3 + 2a_1 a_2\right) y[-1] + \left(a_1^2 a_2 + a_2^2\right) y[-2]\}\\
&\quad + a_2\{x[1] + a_1 x[0] + \left(a_1^2 + a_2\right) y[-1] + a_1 a_2 y[-2]\}\\
&= x[3] + a_1 x[2] + \left(a_1^2 + a_2\right) x[1] + \left(a_1^3 + 2a_1 a_2\right) x[0] +\\
&\quad + \left(a_1^4 + 3a_1^2 a_2 + a_2^2\right) y[-1] + \left(a_1^3 a_2 + 2a_1 a_2^2\right) y[-2]
\end{aligned}
$$

We can continue this way to develop the solution to whatever sample number we want. When we examine the solution carefully, we see that the solution is of a very special nature. Notice that the first four terms in $y[3]$ are all dependent only on the input $x[n]$; this part of the solution is known as the forced solution or the zero state response. This part of the solution depends only on the input that is present in the system.

The last two terms in $y[3]$ depend only on the initial conditions that are present on the system. This is known as the natural response or the zero input solution. This part of the solution depends only on the initial condition that is present in the system. The total solution of the constant coefficient difference equation then is the sum of the two solutions as shown in Eq. (2.64).

$$y[n] = y_{zi}[n] + y_{zs}[n] \tag{2.64}$$

The total solution shown in Eq. (2.64) consists of the zero input response that entirely depends on the initial condition that is present in the system. If the system is completely relaxed than this part of the solution will be zero. The zero state response depends only on the input that is present on the system. This part of the solution is independent of the initial conditions that may be present when the input is applied. If the input is zero, then this solution will be zero. This then suggests to us how we can determine the two different solutions that make up the total solution. In general, we can write the solution to a system that is described by constant coefficient difference equation given in Eq. (2.65).

$$y[n] = -\sum_{k=1}^{N} a_k y[n-k] + \sum_{k=0}^{M} b_k x[n-k] \tag{2.65}$$

In Eq. (2.65), the upper limit N represents the order of the system. We need to remember N previous output values. The parameter M represents the number of past inputs that must be remembered to determine the complete solution. Remember that this is a linear system, so the principle of superposition applies both to the total solution and to the zero input response and the zero state response. So, if the initial condition is doubled, then the zero input response will double, in the same way as we double the input the zero state response will double.

2.9.2
Solution of Linear Constant Coefficient Difference Equation

We have seen in the previous section in Eq. (2.64) that the total solution is the sum of the zero input and the zero state response. Since the two solutions are independent of each other, we can determine the two solutions individually. At present, we are working in the time domain so we will determine a time domain solution. Later, after we study what a Z-transform is, we will approach this same subject and determine the solution using the Z-transform.

2.9.3
The Zero Input Response

This response is also known as the homogeneous response. To determine this response, we set the input to zero and our difference equation given in Eq. (2.65) can be written as shown in Eq. (2.66).

$$\sum_{k=0}^{N} a_k y[n-k] = y[n] \tag{2.66}$$

The solution of this equation is similar to the homogeneous solution of the analog differential equation. Here, just as in the analog case, we assume that the solution is in the form of the exponential as shown in Eq. (2.67).

$$y_{zi}[n] = \lambda^n \tag{2.67}$$

So the zero input or the homogeneous solution is as shown in Eq. (2.67). By substituting this solution in Eq. (2.66), we get a polynomial that is shown in Eq. (2.68) and this equation is known as the characteristic equation.

$$\lambda^{n-N}\left(\lambda^N + a_1\lambda^{N-1} + \cdots + a_{N-1}\lambda^1 + a_N\right) = 0 \tag{2.68}$$

Table 2.1 General form of the particular solution for different inputs.

Input signal $x[n]$	Particular solution $y_{zs}[n]$
A (constant)	K
AM^n	KM^n
An^M	$K_0 n^M + K_1 n^{M-1} + \ldots + K_M$
$A^n n^M$	$A^n (K_0 n^M + K_1 n^{M-1} + \ldots + K_M)$
$A \cos(\omega_0 n)$	$K_1 \cos(\omega_0 n) + K_2 \sin(\omega_0 n)$
$A \sin(\omega_0 n)$	

In general, the characteristic polynomial has N roots, $\lambda_1, \lambda_2, \ldots, \lambda_N$ and these roots can be either real or complex. If all the a_k in Eq. (2.66) are real, then the complex roots of the characteristic equation will occur in pairs. The most general solution of the characteristic equation, when there are no repeated roots in the characteristic equation, is shown in Eq. (2.69).

$$y_{zi}[n] = B_1 \lambda_1^n + B_2 \lambda_2^n + \cdots + B_N \lambda_N^n \tag{2.69}$$

In Eq. (2.69), the coefficients B_i are weighting coefficients. We will need to determine these coefficients and we will use the initial conditions later to determine these weighting constants.

2.9.4
The Zero State Response

The zero state response is the response to the system because of the input. The response because of the input has a form that is similar to the input function itself. The general forms of the zero state response for some typical inputs are given in Table 2.1. To use Table 2.1, look at the type of input that is present in the system and across from it you will find a corresponding particular or the zero state solution. Assume that the particular or the zero state solution is of that type and determine the constants in the particular solution. The general procedure is as follows. Once we assume the solution for the zero state response, we substitute the solution in the difference equation. This solution contains unknown weighting constants that are determined as shown in the example in Section 2.9.5.

2.9.5
Complete Solution of the Difference Equation

Determine the zero input and the zero state solution for the difference Eq. (2.70)

$$y[n] = 0.8y[n-1] - 0.1y[n-2] + 1.2x[n] \tag{2.70}$$

when the input $x[n] = (2.5)^n u[n]$ and zero everywhere with initial condition $y[1] = 1$ and $y[2] = 2$.

Zero input solution: We first determine the zero input solution. The characteristic equation of this system is given in Eq. (2.71).

$$\lambda^2 - 0.8\lambda + 0.1 = 0$$
$$\lambda_{1,2} \qquad = 0.1555 \quad \text{and} \quad 0.6445 \tag{2.71}$$

With this solution to the characteristic equation, the zero input solution is given by $y_{zi}[n] = B_1(0.1555)^n + B_2(0.6445)^n$. We will determine the constants B_1 and B_2 after we have determined the zero state solution.

Zero state solution: For this system, since the input function is of the type AM^n, the solution that we will assume is of the type KM^n. So $y_{zs}[n] = K(1.5)^n$, for $n \geq 0$. When we substitute this solution in the difference equation of the system, we get Eq. (2.72).

$$K(1.5)^n \mu[n] = 0.8K(1.5)^{n-1}\mu[n-1] - 0.1K(1.5)^{n-2}\mu[n-2] + 1.2(1.5)^n \mu[n] \tag{2.72}$$

Equation (2.72) is the zero state solution, so it holds for all sampling instants. So if we evaluate this solution at the value of $n = 2$, we get Eq. (2.73) after we have eliminated identical terms on both sides of the equation.

$$(1.5)^2 K = 0.8(1.5)K - 0.1K + 1.2(1.5)^2$$
$$2.25K = 1.2K - 0.1K + 2.7 \tag{2.73}$$

By solving Eq. (2.73), we get $K = 2.3478$. Now that we have determined the weighting constant, we know the zero state solution and this is $y_{zs}[n] = 2.3478(1.5)^n \mu[n]$.

Now we can determine the constants B_1 and B_2. We do this by using the initial conditions. To do this, we write the complete solution to the system as shown in Eq. (2.74).

$$y[n] = y_{zs}[n] + y_{zi}[n]$$
$$y[n] = B_1(0.1555)^n + B_2(0.6445)^n + 2.2478(1.5)^n \tag{2.74}$$

Since Eq. (2.74) is a solution to the difference equation, this must be a solution also at the instants of the initial conditions. In this example, we know that the output $y[n]$ at $n = -1$ is 1 and at $n = -2$ is 2, we can write Eq. (2.74) as shown in Eq. (2.75).

$$y[1] = 1 = B_1(0.1555)^1 + B_2(0.6445)^1 + 2.3478(1.5)^1$$
$$y[2] = 2 = B_1(0.1555)^2 + B_2(0.6445)^2 + 2.3478(1.5)^2 \tag{2.75}$$

In Eq. (2.75), we have two simultaneous equations in B_1 and B_2 that can be easily solved. By solving these two equations, we get $B_1 = 21.7789$ and $B_2 = -9.169$. Now that we have determined the constants, we are able to write the complete solution as shown in Eq. (2.76).

$$y[n] = y_{zi}[n] + y_{zs}[n]$$
$$y[n] = (21.7789(1.555)^n - 9.169(0.6445)^n + 2.3478(1.5)^n)\mu[n] \tag{2.76}$$

Sometimes, we run into special situations where the method above needs to be modified slightly. This occurs when the zero input solution and the zero state solution are the same.

To see when this comes about and how we modify the solution, consider a system described by the following difference equation:

$$y[n] = 0.5y[n-1] + 1.5y[n-2] + 2x[n]$$

when the input $x[n] = (1.0)^n u[n]$ and zero everywhere, with initial condition $y[1] = 1$ and $y[2] = 2$.

Zero input solution: The characteristic equation of this system is given in Eq. (2.71).

$$\lambda^2 - 0.5\lambda - 1.5 = 0$$

$$\lambda_{1,2} \qquad = -1.5 \quad \text{and} \quad 1.0 \tag{2.77}$$

With this solution to the characteristic equation, the zero input solution, if we follow the method described above, would be given by $y_{zi}[n] = B_1(-1.5)^n + B_2(1.0)^n$.

Zero state solution: For this system, since the input function is of the type AM^n, the solution that we will assume is of the type KM^n. So $y_{zs}[n] - K(1.0)^n$, for $n \geq 0$. Here, we find that the zero input solution and the zero state solution are identical. So the zero input solution already contains the zero state solution, hence the zero state solution is redundant. To have an independent zero state solution, we will modify the zero state solution and write it as shown in Eq. (2.78) and we will keep the zero input solution as it is.

$$y_{zs}[n] = Kn(1.0)^n u[n] \tag{2.78}$$

With the modification we will next determine the constant K. The value of K is determined in Eq. (2.79). In Eq. (2.79), we determine the constant K for $n = 2$.

$$(1.0)^n nK = 0.5(1.0)^{(n-1)}[n-1]K + 1.5(1.0)^{(n-2)}[n-2]K + 2(1.0)^n$$
$$(1.0)^2 2K = 0.5(1.0)^1 1K + 1.5(1.0)^0(0)K + 2(1.0)^2 \tag{2.79}$$
$$2K \qquad = 0.5K + 2$$

By solving Eq. (2.79), we get $K = 1.33$. With this value for the weighting function K, we can next determine the coefficients B_1 and B_2 using the initial conditions. To do this, we write the complete solution to the system as shown in Eq. (2.80).

$$y[n] = \{B_1(-1.5)^n + B_2(1.0)^n + 1.333n(1.0)^n\}u[n] \tag{2.80}$$

Next using the available initial conditions, as shown in Eq. (2.81), we determine the coefficients B_1 and B_2.

$$y[1] = 1 = B_1(-1.5)^1 + B_2(1.0)^1 + 1.333 * (1) * (1.0)^1$$
$$y[2] = 2 = B_1(-1.5)^2 + B_2(1.0)^2 + 1.333 * (2) * (1.0)^2 \tag{2.81}$$

By solving the two simultaneous equations in Eq. (2.81), we get $B_1 = -0.0888$ and $B_2 = -0.4666$. With these constants known, we can write the complete solution as shown in Eq. (2.82).

$$y[n] = y_{zi}[n] + y_{zs}[n]$$
$$y[n] = (-0.0888(-1.5)^n - 0.4666(1.0)^n + 1.3333n(1.0)^n)u[n] \tag{2.82}$$

2.10
Impulse Response to a System

Now that we know how to determine the response to a system described by a difference equation, we can determine the impulse response of a system. Earlier, we have identified the impulse response as $h[n]$, we have also defined the impulse response of a system as the response when the input to the system is an impulse at $n = 0$. This is shown in Figure 2.18. In IIR systems, the impulse response is simply the zero state solution to the system. To examine how to determine the impulse response, we will use the system described in Section 2.8.5 and in Eq. (2.70). The impulse response to a system is defined as the zero state response to the system when the system is in a relaxed state (initial conditions $= 0$) and the input is the impulse function $x[n] = \delta[n]$. Applying this definition to the system described in Eq. (2.70) and repeated here

$$y[n] = 0.8y[n-1] - 0.1y[n-2] + 1.2x[n]$$

we see that that the zero state response can be written as shown in Eq. (2.83).

$$y_{zs}[n] = (B_1(0.1555)^n + B_2(0.6445)^n)\mu[n] \tag{2.83}$$

Since the zero input solution is zero (since the system is relaxed) when $x[n] = \delta[n]$, the impulse response of the system is the zero state response and is known once we determine the coefficients B_1 and B_2, which is done using the given initial conditions, $y[1] = 1$ and $y[2] = 2$. By using these initial conditions, we get Eq. (2.84).

$$y_{zs}[1] = 1 = \left(B_1(0.1555)^1 + B_2(0.6445)^1\right)\mu[n]$$

$$y_{zs}[2] = 2 = \left(B_1(0.1555)^2 + B_2(0.6445)^2\right)\mu[n] \tag{2.84}$$

$$B_1 = -17.825 \quad and \quad B_2 = 5.8535$$

So with these constants determined, the impulse response is known and it is given in Eq. (2.85).

$$h[n] = [-17.825(0.1555)^n + 5.8535(0.6445)^n]\mu[n] \tag{2.85}$$

The impulse response of the system lasts for all values of $n \geq 0$. The value of n is unlimited and keeps on going to infinity. This is what we would expect as the response for the system is an infinite impulse response. The impulse response lasts for an infinite time since the

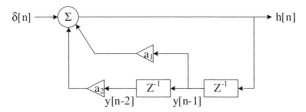

Figure 2.18 Block diagram of a system to determine the impulse response of a system.

system that we are examining is a recursive system and has feedback. From the method that we have used for the solution of the difference equation and the method used to determine the impulse response, we can write a general form of the impulse response as shown in Eq. (2.86).

$$h[n] = \sum_{k=0}^{M} B_k \lambda_k^n \qquad (2.86)$$

where λ_k are the roots of the characteristic equation of the difference equation that describes the IIR system and the coefficients B_k need to be determined from the initial condition. This solution is particularly useful since it allows us to determine if the system is stable or not. Remember, we saw earlier in Eq. (2.51) and repeat here the condition for stability given as

$$\left| \sum_{n=0}^{N} h[n] \right| < \sum_{n=0}^{N} |h[n]| < \infty$$

Now that we know a little more about the impulse response and what it is made of, we can get a better insight into the stability condition. We can rewrite the impulse response as shown in Eq. (2.87).

$$\sum_{n=0}^{N} |h[n]| = \sum_{n=0}^{N} \left| \sum_{k=0}^{M} |B_k \lambda_k^n| \right| \leq \sum_{k=0}^{M} |B_k| \left| \sum_{n=0}^{N} |\lambda_k^n| \right| \leq \infty \qquad (2.87)$$

In Eq. (2.87), the first summation is always finite since all the B_k are finite and there are only a finite number of them. On the other hand, the second summation could be infinite if any one of the $\lambda_k \geq 1$. The λ_k in Eq. (2.87) are the roots of the characteristic equation. So we can say that the system will be stable if and only if all the roots of the characteristic equation lie between -1 and $+1$ or must be less than 1 in magnitude (Figure 2.19). Since the roots of the characteristic equation can also be complex, we can say that the system will be stable if all the

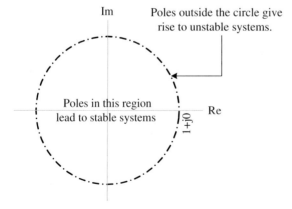

Figure 2.19 Regions showing where poles should be located for stable systems.

roots of the characteristic equation lie inside the unit circle. We will expand on this idea later when we study the Z-transform of a discrete system.

2.11
Examples of Some Discrete Time Systems

The operations that we saw on a sequence in Figure 2.2 can be considered as very simple discrete time systems. The gain, the upsampler, and the downsampler, even though they operate only on a single input sequence while the modulator and the adder operate on two input sequences, are all very simple discrete time systems. Some other discrete time systems are described here.

2.11.1
The Accumulator

A very simple system is one that adds all the inputs and the resulting sum to the output. This system is described in Eq. (2.88).

$$y[n] = \sum_{k=-\infty}^{n} x[k]$$
$$y[n] = \sum_{k=-\infty}^{n-1} x[k] + x[n] \tag{2.88}$$
$$y[n] = y[n-1] + x[n]$$

Equation (2.88) describes a system that is the sum of all the inputs that have arrived till the present instant. Writing the system as shown in the first equation in Eq. (2.88) requires us to remember an infinite number of past inputs. The same system can be written as the third equation in Eq. (2.88) that requires us to remember only the previous output and the present input.

2.11.2
The Moving Average Filter

Very often you are interested in the trend of the input. This is often what a stockbroker sees and tries to predict if a stock's price will rise or fall in the near future.[1] At other times, the input data is changing very slowly, but it is corrupted by noise that is random and changing relatively faster. An example of this may be the temperature of a certain location that is monitored and reported every 20 s. Systems like these can be represented as shown in Eq. (2.89).

$$y[n] = \frac{1}{M} \sum_{k=0}^{M-1} x[n-k] \tag{2.89}$$

---

1) The stockbroker can try but if the stock price is truly random, then nothing can predict a stock's price.

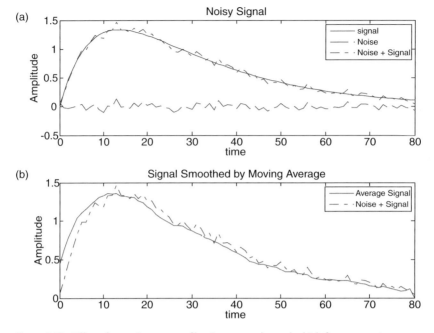

Figure 2.20 Effect of a moving average filter is to smooth out the high-frequency noise.

A two-dimensional moving average filter is often used in image processing; it helps us to smooth out the digital image that could have a lot of noise. In Figure 2.20, we see the effect of a moving average filter on a one-dimensional signal that is corrupted by noise. Using MATLAB we can demonstrate the effect of the moving average filter. Figure 2.20a shows the original signal, the noise signal, and the corrupted signal obtained by adding the noise to the original smooth signal. In Figure 2.20b, we can see the smoothing effect of the moving average filter.

Just as we were able to rewrite the expression for the accumulator in Eq. (2.88) so that we did not need a lot of memory elements, it is possible to rewrite the expression of the moving average filter so that we can use the result of the previous output. This alternative expression for the moving average filter is shown in Eq. (2.90).

$$y[n] = y[n-1] + \frac{1}{M}(x[n]-x[n-M]) \tag{2.90}$$

The expression for the moving average filter in Eq. (2.90) requires only two additions and one division, while the expression for the moving average filter in Eq. (2.89) requires $M-1$ additions and one division. In general, we try to have the least number of operations in the discrete time system since each operation requires time in a digital computer. More operations imply more time and hence the system operates slower. You are asked to show that Eq. (2.90) is the same as Eq. (2.90) in the chapter problems. There is also a hint provided there to steer you in the right direction.

2.11.3
Median Filter

The median value of a sequence of numbers is a value such that half the numbers are greater than the median value and the other half are less than the median value. In the same way, a median filter is a filter that assigns a value to the sample from $(2K + 1)$ adjacent samples in such a way that half the sample values are more and the other half are less than the value of the current sample. So a median value is selected by arranging all the $(2K + 1)$ values in increasing (decreasing) order and then choosing the middle value. To implement a median value filter, we define a sliding window, just like we did in the moving average filter. Then, we arrange the samples within the window in an increasing order and choose the central value. This is shown in Eq. (2.91). In Eq. (2.91), the shaded area represents the sliding filter.

$$\mathrm{med}(2, 4, 1, 5, 7, 3, 1, 6, 3, 5, 7, 3) = 5 \tag{2.91}$$

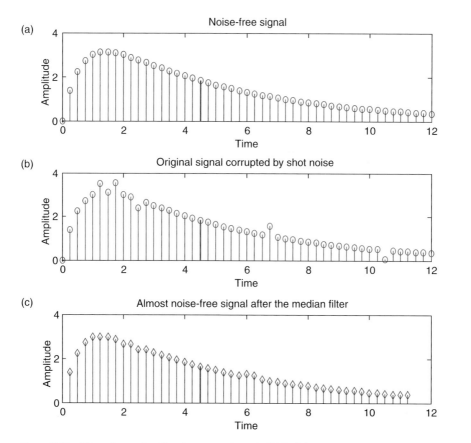

Figure 2.21 Effect of a median filter is to smooth out individual shot noise.

Within the filter the five values, when arranged in order, are [1, 3, 5, 6, 7] and the center value will be 5, so the median value is 5. The median filter is most often used in denoising a digital image. The use of the median filter is demonstrated in Figure 2.21.

Figure 2.21a shows the signal that has no noise associated with it. In Figure 2.21b, we see the signal after it has been corrupted with shot noise. Shot noise is noise with high amplitude, occurring very infrequently. It resembles impulses placed at random locations. In Figure 2.21b, we see only a few samples that are corrupted with this shot noise. In Figure 2.21c, we see the signal after it is passed through a median filter that determines the median value from five samples.

2.11.4
Linear Interpolator

This discrete time system tries to estimate the values of the signal in between the actual samples. To interpolate the values at the in-between locations of a signal, we first create a placeholder for the value to be interpolated. We do this by passing the signal through an upsampler. An upsampler is shown in Figures 2.2 and 2.4a. Figure 2.22a shows the signal that has to be interpolated. Figure 2.22b shows the signal after it is passed through an upsampler. Now that we have created placeholders with zero-value samples, we will interpolate the values at these zero-value sample locations. In Figure 2.22c, we have

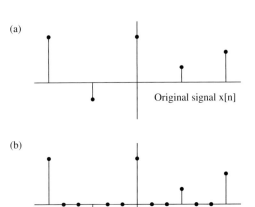

(a)

Original signal x[n]

(b)

Original signal upsampled

(c)

Original signal interpolated

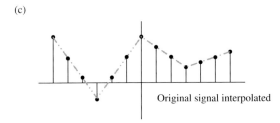

Figure 2.22 Signal after it is interpolated.

interpolated the signal by 3; other values are also possible; for example, interpolating a signal by 2 is done by using Eq. (2.92).

$$y[n] = x_u[n] + \frac{1}{2}[x_u[n-1] + x_u[n+1]] \tag{2.92}$$

In Eq. (2.92), $x_u[n]$ is the signal to be interpolated, $x_u[n+1]$ and $x_u[n-1]$ are the samples in the original signal. Interpolating as shown in Figure 2.22 requires a modified equation since there are two zero values to be interpolated between two sample values. This is done by Eq. (2.93) as follows:

$$y_1[n] = x[n] + \frac{2}{3}x[n-1] + \frac{1}{3}x[n+1]$$

$$y_2[n] = x[n] + \frac{1}{3}x[n-1] + \frac{2}{3}x[n+1] \tag{2.93}$$

In Eq. (2.93), $y_1[n]$ is the interpolated value that is close to the $x[n-1]$ sample point, while $y_2[n]$ is the interpolated value that is close to the $x[n+1]$ sample point.

2.12
Chapter Summary

We began this chapter with the study of how an analog or a continuous time signal is sampled and converted to a discrete time signal. We saw that the impulse function sampled the analog signal and made it into a discrete time signal. Next, we saw that we could activate the impulse anywhere along the time axis by changing the argument of the function. The function is activated when the argument of the function is zero.

The sampled signals that we got can have various operations performed on them to get different sampled signals. The operations that we studied were chiefly those that we encounter in the study of discrete time systems. Among them, we studied the time shift, addition, scaling, upsampling, and downsampling. The time shift, addition, and scaling in general do not affect the frequency content of the signal, but the upsampling and the downsampling do affect the frequency content of the signal as they change the number of sample points in a sequence.

The discrete signals have symmetry of many different types. In this chapter, we studied the odd and the even symmetry. A signal could be conjugate symmetric or conjugate antisymmetric. This is similar to odd and even symmetry for complex signals.

The signals can be classified as either energy or as power signals. If a signal has infinite energy, then it is most probably a power signal. An infinite signal that is a power signal has an average value that is not zero over the entire timescale. We extended this idea of energy and power signals to periodic signals also.

Frequency of a signal is a concept that we discussed next. In this discussion, we paid particular attention to the frequency of a discrete signal. Here, we found that the frequency of a continuous time signal and its sampled version do not have to be same. The sampled signal can have a higher or a lower frequency. It is also possible that the sampled signal may turn out to be periodic and hence have no frequency.

We use these signals in systems, so we next described the various characteristics that we expect to see in a discrete time system. The characteristics we discussed consisted of static versus dynamic, time variant or time invariant, linear versus nonlinear, causal versus noncausal, and bounded input–bounded output stable. Here, we also saw how we determine if a system is linear (by using the superposition principle). We also determined a test for BIBO stability (system impulse response is absolutely summable).

To analyze the shift-invariant systems, we first described the systems by the use of difference equations. By solving the difference equations, we were able to determine the response of the shift-invariant systems. The solution consisted of two separate and distinct parts. We called them the zero input and the zero state response.

The analysis of a shift invariant system is greatly simplified by understanding the convolution and the convolution sum. We examined how the convolution of a discrete sequence with a discrete time system is performed. Convolution as an operation has several characteristics. These properties of the convolution sum were discussed.

Finally, the chapter was closed with several examples of discrete systems. All these systems tried to show where and how some discrete systems can be and are used.

3
Discrete Time Systems in the Frequency Domain

3.1
Introduction

In Chapter 1, we introduced a sequence as a sum of weighted shifted impulses. With this representation of the sequence, we were able to determine the output of the sequence through an LTI system by a convolution sum. To do this, in Chapter 1 we worked entirely in the time domain. We did not have any idea about what the frequency content of either the input or the output was. Sometimes, we are very interested in knowing the frequency characteristics of the system. In this chapter, we discuss the frequency representation of a sequence and the frequency response of the system to the sequence.

We begin this chapter with first relating an analog signal to its sampled representation. Here, we will first see what the relation between the analog signal and the sampling frequency has to be so that all the information content in the analog signal is not compromised.

As the chapter develops, we will see that the frequency representation of the sequence depends on the sequence, the sequence can be periodic or aperiodic, the sequence can be finite or infinite. The various sequences will give rise to slightly different frequency domain representations.

We will make extensive use of MATLAB to demonstrate the frequency domain properties and how the frequency domain and the time domain are just two different ways of looking at the same information.

3.2
Continuous Time Fourier Transform

We first review the continuous time Fourier transform. This transform will tell us the relation between the time domain and the frequency domain properties of the analog signal. The review will also provide us with a basis to compare the properties of the continuous time and the discrete time signals.

Digital Filters: Theory, Application and Design of Modern Filters, First Edition. Rajiv J. Kapadia.
© 2012 Wiley-VCH Verlag GmbH & Co. KGaA. Published 2012 by Wiley-VCH Verlag GmbH & Co. KGaA.

3.2.1
Definition of the Fourier Transform

The continuous time Fourier transform and its inverse are given in Eq. (3.1).

$$x_a(t) = \frac{1}{2\pi} \int_{-\infty}^{+\infty} X_a(j\Omega) e^{j\Omega t} d\Omega$$

$$X_a(j\Omega) = \int_{-\infty}^{+\infty} x_a(t) e^{-j\Omega t} dt$$

(3.1)

The two equations in Eq. (3.1) show the time domain and the frequency domain representation of the continuous time signal $x(t)$. In the Fourier transform, we can interpret $X_a(j\Omega)$ as the amplitude of exponentials $e^{j\Omega t} d\Omega$. There are infinitely many exponentials since the transform of a continuous time signal is a continuous time function so the exponentials extend from $+\infty$ to $-\infty$.

In general, for the continuous time Fourier transform to exist, it has to satisfy the Dirichlet conditions. The conditions written in a simple form state that

a) the time domain function will have only a finite number of maxima or minima in a finite interval of time;
b) all the maxima and the minima will be finite;
c) the signal has only a finite number of discontinuities;
d) the absolute value of the signal $x(t)$ is integrable. $\int_{-\infty}^{+\infty} |x_a(t)| dt < \infty$.

If a signal $x_a(t)$ satisfies the Dirichlet conditions, then the Fourier transform equation given in Eq. (3.1) converges and the Fourier transform exists. In general, the Fourier transform of a time domain signal, if it exists, can be separated in its real and imaginary parts or into its magnitude and phase as shown in Example 3.1.

Example 3.1

Determine the Fourier transform of a real signal $x_a(t) = e^{-\alpha t} u(t)$ when $\alpha > 0$. Separate the Fourier transform into its real and its imaginary parts, represent the Fourier transform as its magnitude and phase.

Solution 3.1: Since this function satisfies all the Dirichlet conditions (verify on your own that the integral of the absolute value is finite) the Fourier transform exists. We determine the transform in Eq. (3.2).

$$X_a(j\Omega) = \int_{-\infty}^{+\infty} x_a(t) e^{-j\Omega t} dt$$

$$X_a(j\Omega) = \int_{-\infty}^{+\infty} e^{-\alpha t} u(t) e^{-j\Omega t} dt$$

$$X_a(j\Omega) = \int_0^{+\infty} e^{-(\alpha + j\Omega)t} dt$$

(3.2)

$$X_a(j\Omega) = \frac{-1}{\alpha + j\Omega} e^{-\alpha + j\Omega t} \Big|_0^\infty = \frac{1}{\alpha + j\Omega}$$

Next, we determine the real and the imaginary parts of the Fourier transform as shown in Eq. (3.3).

$$\frac{1}{\alpha + j\Omega} = \frac{\alpha - j\Omega}{\alpha^2 + \Omega^2}$$

$$\text{Re}\{X_a(j\Omega)\} = \frac{\alpha}{\alpha^2 + \Omega^2}; \quad \text{Im}\{X_a(j\Omega)\} = \frac{-j\Omega}{\alpha^2 + \Omega^2} \tag{3.3}$$

The magnitude and the phase functions can also be determined similarly as shown in the following equation:

$$\frac{1}{\alpha + j\Omega} = \underbrace{\frac{1}{\sqrt{\alpha^2 + \Omega^2}}}_{\text{Magnitude}} \underbrace{e^{-j \tan^{-1}(\Omega/\alpha)}}_{\text{Phase}} \tag{3.4}$$

3.3
Sampling an Analog Signal

In Chapter 1, we saw some different discrete time signals and saw how to work with them. Often, there are questions as to where these signals come from. Signals such as speech, biological signals, seismic signals, and many other types of signals are all analog signals. To process these signals, we need to first convert them to discrete signals. The process of converting an analog signal to discrete signal is known as *analog-to-digital(A/D) conversion*.

The process of converting an analog signal to a digital is explained in Figure 3.1. To convert an analog signal to a discrete signal, we must first hold the signal so that it does not change while the conversion takes place. The conversion is very quick but it is not instantaneous, so the analog level at the instant when the conversion starts must be maintained so that it is converted correctly. This is done by a device known as the sample and hold device. The next process is the process of quantization. This process is necessary because the analog signal has infinite values, while the digital signal will have only a finite number of values. The quantizer represents the present sampled value to one of the levels that is closest to the present sampled value. It will be this value that will be converted to a digital code that we will be using. The last step is that of coding the quantized value, this is just converting the quantized value to its binary representation. In a practical A/D converter, all the three processes are built into one integrated circuit, so you may not see the three processes taking place separately. They are all contained into a single circuit. The sampling of the analog signal is usually done periodically, the sampling period is T_s, and since we are holding the analog signal at a particular instant, we can make the assumption that this is impulse sampling.

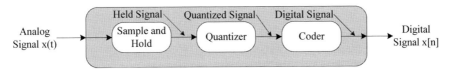

Figure 3.1 Various processes involved in converting an analog signal to a digital signal.

3.4
Discrete Time Fourier Transform

In Section 3.1, we saw the Fourier transform of a continuous time infinite signal. If we sample this signal using periodic sampling, then we will have a discrete signal that is infinite in length. In this section, we examine what is the Fourier transform of this infinite discrete time sequence.

3.4.1
Definition of the Discrete Time Fourier Transform

The discrete time Fourier transform (DTFT) of a sequence is a representation of the sequence in terms of the complex exponentials $e^{j\omega n}$. Here, we have used ω to represent the real frequency. There are only a finite number of complex exponentials in contrast to the infinite number of complex exponentials for the continuous time Fourier transform that we saw in Eq. (3.1). The DTFT of a sequence is given in Eq. (3.5).

$$X(e^{j\omega}) = \sum_{n=-\infty}^{\infty} x[n]e^{-j\omega n} \tag{3.5}$$

The transform in Eq. (3.5) exists if the infinite series $x[n]e^{-j\omega n}$ converges. This series will converge if the sequence $x[n]$ is absolutely summable. This is shown in Eq. (3.6). In Eq. (3.6), we have made use of the fact that $\left|e^{-j\omega n}\right|$ is one for all values of ω. Thus, the existence of the DTFT is guaranteed if the sequence $x[n]$ is absolutely summable.

$$\left|X(e^{j\omega})\right| = \left|\sum_{n=-\infty}^{\infty} x[n]e^{-j\omega n}\right| \leq \sum_{n=-\infty}^{\infty} |x[n]|\left|e^{-j\omega n}\right|$$

$$\left|X(e^{j\omega})\right| \leq \sum_{n=-\infty}^{\infty} |x[n]| \leq \infty \tag{3.6}$$

Computing the DTFT is similar to computing the Fourier transform and it is demonstrated in Example 3.2.

Example 3.2

Determine the DTFT of a real signal $x_a(t) = e^{-\alpha n T}u(t)$, when $\alpha > 0$. Separate the DTFT into its real and its imaginary parts, and also represent the DTFT as its magnitude and phase.
   Solution 3.2: Since this function is absolutely summable (verify on your own that the function is absolutely summable), the DTFT exists. We determine the transform in Eq. (3.7)

$$X[e^{j\omega}] = \sum_{n=-\infty}^{\infty} x[n]e^{-j\omega n} = \sum_{n=-\infty}^{\infty} e^{-\alpha n T}u[n]e^{-j\omega n}$$

$$X[e^{j\omega}] = \sum_{n=0}^{\infty} e^{-\alpha n T}e^{-j\omega n} = \sum_{n=0}^{\infty} e^{-(\alpha T + j\omega)n} = \frac{1}{1 - e^{-(\alpha T + j\omega)}} \tag{3.7}$$

Next, we determine the real and the imaginary parts of the DTFT as shown in Eq. (3.8)

$$\frac{1}{1-e^{-aT}e^{-j\omega}} = \frac{1}{1-e^{-aT}\cos(\omega)+je^{-aT}\sin(\omega)}$$

$$\frac{1}{1-e^{-aT}e^{-j\omega}} = \frac{1-e^{-aT}\cos(\omega)-je^{-aT}\sin(\omega)}{[1-e^{-aT}\cos(\omega)]^2 + [e^{-aT}\sin(\omega)]^2}$$

$$\text{Re}\{X_a(j\Omega)\} = \frac{1-e^{-aT}\cos(\omega)}{[1-e^{-aT}\cos(\omega)]^2 + [e^{-aT}\sin(\omega)]^2}$$
(3.8)

$$\text{Im}\{X_a(j\Omega)\} = \frac{-je^{-aT}\sin(\omega)}{[1-e^{-aT}\cos(\omega)]^2 + [e^{-aT}\sin(\omega)]^2}$$

The magnitude and the phase functions can also be determined similarly as shown in the following equation:

$$\frac{1}{1-e^{-aT}e^{-j\omega}} = \underbrace{\frac{1}{\sqrt{\left[[1-e^{-aT}\cos(\omega)]^2 + [e^{-aT}\sin(\omega)]^2\right]}}}_{\text{Magnitude}} \underbrace{\tan^{-1}\left[\frac{-e^{aT}\sin(\omega)}{[1-e^{aT}\cos(\omega)]}\right]}_{\text{Phase}}$$
(3.9)

Now that we know what the DTFT is and how to determine the DTFT of a continuous time signal, we can look at it in more detail. First, note that the DTFT is periodic and the period is 2π. This can be easily shown by substituting $\omega + 2\pi$ instead of ω in Eq. (3.5) and it is shown in Eq. (3.10)

$$X\left(e^{j(\omega+2\pi)}\right) = \sum_{n=-\infty}^{\infty} x[n]e^{-j(\omega+2\pi)n} = \sum_{n=-\infty}^{\infty} x[n]e^{-j\omega n}e^{-j2\pi n}$$
(3.10)

$$X\left(e^{j(\omega+2\pi)}\right) = \sum_{n=-\infty}^{\infty} x[n]e^{-j\omega n} = X\left(e^{j\omega}\right)$$

So the DTFT of a discrete sequence is a continuous function that is periodic with a period of 2π. We can think of $x[n]$ as the Fourier series coefficients of the continuous, periodic function $X(e^{j\omega})$. So the relation between $x[n]$ and $X(e^{j\omega})$ can be written as the Fourier series of $X(e^{j\omega})$ and this will represent the inverse DTFT as shown in Eq. (3.11).

$$x[n] = \frac{1}{2\pi} \int_{-\pi}^{\pi} X\left(e^{j\omega}\right)e^{j\omega n}d\omega$$
(3.11)

To verify that Eqs. (3.5) and (3.11) represent a transform pair, we can substitute one into the other and verify that we end up with an identity as shown in Eq. (3.12)

$$x[n] = \frac{1}{2\pi} \int_{-\pi}^{\pi} X(e^{j\omega}) e^{j\omega n} d\omega$$

$$x[n] = \frac{1}{2\pi} \int_{-\pi}^{\pi} \left[\sum_{k=-\infty}^{\infty} x[k] e^{-j\omega k} \right] e^{j\omega n} d\omega$$

$$x[n] = \sum_{k=-\infty}^{\infty} x[k] \left[\frac{1}{2\pi} \int_{-\pi}^{\pi} e^{-j\omega k} e^{j\omega n} d\omega \right] \qquad (3.12)$$

$$x[n] = \sum_{k=-\infty}^{\infty} x[k] \left[\frac{1}{2\pi} \int_{-\pi}^{\pi} e^{-j\omega(k-n)} d\omega \right]$$

In Eq. (3.12), the last equation has a very special integral that we evaluate in Eq. (3.13).

$$\frac{1}{2\pi} \int_{-\pi}^{\pi} e^{-j\omega(k-n)} d\omega = \frac{1}{2\pi} \left[\frac{e^{-j\omega(k-n)}}{-j(k-n)} \right]_{-\pi}^{\pi} = \frac{\sin(\pi(k-n))}{\pi(k-n)} \qquad (3.13)$$

So the last part of Eq. (3.12) can be modified by substituting from Eq. (3.13), but before we do that let us examine what Eq. (3.13) represents. We first note that the variable k changes from $-\infty$ to $+\infty$, while the variable n is fixed. So, we have two unique cases, one when $n = k$ and the other when $n \neq k$. These two cases are evaluated in Eq. (3.14).

$$\frac{\sin(\pi(k-n))}{\pi(k-n)} = \text{sinc}(\pi(n-k)) = \begin{cases} \delta(n-k) & n = k \\ 0 & n \neq k \end{cases} \qquad (3.14)$$

We get the last equality of Eq. (3.14) by noting that the sinc function is 1 at $n = k$ and the unity value is attained as $k \rightarrow n$. With all these substitutions, the last part of Eq. (3.12) becomes

$$x[n] = \sum_{k=-\infty}^{\infty} x[k] \left[\frac{1}{2\pi} \int_{-\pi}^{\pi} e^{-j\omega(k-n)} d\omega \right] = \sum_{k=-\infty}^{\infty} x[k] \delta[n-k] = x[n] \qquad (3.15)$$

Equation (3.15) tells us that the two equations in Eq. (3.5) and in Eq. (3.11) indeed represent DTFT equations.

3.4.2
Properties of the DTFT

The DTFT has some very basic properties that we have seen in other transforms.

3.4.2.1 Linearity

The DTFT is a linear transform. This implies that if $x_1[n] \Leftrightarrow X_1(e^{j\omega})$ and $x_2[n] \Leftrightarrow X_2(e^{j\omega})$ are DTFT pairs, then the DTFT of $(x_1[n] + x_2[n])$ is $(X_1(e^{j\omega}) + X_2(e^{j\omega}))$. This is proved in Eq. (3.16).

$$\mathcal{D}(x_1[n] + x_2[n]) = \sum_{n=-\infty}^{\infty} (x_1[n] + x_2[n])e^{-j\omega n}$$

$$= \sum_{n=-\infty}^{\infty} x_1[n]e^{-j\omega n} + \sum_{n=-\infty}^{\infty} x_2[n]e^{-j\omega n} \qquad (3.16)$$

$$= X_1(e^{j\omega}) + X_2(e^{j\omega})$$

3.4.2.2 Time Reversal

This property tells us how the DTFT is altered if we look at the time domain sequence in a reverse order. This property is proved in Eq. (3.17).

$$\mathcal{D}(x[-n]) = \sum_{n=-\infty}^{\infty} (x[-n])e^{-j\omega n} \qquad \text{let} -n = m$$

$$= \sum_{m=\infty}^{-\infty} x[m]e^{j\omega m} \qquad (3.17)$$

$$= \sum_{m=\infty}^{-\infty} x[m]e^{-j(-\omega)m} = X(e^{-j\omega})$$

To understand how we interpret this property, consider Example 3.3.

Example 3.3

We determined the DTFT of a real signal $x[n] = x_a(t) = e^{-\alpha nT}u(t)$ when $\alpha > 0$ in Example 3.2. Here, we will determine the DTFT of the time-reversed sequence $x[-n]$.

Solution 3.3: Using the definition of the DTFT, we can determine the DTFT of the time-reversed sequence as shown in the following Eq. (3.18):

$$\mathcal{D}[x[-n]] = \sum_{n=-\infty}^{\infty} x[-n]e^{-j\omega n} = \sum_{n=-\infty}^{\infty} e^{-\alpha(-n)T}u(-nT)e^{-j\omega n}$$

$$= \sum_{n=-\infty}^{0} e^{-\alpha(-n)T}e^{-j\omega n} \qquad \text{let } n = -m$$

$$= \sum_{m=0}^{\infty} e^{-\alpha(m)T}e^{j\omega m} = \sum_{m=0}^{\infty} e^{-\alpha(m)T}e^{-j(-\omega)m} \qquad (3.18)$$

$$= \sum_{m=0}^{\infty} e^{-m(\alpha T - j\omega)} = \frac{1}{1 - e^{-(\alpha T - j\omega)}} = X(e^{-j\omega})$$

We will get a much better idea of what happens when we time reverse the time domain sequence if we separate the DTFT in its magnitude and phase as we did in Example 3.2. So we separate the DTFT of the time-reversed sequence in Eq. (3.19).

$$\frac{1}{1-e^{-aT}e^{j\omega}} = \frac{1}{1-e^{-aT}(\cos(\omega)+j\sin(\omega))}$$

$$\frac{1}{1-e^{-aT}e^{j\omega}} = \underbrace{\frac{1}{\sqrt{\left[1-e^{-aT}\cos(\omega)\right]^2 + \left[e^{-aT}\sin(\omega)\right]^2}}}_{\text{Magnitude}} \underbrace{\tan^{-1}\left[\frac{e^{aT}\sin(\omega)}{\left[1-e^{aT}\cos(\omega)\right]}\right]}_{\text{Phase}}$$

(3.19)

When we compare Eq. (3.19) representing the DTFT of the time-reversed sequence with Eq. (3.9), the DTFT of the original sequence, we see that the magnitude function is the same. The magnitude does not change. The phase function is the only part that changes. The phase of the time-reversed function is negative of the phase of the original function.

3.4.2.3 Shift in the Time Domain

This property tells us how the DTFT is altered if we keep the time domain sequence as it is but shift it to a different time origin. This property is proved in Eq. (3.20) as follows:

$$\mathcal{D}(x[n-k]) = \sum_{n=-\infty}^{\infty} x[n-k]e^{-j\omega n}$$

$$= \sum_{m=-\infty}^{\infty} x[m]e^{-j\omega(m+k)}$$

(3.20)

$$= e^{-j\omega k}\sum_{m=-\infty}^{\infty} x[m]e^{-j\omega m}$$

$$= e^{-j\omega k}X\left(e^{j\omega}\right)$$

The last part of Eq. (3.20) tells us that the DTFT of the shifted time domain sequence undergoes an extra phase shift compared to the original sequence. This rotation of the phase of the DTFT is demonstrated in Figure 3.2, where we take a periodic signal and determine its DTFT and then shift the signal by seven samples and determine its DTFT. As you can see in Figure 3.2, the two magnitude plots are identical, while there is a change in the phase plot. The MATLAB script to draw the plot is given below and the plot is included in Figure 3.2.

MATLAB 3.1

```
n = 0:59; %Define a limit on number of samples.
x = cos(7*pi*n/20)+cos(11*pi*n/20); %Define a sampled signal in
%time domain.
x2 =[ x(53:60) x(1:52)] ; %Shift the signal by 7 samples.
X = fft(x); %Determine the Fourier Transform of the
X2 = fft(x2); %original and the shifted signals.
```

```
subplot(411);stem(n,abs(X)); %Plot the Magnitude Spectrum.
title('Magnitude plot original signal'); %of the original signal
subplot(412);stem(n,abs(X2)); %and of the shifted signal.
title('Magnitude plot shifted signal');
subplot(413);stem(n,angle(X)); %Plot the Phase Spectrum.
title('Phase plot original signal'); %of the original signal
subplot(414);stem(n,angle(X2)); %and of the shifted signal.
title('Phase plot shifted signal');
```

Example 3.4

We determined the DTFT of a real signal $x[n] = x_a(t) = e^{-anT}u(t)$ when $\alpha > 0$ in Example 3.2. Here, we will determine the DTFT of the time-shifted sequence $x[n - n_0]$.

Solution 3.4: By using the definition of the DTFT, we can determine the DTFT of the time-shifted sequence as shown in Eq. (3.21).

$$\mathcal{D}(x[n-n_0]) = \sum_{n=-\infty}^{\infty} x[n-n_0]e^{-j\omega n}$$

$$\mathcal{D}(x[n-n_0]) = e^{-j\omega n_0}X(e^{j\omega}) = \frac{e^{-j\omega n_0}}{1-e^{-(\alpha T+j\omega)}}$$

$$(3.21)$$

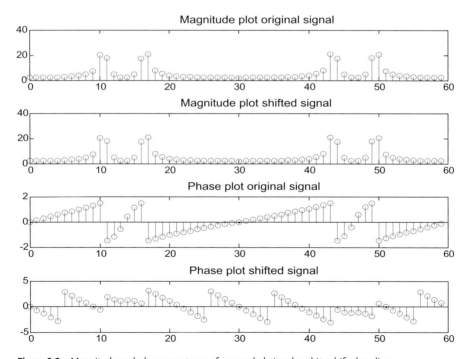

Figure 3.2 Magnitude and phase spectrum of a sampled signal and its shifted replica.

And now from Eq. (3.21) we can say that the magnitude response of the shifted sequence is unaltered; however, the phase undergoes a rotation if the amount of the shift specifically the phase of $e^{-j\omega n_0}$ is added to every component of the DTFT.

3.4.2.4 Shift in the Frequency Domain

When the DTFT of a sequence is shifted to a different frequency region, it affects the time domain sequence in a manner that is similar to that seen in Section 3.3.1.3. This is shown in Eq. (3.22).

$$\mathcal{D}\left(e^{j\omega_0 n}x[n]\right) = \sum_{n=-\infty}^{\infty} e^{j\omega_0 n}x[n]e^{-j\omega n} = \sum_{n=-\infty}^{\infty} x[n]e^{-j(\omega-\omega_0)n} = X(\omega-\omega_0) \qquad (3.22)$$

Equation (3.22) tells us that shifting the entire spectrum of the DTFT by an amount ω_0 causes the time domain sequence to undergo a rotation (phase gets added to the time domain sequence). Remember the shape of the frequency spectrum is not altered, it is only shifted.

Relation between the Time and the Frequency Domain When we compare the two properties, shift in time and shift in frequency, we see a rather special relation between the time domain and the frequency domain. The relation can be summed up as follows:

> When you perform a certain action on the sequence in the time domain, it will have some effect on the frequency domain. Now, if you were to perform the same action in the frequency domain as we performed in the time domain (shifting in the two properties above), then the effect on the other domain will be very similar (rotation in the two properties above).

3.4.2.5 Multiplication by a Linear Ramp

When the time domain sequence is multiplied by a linear ramp, the DTFT of the modified function is the derivative of the original time domain sequence; this is shown in Eq. (3.23). In the first equation in Eq. (3.23), we determine the derivative of the DTFT. We use this result in the second equation in Eq. (3.23) to get the result that we want.

$$\frac{dX\left(e^{j\omega}\right)}{d\omega} = \frac{d\sum_{n=-\infty}^{\infty} x[n]e^{-j\omega n}}{d\omega} = \sum_{n=-\infty}^{\infty} (-jn)x[n]e^{-j\omega n} = -j\sum_{n=-\infty}^{\infty} nx[n]e^{-j\omega n}$$

$$\mathcal{D}(nx[n]) = \sum_{n=-\infty}^{\infty} nx[n]e^{-j\omega n} = \frac{-1}{j}\frac{dX\left(e^{j\omega}\right)}{d\omega} = j\frac{dX\left(e^{j\omega}\right)}{d\omega} \qquad (3.23)$$

Example 3.5

We determined the DTFT of a real signal $x[n] = x_a(t) = e^{-\alpha n T}u(t)$ when $\alpha > 0$ in Example 3.2. Here, we will determine the DTFT of the modified sequence $(n + 5)x[n]$.

Solution 3.5: To determine the required DTFT, we will first rewrite the modified function as a sum of two different functions, $(n+5)x[n] = \{nx[n] + 5x[n]\}$. Next, using the properties that we have just seen we will be able to write the DTFT of the modified function as shown in Eq. (3.24).

$$\mathcal{D}((n+5)x[n]) = D(nx[n]) + D(5x[n])$$

$$\mathcal{D}((n+5)x[n]) = j\frac{d(X(e^{j\omega}))}{d\omega} + 5X(e^{j\omega}) \tag{3.24}$$

Since we know $X(e^{j\omega})$ given in Eq. (3.7), we are able to write the DTFT of the two individual terms of the modified sequence as shown in Eq. (3.25), where we have first used the multiplication by a linear ramp property and then for the second term we have used the linearity property.

$$j\frac{d(X(e^{j\omega}))}{d\omega} = j\frac{d(1/(1-e^{-(\alpha T+j\omega)}))}{d\omega} = \frac{e^{-(\alpha T+j\omega)}}{(1-e^{-(\alpha T+j\omega)})^2} \quad \text{and}$$

$$5X(e^{j\omega}) = \frac{5}{1-e^{-(\alpha T+j\omega)}} \tag{3.25}$$

Finally, by combining the two individual transforms we get the transform of the modified sequence as shown in Eq. (3.26). This example shows you one of the very important ways to determine the DTFT especially of those sequences that are not familiar to you. Begin by breaking up the sequence into its simpler parts, work on the various individual parts, and then combine the transforms to get the transform of the complex composite signal.

$$\mathcal{D}((n+1)x[n]) = \frac{e^{-(\alpha T+j\omega)}}{(1-e^{-(\alpha T+j\omega)})^2} + \frac{5}{1-e^{-(\alpha T+j\omega)}} = \frac{5-4e^{-(\alpha T+j\omega)}}{(1-e^{-(\alpha T+j\omega)})^2} \tag{3.26}$$

3.4.2.6 Convolution in the Time Domain

In Chapter 1, we saw that when a time domain signal goes through a system the result of this operation is obtained by convolving the time domain signal with the impulse response of the system. Here, we show, in Eq. (3.27), that we can evaluate the convolution in the time domain by multiplying the two individual transforms in the frequency domain and then determining the inverse transform of the product to get the convolution in the time domain.

$$\mathcal{D}(x[n]*h[n]) = D\left(\sum_{k=-\infty}^{\infty} x[k]h[n-k]\right)$$

$$\mathcal{D}(x[n]*h[n]) = \sum_{n=-\infty}^{\infty}\left\{\sum_{k=-\infty}^{\infty} x[k]h[n-k]\right\}e^{-j\omega n} \qquad \text{let } n = m+k$$

$$\mathcal{D}(x[n]*h[n]) = \sum_{m=-\infty}^{\infty}\left\{\sum_{k=-\infty}^{\infty} x[k]h[m]\right\}e^{-j\omega(m+k)} \qquad \text{interchange the order of the summation}$$

$$\mathcal{D}(x[n]*h[n]) = \underbrace{\sum_{k=-\infty}^{\infty} x[k]e^{-j\omega k}}_{X(e^{j\omega})}\underbrace{\sum_{m=-\infty}^{\infty} h[m]e^{-j\omega m}}_{H(e^{j\omega})} = X(e^{j\omega})H(e^{j\omega}) \tag{3.27}$$

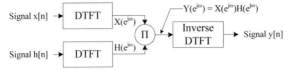

Signal y[n] equals the convolution of x[n] with h[n]

Using the convolution property to determine convolution of two signals

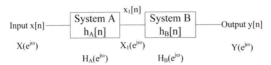

Using the convolution property to determine system output

Figure 3.3 Using the convolution property.

So convolution in the time domain becomes multiplication in the frequency domain. This is indeed very nice since now we can talk about a signal going through several different filters and not have to compute the convolution of the signal at every step of the way. Multiplication of the signal in the frequency domain is definitely a lot easier to understand than convolution.

Figure 3.3 shows two of the many ways we can use the convolution property. The top half of Figure 3.3 shows that to determine the convolution of two signals, we can first determine the DTFT of the two signals, then we multiply the two DTFTs together, this gives us the DTFT of the signal that will be the convolution of the two time domain signals, and finally when we determine the inverse DTFT, we get the result of the convolution.

This is also demonstrated by the MATLAB script in MATLAB 3.2. In this script, we have first defined the two sequences and determined their convolution using the MATLAB command *conv*. Next, we have determined the Fourier transform of the two signals by using a different MATLAB command *fft*. These two Fourier transforms are multiplied together and then we determine the inverse Fourier transform by using a different MATLAB command *ifft*. Finally, after correcting for round-off errors, we subtract the two time domain signals to determine the difference between the two time domain signals. We show that the difference between the two signals, one obtained directly by performing the convolution and the other obtained by multiplying the two Fourier transforms together, is zero for every sample value, hence showing that the two sequences are identical.

To determine the output of the system shown in Figure 3.3, you would have to evaluate the convolution of $x[n]$ with $h_A[n]$ to get $x_1[n]$. Then to get $y[n]$, you would have to convolve $x_1[n]$ with $h_B[n]$ if we worked strictly in the time domain. Working in the frequency domain, we first determine the DTFT of $x[n]$, $h_A[n]$, and $h_B[n]$ that will be $X(e^{j\omega})$, $H_A(e^{j\omega})$, and $H_B(e^{j\omega})$, respectively. When we multiply all the three DTFTs together, we get the DTFT of the output $y[n]$ as $Y(e^{j\omega})$. Now, if we determine the inverse DTFT of $Y(e^{j\omega})$, then we have determined the output from a complex system without evaluating any convolution.

MATLAB 3.2

```
h=[ 0.9 0.7 0.5 0.5 0.7 0.9] ; %Define one sequence.
x=[ 1 0.9 0.8 0.7 0.6 0.5 0.4 0.3 0.2 0.1 0] ; %Another sequence.
y=conv(x,h); %Determine their convolution.
X=fft(x,16); %Determine the Fourier Transform of
H=fft(h,16); %the two time domain signals.
Y=X.*H; %Multiply the Fourier Transform.
y2=ifft(Y); %determine inverse Fourier transform.
error=y-y2; %Determine how close the two
for i=1:16
if abs(error(i))< 0.000001 %Check the Error signal set it to
     error(i)=0; %zero if less than very small value.
end     %This corrects for rounding error.
end
error =
   0   0   0   0   0   0   0   0   0   0   0   0   0   0   0   0
```

3.4.2.7 Parseval's Relation

When two sequences are multiplied together in the time domain by multiplying their corresponding samples together, we get a new sequence. To determine the DFT of this product sequence, we use the Parseval's relation. To prove this relation, we replace one of the sequences in the product by its DTFT representation. Next, we interchange the summation and the integral. The summation will now be the DTFT of the other sequence. By replacing the summation, we get the relation that we want to prove.

$$\sum_{n=-\infty}^{\infty} (x[n]g^*[n]) = \sum_{n=-\infty}^{\infty} \left(x[n] \left(\frac{1}{2\pi} \int_{-\pi}^{\pi} G^*\left(e^{j\omega}\right)e^{-j\omega n}d\omega \right) \right)$$

$$= \frac{1}{2\pi} \int_{-\pi}^{\pi} G^*\left(e^{j\omega}\right) \underbrace{\left(\sum_{n=-\infty}^{\infty} x[n]e^{-j\omega n} \right)}_{X(e^{j\omega})} d\omega \qquad (3.28)$$

$$= \frac{1}{2\pi} \int_{-\pi}^{\pi} G^*\left(e^{j\omega}\right)X\left(e^{j\omega}\right)d\omega$$

All the properties of the DTFT are summarized in Table 3.1 for your convenience.

3.5
Sampling a Continuous Time Signal

In Section 3.1, we have seen that a continuous time signal in the time domain can be represented in the frequency domain by the continuous time Fourier transform. Then in

Table 3.1 Properties of the discrete time Fourier transform.

Theorem	Sequence	Discrete time Fourier transform
Linearity	$Ax[n] \pm Bh[n]$	$AX(e^{j\omega}) \pm BH(e^{j\omega})$
Time reversal	$x[-n]$	$X(e^{-j\omega})$
Shift in time domain	$x[n-k]$	$e^{j\omega k}X(e^{j\omega})$
Shift in frequency domain	$e^{j\omega_0 k}x[n]$	$X(e^{j(\omega-\omega_0)})$
Multiplication by a linear ramp	$nx[n]$	$j\dfrac{dX(e^{j\omega})}{d\omega}$
Convolution	$x[n]*h[n]$	$X(e^{j\omega})\cdot H(e^{j\omega})$
Parseval's theorem	$\displaystyle\sum_{n=-\infty}^{\infty}(x[n]\cdot g^*[n])$	$\dfrac{1}{2\pi}\displaystyle\int_{-\pi}^{\pi}X(e^{j\omega})G^*(e^{j\omega})d\omega$

Section 3.3, we saw that a discrete time sequence in the time domain can be represented in the frequency domain by the DTFT. Suppose, we sample an analog signal and then determine analog signals Fourier transform and the sampled signals DTFT, would there be any similarity in the frequency domain representation of the two signals? Here, we consider the relation between the Fourier representations of a signal when it is a continuous time signal and then when the same continuous time signal is sampled. We now have the Fourier representation of the continuous time signal and the Fourier representation of the sampled signal. Is there a relation between the two Fourier representations of the same signal? We would expect that just because the signal is sampled it does not lose its frequency information and that the frequency information from the continuous time signal and the frequency information from its sampled representation should be same or similar. This indeed is the case provided the sampling is done so that the frequency information is not destroyed. In this section, we examine the relation between the frequency contents of the two signals and then determine the sampling requirements for the continuous time signal so that the frequency content is not destroyed.

3.5.1
Nyquist Sampling Theorem

To see the effect of sampling a continuous time signal, we begin with a continuous time signal $x_a(t)$. We will uniformly sample this signal at an interval of T seconds to give us a sampled signal $x\{n\}$ shown in Eq. (3.29) as follows:

$$x[n] = x_a(t)|_{t=nT} = x_a(nT) \tag{3.29}$$

Also, since $x_a(t)$ is a continuous time signal, it will have its Fourier transform as shown in Eq. (3.1). As shown in Eq. (3.29), we will periodically sample $x_a(t)$ to get the sampled signal $x[n]$ and its DTFT will be obtained as shown in Eq. (3.5). The question that we want answered here is what is the relation between the frequency content of $x_a(t)$ and the frequency content of $x[n]$? To answer this question, let us first determine $X(e^{j\omega})$ the DTFT of the sampled signal $x[n]$.

First, we define the sampled signal $x[n]$ by determining the inverse Fourier transform of the continuous time signal $X_a(j\omega)$ as shown in Eq. (3.30).

$$x[n] = x_a(nT) = \frac{1}{2\pi} \int_{-\infty}^{\infty} X_a(j\Omega)e^{jn\Omega T}d\Omega \qquad (3.30)$$

We can also define the sampled signal $x[n]$ by determining the inverse DTFT of the sampled signal $X(e^{j\omega})$ as shown in Eq. (3.31).

$$x[n] = \frac{1}{2\pi} \int_{-\pi}^{\pi} X(e^{j\omega})e^{j\omega n}d\omega \qquad (3.31)$$

By examining Eqs. (3.30) and (3.31), we see that we have related $x[n]$ the sampled signal to two different frequency domain functions; $X(j\Omega)$ and $X(e^{j\omega})$. What we want to determine is the relation between these two frequency domain functions. This relation can be determined in several different ways; here, we will begin with Eq. (3.30) and break it up first into infinite strips of length $2\pi/T$, so it covers the entire range from $-\infty$ to $+\infty$. This is shown in Eq. (3.32).

$$x[n] = \frac{1}{2\pi} \sum_{k=-\infty}^{\infty} \int_{(2k-1)\pi/T}^{(2k+1)\pi/T} X_a(j\Omega)e^{j\Omega nT}d\Omega \qquad (3.32)$$

Next, we will change the variable of integration in Eq. (3.32) from Ω to $\Omega + (2\pi k/T)$; this will change the limits of the integration so that all the integrals are performed from $-\pi/T$ to π/T, and this is completed in Eq. (3.33).

$$x[n] = \frac{1}{2\pi} \sum_{k=-\infty}^{\infty} \int_{-\pi/T}^{\pi/T} X(j\Omega + (j2\pi k/T))e^{j(\Omega + (2\pi k/T))nT}d\Omega \qquad (3.33)$$

Next, we realize that $e^{j(2\pi k/T)nT} = 1$, so we can eliminate this factor from the integral. Then, we will interchange the order of the summation and the integration in Eq. (3.33) and we get Eq. (3.34).

$$x[n] = \int_{-\pi/T}^{\pi/T} \left[\frac{1}{2\pi} \sum_{k=-\infty}^{\infty} X_a\left(j\Omega + \frac{j2\pi k}{T}\right) \right] e^{j\Omega nT}d\Omega \qquad (3.34)$$

Now, we change the variable one more time from Ω to ω/T; this will also change the limits of integration from $(-\pi/T \rightarrow \pi/T)$ to $(-\pi \rightarrow \pi)$ to get Eq. (3.35).

$$x[n] = \frac{1}{2\pi} \int_{-\pi}^{\pi} \frac{1}{T} \sum_{k=-\infty}^{\infty} X_a\left(\frac{j\omega}{T} + \frac{j2\pi k}{T}\right) e^{j\omega n}d\omega \qquad (3.35)$$

When we compare Eq. (3.31) with Eq. (3.35), we can make the following observations. Since both the equations are the frequency domain representation of the sampled sequence,

$x[n]$ then must be equal to one another, hence we can equate the terms that are different as shown in Eq. (3.36).

$$X\left(e^{j\omega}\right) = \frac{1}{T}\sum_{k=-\infty}^{\infty} X_a\left(\frac{j\omega}{T} + \frac{j2\pi k}{T}\right) \tag{3.36}$$

In Figure 3.4a, we see the spectrum of the continuous time signal. This signal is band limited to $\pm\Omega$. In Figure 3.4b, we see the sampling function; note that we are performing impulse sampling and the sampling frequency is greater than twice the largest frequency present in the continuous time signal. In Figure 3.4c, we see the spectrum of the sampled signal. Notice that the sampled signal has a periodic spectrum and the period is 2π just as the DTFT required.

Figure 3.4 shows us that the sampled spectrum is repeated and the period is the sampling frequency Ω_s. This is what Eq. (3.36) tells us also. In Figure 3.4, we see that there is a possibility that the spectrum of the sampled signal centered at Ω_s could overlap the spectrum of the sampled signal at the origin. To make sure that the two replicas of the frequency spectrum of the sampled signal do not overlap and hence do not destroy the frequency content, we see from the figure that $\Omega_s - \Omega > \Omega$ or that the sampling frequency $\Omega_s > 2\Omega$. This relation is the famous Nyquist sampling theorem.

You should also notice from Figure 3.4 and the derivation that we have just completed that the band limit of the continuous signal that was $\pm\Omega$ is now changed to $\pm\Omega$, the band limit of one of the sampled replica, and that this is equal to ΩT; we used this relation in arriving at Eq. (3.35). Also, the sampling interval of the continuous time signal Ω_s is equal to 2π in the discrete signal, so we have the relation $\Omega_s = 2\pi/T$ or $2\pi = T\Omega_s$. This gives us the relation of the frequencies in the continuous and the discrete time domains.

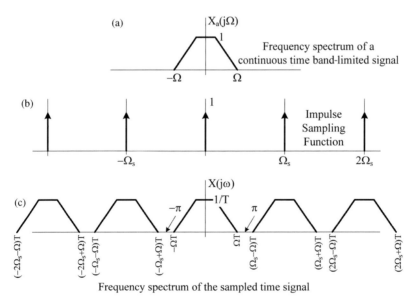

Figure 3.4 Relation between the spectrum of a continuous time band-limited signal and its sampled version.

Nyquist's theorem states that to avoid aliasing (this is only a fancy way of saying that the frequency content is not destroyed) of the frequency of the sampled signal, a band-limited signal must be sampled at least at the rate of twice the highest frequency present in the band-limited signal.

Figure 3.5 shows us the effect of sampling a band-limited signal with a sampling rate that is less than the Nyquist rate. In Figure 3.5, we have $\Omega_s < 2\Omega$, the sampling frequency is less than twice the highest frequency present in the band-limited signal. Now, the spectrum of the sampled signal will suffer from aliasing. This results in the ends of adjacent spectra overlapping each other; this is shown in Figure 3.5c. Sampling at less than the Nyquist rate results in loss due to the overlapping of the spectrum. The loss of the spectrum occurs where the two adjacent spectrum have overlapped and their energies in these frequency bands have added together. The added content is now corrupted as it cannot be separated.

When the Nyquist sampling theorem is not satisfied, we see that aliasing occurs; this is shown in Figure 3.5. In Figure 3.6, we see a specific example of this. Figure 3.6 shows the result of sampling three signals with each signal being sampled at the sampling rate of 10 Hz. The three signals are 4, 6, and 14 Hz. First, notice that all the 20 samples in 2 s are exactly identical to each other for all the 3 signals. So by looking at only the samples, we cannot tell which signal was sampled to get this sampled sequence. From the three signals, only the signal that is a 4 Hz signal satisfies the Nyquist sampling theorem, the other two signals do not satisfy the sampling theorem. The three signals are specially chosen to show us that the sampled values of the 4, 6, and 14 Hz signals are identical.

Since the sampled values of the two higher frequencies are the same as the 4 Hz signal, we can say that the two higher frequency signals resemble the lower frequency signal when

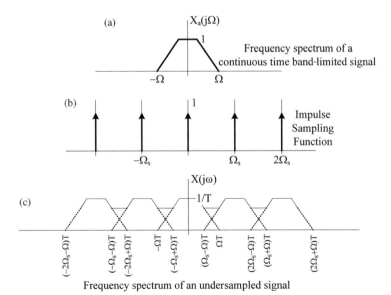

Figure 3.5 Frequency spectrum of an undersampled signal. Where the two spectra overlap, there is a loss of information due to aliasing of the signal.

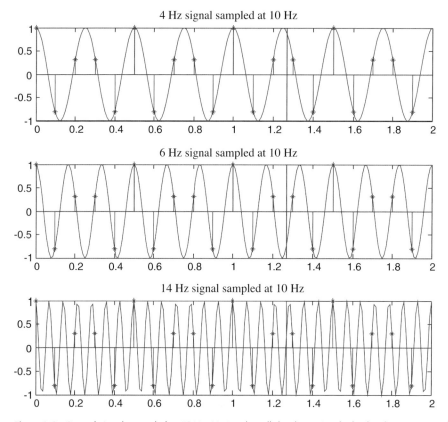

Figure 3.6 Several signals sampled at 10 Hz. Notice that all the three signals display the same sampled values even though the three signals have different frequencies.

sampled at this specific sampling rate. We say that the high-frequency signals are aliased to the lower frequency signal; in our example, the 6 and 14 Hz signals are aliased to the 4 Hz signal. In the figure, we say that sampling the 6 and 14 Hz signals resulted in their aliasing to the 4 Hz signal. So aliasing makes the 6 and 14 Hz signals appear as the 4 Hz signal would appear once the signal is sampled. Since the 6 and 14 Hz signals appear the same as the 4 Hz signal, we would not be able to recover the 6 Hz or the 14 Hz signal from the sampled value of the signal. We can extend this idea a little further and say that if we had a complex signal that is a sum of the 4, 6, and 14 Hz signals, then we would be able to recover only the information content in the 4 Hz signal if we sampled this complex signal at a sampling rate of 10 Hz. So the information present in the 6 and 14 Hz signals would be lost. For this reason, we say that there is loss of information when a signal is aliased.

Example 3.6

Now that we have seen the spectrum of a sampled signal is a periodic signal, we would expect that the DTFT be a periodic function. Determine the period of the DTFT.

Solution 3.6: To determine the period of the DTFT, let us examine the definition of a general signal $x[n]$; this is shown in Eq. (3.5). Also knowing that the period of the sampled signal, as shown in Eq. (3.36), is 2π, we would expect that the period of the signal would be 2π. So assuming that the period of the signal is 2π, we will show that the DTFT of $X(e^{j\omega})$ is the same as the DTFT of $X(e^{j(\omega + 2\pi)})$. In Eq. (3.37), we have made use of the fact that $e^{j2\pi n} = 1$ for all values of n. So the period of the DTFT is 2π.

$$
\begin{aligned}
X\left(e^{j(\omega + 2\pi)}\right) &= \sum_{n=-\infty}^{\infty} x[n] e^{j(\omega + 2\pi)n} \\
&= \sum_{n=-\infty}^{\infty} x[n] e^{j\omega n} e^{j2\pi n} \\
&= \sum_{n=-\infty}^{\infty} x[n] e^{j\omega n} = X\left(e^{j\omega}\right)
\end{aligned}
\tag{3.37}
$$

Example 3.7

In Figure 3.7, we see a frequency domain representation of a continuous time signal. We will sample the signal at a sampling rate of 80 Hz. After the signal is sampled, we wish to filter the signal so that we are able to recover the signal up to point A, which is a frequency of 20 Hz. What should be the filter cutoff frequency of the digital filter if we assume that we can design ideal filters?

Solution 3.7: Point A in Figure 3.7 is at a frequency of 20 Hz. So we need to design a digital filter that will permit all the signal frequencies up to 20 Hz go through and block all the frequencies that are greater than 20 Hz. The point A when the signal is sampled gets transformed to a digital frequency of $\omega = 2^* \pi^* \Omega^* T = 2^* \pi^* 20^* 0.0125 = 0.5^* \pi$. This gives us the cutoff frequency of an ideal low pass filter to recover the signal upto point A in the signal.

Example 3.8

Repeat Problem 3.7 but change the sampling rate to 30 Hz.

Solution 3.8: With the sampling rate being only 30 Hz, the signal will be aliased and there will not be a complete recovery. To find out where the 20 Hz signal maps to in the sampled

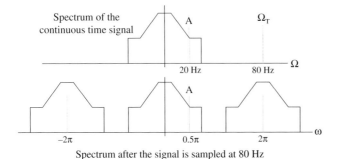

Spectrum of the continuous time signal

Spectrum after the signal is sampled at 80 Hz

Figure 3.7   Frequency spectrum of a continuous time signal and after it is sampled with a sampling rate of 80 Hz.

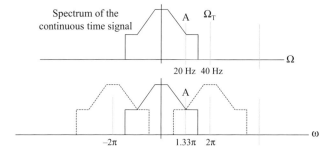

Figure 3.8 Frequency spectrum of a continuous time signal that is aliased so the original is corrupted.

domain, we proceed in the same way as we did in Problem 3.7. Completing the computation, we get a digital frequency of $\omega = 2^*\pi^*\Omega^*T = 2^*\pi^*20^*0.033 = 1.332^*\pi$. Since this digital frequency exceeds π, we know that the signal is aliased and we cannot recover the signal. The effect of sampling at this rate on the signal is shown in Figure 3.8.

3.5.2
Reconstruction of the Sampled Signal

From Figure 3.4, it is clear that if we sample the continuous time signal so that it satisfies the Nyquist theorem, then the original continuous time signal $x_a(t)$ can be recovered by using a low-pass filter that has its pass band equal to ΩT. Here, we examine what the appropriate interpolation function needs to be so that we can recover the original continuous time signal $x_a(t)$ from the samples $x[n]$ of the discrete signal. To determine this interpolation formula, let us first assume that the continuous time signal $x_a(t)$ is a band-limited signal limited to frequency Ω. Also, the sampling frequency Ω_s chosen satisfies the Nyquist criteria, so $\Omega T \leq (\Omega_s T/2) = \pi$ as shown in Figure 3.4. Then, from the continuous time Fourier transform, we get Eq. (3.38) as follows:

$$x_a(t) = \frac{1}{2\pi} \int_{-\pi/T}^{\pi/T} X_a(j\Omega) e^{j\Omega t} d\Omega \tag{3.38}$$

From Figure 3.4, we can see that in the region $-\pi/T < \Omega < \pi/T$ the relation between the frequency spectrum of the discrete time signal and the continuous time signal is given by Eq. (3.39).

$$X(e^{j\omega}) = \frac{1}{T} X_a(j\Omega) \tag{3.39}$$

Combining Eqs. (3.38) and (3.39) we can write the expression for the continuous time signal as shown in Eq. (3.40).

$$x_a(t) = \frac{1}{2\pi} \int_{-\pi/T}^{\pi/T} T X(e^{j\omega}) e^{j\Omega T} d\Omega \tag{3.40}$$

In Eq. (3.40), we will substitute for $X(e^{j\omega})$ from its DTFT as $X(e^{j\omega}) = \sum_{k=-\infty}^{\infty} x_a(kT)e^{-j\omega k}$, then we get Eq. (3.41), and to get the last equality in Eq. (3.41) we substitute $\omega = \Omega T$.

$$x_a(t) = \frac{T}{2\pi}\int_{-\pi/T}^{\pi/T}\left[\sum_{k=-\infty}^{\infty} x_a(kT)e^{-j\omega k}\right]e^{j\Omega t}d\Omega = \frac{T}{2\pi}\int_{-\pi/T}^{\pi/T}\left[\sum_{k=-\infty}^{\infty} x_a(kT)e^{-j(\Omega T)k}\right]e^{j\Omega t}d\Omega$$

(3.41)

When we interchange the order of the summation and the integration, we get Eq. (3.42) as follows:

$$x_a(t) = \frac{T}{2\pi}\sum_{-\infty}^{\infty} x_a(kT)\int_{-\pi/T}^{\pi/T} e^{-j\Omega kT}e^{j\Omega t}d\Omega = \frac{T}{2\pi}\sum_{-\infty}^{\infty} x_a(kT)\int_{-\pi/T}^{\pi/T} e^{j\Omega(t-kT)}d\Omega$$

(3.42)

The integral in Eq. (3.42) is a special integral. It is evaluated in Eq. (3.43).

$$\frac{T}{2\pi}\int_{-\pi/T}^{\pi/T} e^{j\Omega(t-kT)}d\Omega = \frac{T}{2\pi}\left[\frac{e^{j\Omega(t-kT)}}{j(t-kT)}\Big|_{-\pi/T}^{\pi/T}\right] = \frac{\mathrm{Sin}\left((\pi/T)(t-kT)\right)}{(\pi/T)(t-kT)}$$

(3.43)

Using the result of Eq. (3.43) in Eq. (3.42), we get the required interpolation formula as shown in Eq. (3.44). Remember that this interpolation formula has assumed that the continuous time signal was band limited to begin with and that the sampling interval was chosen so that no aliasing occurs, that is, the signal is sampled at, at least, the Nyquist sampling rate.

$$x_a(t) = \sum_{-\infty}^{\infty} x_a(kT)\frac{\sin((\pi/T)(t-kT))}{(\pi/T)(t-kT)}$$

(3.44)

Equation (3.44) tells us how to reconstruct the original signal from the sampled version of the signal. Each sample at location kT is multiplied by the sinc function that is shifted to the location kT. The sinc function from any other location does not interfere with the sinc function at location kT since the sinc functions from all the other locations are exactly zero at this location as shown in Figure 3.9. So say, for example, we have a vector of discrete values $y = [1.3\ 1.5\ 1.7\ 1.8\ 1.85\ 1.7\ 1.6\ 1.45\ 1.2\ 1.0]$ and we use the reconstruction/interpolation formula on these values, then we will get a graph as shown in Figure 3.9. Figure 3.9 was produced by using MATLAB 3.3 as shown below. In the MATLAB script in the for loop, we compute the sinc functions for each value in the vector y and add them up together. They are then plotted in the last command to get the plot of Figure 3.9.

MATLAB 3.3

```
T=0.2;
y=[ 1.3 1.5 1.7 1.8 1.85 1.7 1.6 1.45 1.2 1.0] ;
t=-2.5:0.01:2.5;
```

```
x=zeros(10,length(t));
for k=-4:5,
    x(k+5,:)=y(k+5)*sin(pi/T*(t-k*T))./(pi/T*(t-k*T));
end;
plot(t,x(1:10,:))
```

Reconstruct the analog signal that is represented by the following
sampled points. y=[1.3 1.5 1.7 1.8 1.85 1.7 1.6 1.45 1.2 1.0]

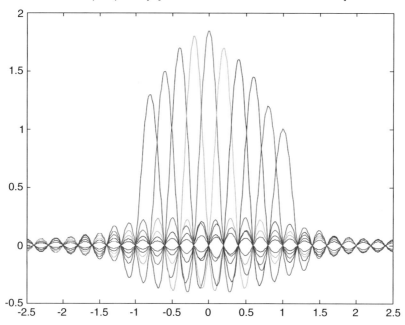

Figure 3.9 Using the interpolation formula to reconstruct the analog signal from a vector of discrete values.

The MATLAB script draws several weighted shifted sinc functions, each shifted sinc function has an amplitude exactly equal to the sample value at that location. Also, notice that the sinc function becomes active when $t = kT$ value. For any particular value of k when the function is active, all the other sinc functions have their values at zero, so they do not contribute, and only the active function has any value. Now, if we connect all the tops of the various sinc functions, we will have recovered the original analog signal provided it was band limited and that we sampled it at a sampling rate exceeding the Nyquist rate.

3.6
Discrete Fourier Transform

We began this chapter with the Fourier transform of a continuous time signal. We found that this transform was also a continuous time function. Next, we determined the Fourier transform of an infinite sampled sequence. We found that its Fourier transform was a continuous function of ω. The DTFT is useful in only a limited way because one of its requirements is that the sampled time domain signal has to be infinite in length. We do not have any signal that is infinite in length that we can receive or have received and can transmit or have transmitted. All the signals that we have, have a beginning and an end so they are all finite length signals. Next, in this section, we examine the Fourier transform of a finite length sequence. All the sequences that we have in practice begin and end at a finite time, so they out of necessity are of finite length. To obtain the transform of a finite length sequence, we examine a slightly modified version of the DTFT as shown in this section. To distinguish this transform from the DTFT, we will call this transform simply *discrete Fourier transform* (DFT). We will find that this transform, the DFT itself, is a finite length sequence rather than a continuous function. The DFT has gained in importance chiefly because of the construction of cheap digital computing stations and the discovery of an algorithm that is able to take advantage of the symmetry and the periodicity of the DFT.

3.6.1
Definition

Before we define the transform, we would like to discuss several different interpretations that are present in viewing this transform. First, we can think of this transform as a transform of a finite length sequence. Second, we can think of this transform as a transform of one period of a periodic sequence. As we study the transform, we will find that both these interpretations are very consistent and that there is no confusion involved. Most of the time we will treat the finite length sequence as one period of a periodic sequence. This is shown in Figure 3.10. This interpretation will permit us to fully understand the circular shift and circular convolution, the two very important operations in the DFT domain.

Consider a sequence $x[n]$ that is periodic with period N so that $x[n] = x[n + N]$ as shown in Figure 3.10. Visualizing the sequence as a periodic sequence, we can compute the Fourier series representation of this sequence. To determine the Fourier series representation, we will use the exponential Fourier series as it will give us an insight into what a DFT represents physically. The complex exponentials $e_k = e^{j2\pi nk/N}$ that we will use to determine the Fourier series are periodic with period N since $e^{j2\pi n} = 1$; $e_{k+N} = e^{j2\pi n(k+N)/N} = e^{j2\pi nk/N}e^{j2\pi nN/N} = e_k$. Knowing that the complex exponentials are periodic, we can summarize that there are only N unique exponentials that are multiples of the fundamental frequency $2\pi/N$. This implies that the DFT will also have only N coefficients and that these N

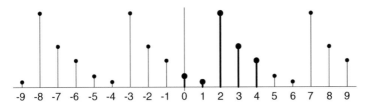

Figure 3.10 A discrete sequence that is a periodic sequence. The period of this sequence is five samples.

coefficients are discrete. With this we can write the Fourier series expansion of the sequence $x[n]$ as shown in Eq. (3.45).

$$x[n] = \frac{1}{N} \sum_{k=0}^{N-1} X[k] e^{j2\pi nk/N} \tag{3.45}$$

The term $1/N$ is present for convenience and has no effect on the transform itself. The inverse transform is given by Eq. (3.46) that is the inverse Fourier series representation.

$$X[k] = \sum_{n=0}^{N-1} x[n] e^{-j2\pi nk/N} \tag{3.46}$$

Note that the sequence $x[n]$ and $X[k]$ are both finite length sequences of length N and both are periodic and both have a period N and that there are only N distinct complex exponentials.

To verify that Eqs. (3.45) and (3.46) represent a transform pair, we can substitute one equation in the other equation and see if we get an identity as the result. When we attempt to do this, we will have to make use of the fact that the complex exponential

$$\frac{1}{N} \sum_{n=0}^{N-1} e^{j2\pi n(k-r)/N} = \begin{cases} 1 & k = r \\ 0 & k \neq r \end{cases} \tag{3.47}$$

when r is equal to k, the summation is equal to 1, and when k is not equal to r, the summation is equal to zero. This can be easily proved and you are asked to do so as an exercise in the chapter. To verify the transform pair, we begin by substituting Eq. (3.46) into Eq. (3.45) as shown in Eq. (3.48).

$$x[n] = \frac{1}{N} \sum_{k=0}^{N-1} X[k] e^{j2\pi nk/N} = \frac{1}{N} \sum_{k=0}^{N-1} \left[\sum_{r=0}^{N-1} x[r] e^{-j2\pi rk/N} \right] e^{j2\pi nk/N} \tag{3.48}$$

Interchange the order of the two summations and rearrange to get Eq. (3.49).

$$x[n] = \sum_{r=0}^{N-1} x[r] \left[\frac{1}{N} \sum_{k=0}^{N-1} e^{j2\pi(n-r)k/N} \right] \tag{3.49}$$

Now, when we utilize Eq. (3.47), we find that the term in the square bracket is 1 when $r = n$ and zero every other time, and letting $r = n$ in Eq. (3.49) we get an identity that $x[n] = x[n]$. Thus, the transform pair is verified. Very often as a matter of convenience, we write the complex exponential as $e^{-j2\pi/N} = W_N$. With this substitution, the DFT pair of Eq. (3.46) into Eq. (3.45) will become

$$X[k] = \sum_{n=0}^{N-1} x[n] W_N^{nk}$$
$$x[n] = \frac{1}{N} \sum_{k=0}^{N-1} X[k] W_N^{-nk} \tag{3.50}$$

As an example of determining the DFT of a periodic sequence, consider the sequence given by $x[n]$ in Figure 3.11.

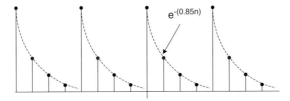

Figure 3.11 Periodic sequence for which DFT is to be computed.

Example 3.9

You are shown a periodic sequence in Figure 3.11. The sequence is obtained by sampling the continuous time function $e^{-0.85t}$ with a sampling interval of 1 s. Determine the DFT of this sequence.

Solution 3.9: Since we need to find the DFT, we begin with the first equation in Eq. (3.50) with $N=4$ as

$$X[k] = \sum_{k=0}^{N-1} e^{(-0.85n)} W_N^{nk} = \sum_{k=0}^{3} e^{(-0.85n)} e^{((-j2\pi nk/3))}$$

$$X[k] = \sum_{n=0}^{3} e^{-n((j2\pi k/3) + 0.85)} \tag{3.51}$$

$$X[k] = e^0 + e^{-1((j2\pi k/3) + 0.85)} + e^{-2((j2\pi k/3) + 0.85)} + e^{-3((j2\pi k/3) + 0.85)}$$

Now, all the values of $X[k]$ can be computed for $k=0$ to 3.

Example 3.10

As another example, we will determine the DFT of the sequence shown in Figure 3.12. This time the period is 9 but only four samples have a nonzero value.

Solution 3.10: Using the first equation in Eq. (3.50) with $N=9$, we get

$$X[k] = \sum_{n=0}^{8} x[n] W_9^{nk} = \sum_{n=0}^{3} W_9^{nk} = \sum_{n=0}^{3} e^{-j(2\pi nk/9)}$$

$$X[k] = e^{-j(3\pi k/9)} \left[\frac{\sin(4\pi k/9)}{\sin(\pi k/9)} \right] \tag{3.52}$$

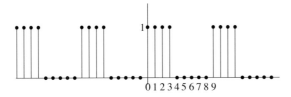

Figure 3.12 Periodic sequence for which DFT is to be computed.

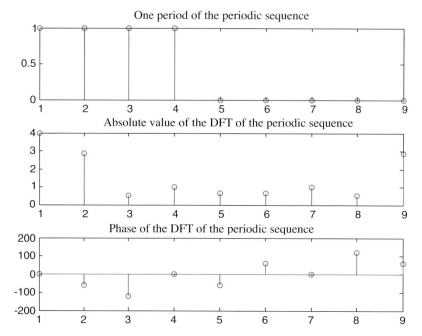

Figure 3.13 Periodic sequence and its DFT.

The plot of the DFT coefficients is also drawn in Figure 3.13. Compare Figure 3.12 with Figure 3.13. The time domain sequence in Figure 3.12 is a periodic sequence. We took one period of the periodic sequence as shown in Figure 3.13 and obtained its DFT. The absolute value and the phase of the DFT are also drawn in Figure 3.13.

3.6.1.1 Computing the DFT from the Definition

To get an idea of why the DFT did not become a popular tool with the engineer and designer, even though the DFT was discovered in the early twentieth century, we need to see how much difficulty is involved in computing the DFT. This will also enable us to determine importance of a later development that led to the many different fast Fourier transform (FFT) algorithms. In MATLAB 3.4, we have listed the MATLAB script that will compute the DFT using the definition of the DFT. In the script of MATLAB 3.4, we compute the DFT of the sequence given in Figure 3.12. Notice that we have used two loops to compute the DFT. The outer loop computes each value of the DFT and to compute each value of the DFT we need to execute the inner loop completely. This means that the inner loop executes N times for each count on the outer loop and there are N counts on the outer loop. So to compute the DFT from the definition, we need about N^2 computations approximately. This number can become quite large when the number of samples N is a large number. It was this, the number of complex computations involved, that kept the DFT from being a very frequently used tool till the discovery of the FFT algorithm.

MATLAB 3.4

```
x=[ 1 1 1 1 0 0 0 0 0];
N=length(x);
X=zeros(1,N);
for k=0:(N-1)
     for n=0:(N-1)
         X(k+1)=X(k+1)+x(n+1)*cos(2*pi*n*k/N)-
             j*x(n+1)*sin(2*pi*n*k/N);
     end
end
subplot(311),stem(x);
title('One period of the periodic sequence.');
subplot(312),stem(abs(X));
title('Absolute value of the DFT of the periodic sequence.');
subplot(313),stem(angle(X)*360/2/pi);
title('Phase of the DFT of the periodic sequence.');
```

Examine Figure 3.14. Notice that it is exactly identical to Figure 3.13. This you would expect since the two figures represent the time domain sequence that are the same and its DFT that should also be the same. The only difference between the two is the way that the DFT is computed. In Figure 3.13, we used the FFT algorithm, while to compute and draw the Figure 3.14 we used the DFT definition.

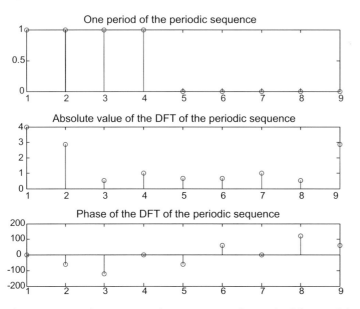

Figure 3.14 Periodic sequence and its DFT computed using the definition of the DFT.

3.6.2
Relation between the DTFT and the DFT

In Section 3.3, we studied the DTFT and in Section 3.5.1 we have studied the DFT. Often there is a question: Is there a relation between the DTFT and the DFT. We answer this question here. We have seen that the DTFT is a continuous function of ω and is periodic with period 2π. The DFT is a discrete function and is also periodic with a period N. We have seen that the DFT is a Fourier representation with the fundamental frequency of $2\pi/N$. So we can say that there are N unique frequencies represented from 0 to 2π in the DFT. When we examine the two Fourier transforms (the DTFT and the DFT) in the range 0–2π, we should find some relation between them.

Since the DFT is a transform of a finite length sequence, we will begin with a finite length sequence, determine its DTFT, and relate that to the DFT; in Eq. (3.53), we determine the DTFT of a finite length sequence.

$$X\left(e^{j\omega}\right) = \sum_{n=0}^{\infty} x[n]e^{-j\omega n} = \sum_{n=0}^{N-1} x[n]e^{-j\omega n} \tag{3.53}$$

Next, we sample the DTFT at N equidistant points in the region 0–2π. This is shown in Eq. (3.54) as follows:

$$|X\left(e^{j\omega}\right)|_{\omega=2\pi k/N} = \sum_{n=0}^{N-1} x[n]e^{-j(2\pi k/N)n} \quad 0 \leq k \leq N \tag{3.54}$$

Comparing the first equation in Eq. (3.50) or Eq. (3.46) with Eq. (3.54), we see that they are both exactly the same equations. So we can say that the DFT is the sampled version of the DTFT when we take N samples of the DTFT from 0 to 2π. Example 3.11 shows a different point of view of this same relation.

Example 3.11

Let $X(e^{j\omega})$ be the DTFT of a sequence $x[n] = (1/2)^n \mu[n]$. Let another sequence $Y[K]$ be a finite length sequence of length 10. The 10-point sequence $Y[k]$ is obtained by sampling the transform $X(e^{j\omega})$, that is, $Y[k] = X(e^{j2\pi k/10})$. Now, determine the sequence $y[n]$ by determining the inverse DFT of the sequence $Y[k]$. Compare the sequence $y[n]$ with the sequence $x[n]$ and determine the relation between the two sequences.

Solution 3.11: We begin by obtaining the sequence $Y[k]$ by sampling the DTFT $X(e^{j\omega})$.

$$Y[k] = X\left(e^{j2\pi k/10}\right)$$

Using the sequence and the definition of the DFT, we determine the sequence $y[n]$

$$y[n] = \sum_{k=0}^{N-1} Y[k]e^{-j2\pi nk/N} = \sum_{k=0}^{9} Y[k]e^{-j2\pi nk/10}$$

Replacing for $Y[k]$, we get Eq. (3.55).

$$y[n] = \frac{1}{10} \sum_{n=0}^{9} X\left(e^{j2\pi k/10}\right) e^{j2\pi nk/10} \tag{3.55}$$

But we know that $X(e^{j\omega}) = \sum_{m=0}^{\infty} x[m] e^{-j\omega m} = \sum_{m=0}^{\infty} (1/2)^m e^{-j2\pi mk/10}$, so substituting for $X(e^{j\omega})$ in Eq. (3.55), we get (using $N = 10$)

$$y[n] = \frac{1}{10} \sum_{k=0}^{N-1} \sum_{m=0}^{\infty} (1/2)^m e^{(-j2\pi mk/N)} e^{(j2\pi nk/N)}$$

$$y[n] = \sum_{m=0}^{\infty} (1/2)^m \frac{1}{10} \sum_{k=0}^{N-1} e^{(j2\pi(n-m)k/N)}$$

The second summation, we have already seen this summation in Eq. (3.47), is equal to 1 only when $n = m + rN$ and it is zero everywhere else (r is an integer). So, we can write $y[n] = (^1/_2)^n \, 0 < n < 9$. Therefore, we can write the relation between $x[n]$ and $y[n]$ in Eq. (3.56) as follows:

$$y[n] = \sum_{r=0}^{9} x[n] \tag{3.56}$$

Figure 3.15 shows the relation between the DTFT and the DFT of the two sequences. Notice that the DFT is just the sampled version of the DTFT. Now that we know what the two

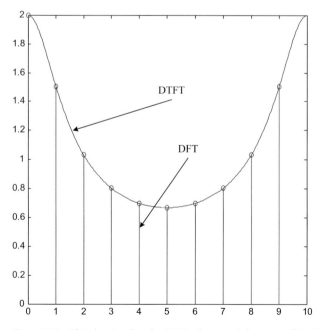

Figure 3.15 Plot showing that the DFT is the sampled version of the DTFT.

sequences $x[n]$ and $y[n]$ in the example are, we can compute their transforms directly as shown in Eq. (3.57) for the DTFT

$$X(e^{j\omega}) = \sum_{n=0}^{\infty} (1/2)^n e^{-j\omega n} = \sum_{n=0}^{\infty} ((1/2)e^{-j\omega})^n = \frac{1}{1-(1/2)e^{-j\omega}} \tag{3.57}$$

and we can evaluate the DFT directly as shown in Eq. (3.58)

$$y[k] = \sum_{n=0}^{9} (1/2)^n e^{-j2\pi n k/10} = \sum_{n=0}^{9} ((1/2)e^{-j2\pi k/10})^n = \frac{1-(1/2)e^{-j2\pi k}}{1-(1/2)e^{-j2\pi k/10}} \tag{3.58}$$

Again, if we substitute $\omega = (j2\pi k/10)$ in Eq. (3.57), we get back Eq. (3.58) except the term $(-1/2\,e^{-j2\pi k})$ in the numerator. This term represents the loss we have in taking only 10 samples instead of infinite samples.

3.6.3
Relation between the Fourier Transform and the DFT

Just as we were able to determine the relation between the DTFT and the DFT, we can also determine the relation between the Fourier transform and the DFT. We begin with a Fourier transform pair $X(j\omega) \leftrightarrow x(t)$. We next sample $X(j\omega)$ at N equally spaced points in the frequency range $0-2\pi$. This gives us a finite length sequence $Y[k]$. We will then determine the inverse DFT of the sequence $Y[k]$ to determine the relation between $y[n]$ and $x[n]$, the sampled version of $x(t)$. This is shown in Eq. (3.59)

$$Y[k] = X(j\omega)|_{\omega=2\pi k/N} = \sum_{n=-\infty}^{\infty} x[n]e^{-j2\pi n k/N} = \sum_{n=-\infty}^{\infty} x[n]W_N^{nk} \tag{3.59}$$

Now, the sequence $y[n]$ can be determined from the inverse DFT of the sequence $Y[k]$ as shown in Eq. (3.60).

$$y[n] = \frac{1}{N} \sum_{k=0}^{N-1} Y[k]W_N^{-nk} \tag{3.60}$$

Substitute for $Y[k]$ from Eq. (3.59) to get

$$y[n] = \frac{1}{N} \sum_{k=0}^{N-1} \left[\sum_{l=-\infty}^{\infty} x[l]W_N^{lk} \right] W_N^{-nk} \tag{3.61}$$

Interchange the order of the summations and after some simple algebra (noting that $(1/N)\sum_{k=0}^{N-1} W_N^{-(n-l)k} = 1$, when $n = l \pm rN$), we get Eq. (3.62).

$$y[n] = \sum_{l=-\infty}^{\infty} x[l]\frac{1}{N}\sum_{k=0}^{N-1} W_N^{lk} W_N^{-nk} = \sum_{l=-\infty}^{\infty} x[l]\underbrace{\frac{1}{N}\sum_{k=0}^{N-1} W_N^{-(n-l)k}}_{\substack{=1 \text{ when } n=l\pm rN \\ =0 \text{ otherwise}}} \tag{3.62}$$

So we get the final relation between $y[n]$ and $x[n]$ as $y[n] = \sum_{r=-\infty}^{\infty} x\left[\langle n + rN \rangle_N\right]$. This last relation tells us that if the sequence $x[n]$ is longer than N terms, then it wraps around to the beginning and starts modifying the term $y[n]$. If the sequence $x[n]$ is smaller than N samples, then there will be no error in using the DFT. This has a very interesting interpretation in the sense that when we want to determine the DFT of any sequence, the DFT must contain at least N terms. It can have more terms but not less. If it has fewer terms, then an error results. One of the homework problems in the exercises makes this point clear to the reader.

3.7
Properties of the DFT

Before we study the properties of the DFT, we will find it is most instructive to study some of the operations on finite length sequences. While we are examining these operations, we should keep in mind both the interpretations we have adopted for the finite length sequences. One interpretation is that the sequence is of finite length, implying that there is no sequence either before $n = 0$ or after $n = (N - 1)$. The other interpretation that we have adopted is that the sequence that we are looking at is one period of the periodic sequence. We will find that both these interpretations are very consistent, but sometime it will be easier to visualize the concept being discussed with one interpretation, while at other times it will be easier with the other interpretation. You should, however, try to adopt both the interpretations for all the properties and operations that we will consider here.

3.7.1
Circular Shift

The shifting operations that we have seen earlier take on a special meaning here. Examine Figure 3.16 to visualize the shift operations. If we visualize the sequence as a finite length sequence, then we know that the sequence exists between samples 0 and $N - 1$. So after the shifting operation, the sequence must exist between samples 0 and $N - 1$, and this can happen after the shift only if the shift is visualized as a circular shift. If we visualize

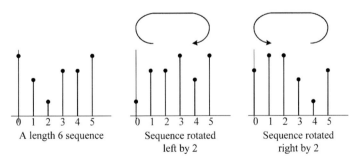

| A length 6 sequence | Sequence rotated left by 2 | Sequence rotated right by 2 |

Figure 3.16 Circular operations on a finite length sequence.

the sequence as a periodic sequence, then when the shift takes place the sample values from the next cycle shift into the current cycle (left shift), while the sample values of the current cycle shift into the previous cycle, thus once again representing a circular shift. This still leaves behind a periodic sequence and one period of the periodic sequence gives us the same result with both the visualizations.

We refer to this shifting operation as modulo shifting operation. To determine n modulo N, we perform modulo arithmetic. To perform modulo arithmetic, we divide n by N and concentrate only on the remainder of the division. The result of the modulo operation is the remainder of the division. So 13 modulo 4 is 1, since $13/4 = 3$ with a remainder of 1. We throw away the 3 and keep only the 1, which is the result of modulo arithmetic. Similarly, 13 modulo 3 is also 1 since $13/3 = 4$ with a remainder of 1. In general terms, we can write m modulo $N = r$, then $m = r + 1^*N$. We usually specify the modulo operation as shown in Eq. (3.63).

$$((m))_N = (m \text{ mudulo } N)$$
$$\text{if } r = ((m))_N \text{ then } m = r + lN$$

(3.63)

By using the modulo notation, we can define the circular shift of a length N sequence as shown in Eq. (3.64).

$$x_c[n] = x\big[((n-n_0))_N\big]$$

(3.64)

In Eq. (3.64), the original sequence $x[n]$ and the rotated sequence $x_c[n]$ are both length N sequences. Both the sequences have the same samples, but they are shifted as shown in Figure 3.16. In Eq. (3.64) if $n_0 > 0$, we have a right shift and if $n_0 < 0$, we have a left shift, so when $n_0 > 0$ we have, as shown in Eq. (3.65),

$$x_c[n] = \begin{cases} x\big[((n-n_0))_N\big] & \text{for } n_0 \le n \le (N-1) \\ x\big[((n+N-n_0))_N\big] & \text{for } 0 \le n \le n_0 \end{cases}$$

(3.65)

Another important concept that we should consider in this section is time reversal (Figure 3.17). When we time reverse the sequence, we get a sequence $x[((-n))_N]$, this sequence can be written as shown in Eq. (3.66). To understand time reversal, think of the sequence being flipped around the zero sample and think of the sequence as a periodic sequence. Then, the zero sample stays as it is and the rest of the samples take their places as

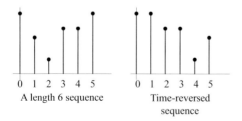

0 1 2 3 4 5
A length 6 sequence

0 1 2 3 4 5
Time-reversed
sequence

Figure 3.17 Time reversal of a finite length sequence.

$N - n$. That is to say that the new sample n will be the original sample $N - n$. We will use the shifting and the time reversal properties in proving the other properties of the DFT.

$$x\left[((-n))_N\right] = \begin{cases} x[N{-}n] & \text{for } 1 \leq n \leq (N{-}1) \\ x[n] & n = 0 \end{cases} \tag{3.66}$$

Example 3.12

Consider a sampled sequence $x[n] = [1\ 2\ -1\ 3\ 4\ -3\ 2\ 1\ 0\ 2]$ of length 10. Determine the resulting sequence after the following operations of the original sequence.

a) Shift the sequence right three places.
b) Shift the sequence left 22 places.
c) Time reverse the sequence.

Solution 3.12: The shifting process is executed according to Eq. (3.65) keeping in mind that we have to perform modulo operation every time.

a) When we have to shift three places to the right, $n_0 = 3$. So for $n = 0, 1, 2, 3$, we will use the bottom half of Eq. (3.65) and for the other value of n, we will use the top half of Eq. (3.65). Doing this we get $x[n - 3] = [1\ 0\ 2\ 1\ 2 - 1\ 3\ 4 - 3\ 2]$.
b) This time we have to shift right by 22. Performing modulo arithmetic, this is equivalent to (22 mod 10 = 2) or a right shift by 2. Once again using Eq. (3.65), we get the shifted sequence as $x[n - 2] = x[n - 22] = [0\ 2\ 1\ 2 - 1\ 3\ 4 - 3\ 2\ 1]$.
c) This time we have to time reverse the sequence. For time reversing, we use Eq. (3.66). Remember we keep the zero sample as it is and adjust the rest of the samples by replacing sample n by sample $(N - n)$. Using this equation, we get the time-reversed sequence as $x = [1\ 2\ 0\ 1\ 2 - 3\ 4\ 3 - 1\ 2]$.

3.7.2
Symmetry Relations in the DFT

As in the DTFT, we can separate the DFT in its real and imaginary parts, in its even and its odd parts, or in its conjugate symmetric and conjugate antisymmetric parts. We do this here.

3.7.2.1 Real and Imaginary Parts of the DFT
The sequence $X[k]$ can be separated into its real and its imaginary parts as shown in Eq. (3.67). Here, we determine what makes up the real part of the DFT and the imaginary part of the DFT. To get the real and the imaginary parts, we have made use of the fact that in general $x[n]$ itself will be complex and that $W_N^{nk} = e^{-j2\pi nk/N} = \cos(2\pi nk/N) - j\sin(2\pi nk/N)$.

$$X[k] = X_{\text{Re}}[k] + jX_{\text{Im}}[k] = \sum_{n=0}^{N-1} x[n]\, W_N^{nk}$$

$$X_{\text{Re}}[k] = \sum_{n=0}^{N-1} x_{\text{Re}}[n]\cos(2\pi nkN) + x_{\text{Im}}[n]\sin(2\pi nkN) \tag{3.67}$$

$$X_{\text{Im}}[k] = \sum_{n=0}^{N-1} x_{\text{Im}}[n]\cos(2\pi nkN) - x_{\text{Re}}[n]\sin(2\pi nkN)$$

3.7.2.2 DFT of the Conjugate of x[n]

We use the definition to determine the transform of the conjugate of $x[n]$ as shown in Eq. (3.68). In Eq. (3.68), we have made use of the fact that the conjugation process of a complex number is the same as changing the sign of the imaginary part.

$$\mathcal{D}\{x^*[n]\} = \sum_{n=0}^{N-1} x^*[n]\, W_N^{nk} = \left\{ \sum_{n=0}^{N-1} x[n]\, W_N^{n\langle -k\rangle_N} \right\}^* = X^*\big[\langle -k\rangle_N\big] \tag{3.68}$$

3.7.2.3 DFT of the Time-Reversed Sequence

Here also, we use the definition of the DFT to determine the required transform as shown in Eq. (3.69).

$$\mathcal{D}[x^*[-n]] = \sum_{n=0}^{N-1} x^*[-n]\, W_N^{nk} \qquad\qquad \text{let } m = N-n$$

$$\mathcal{D}[x^*[-n]] \quad \sum_{m=N}^{1} x^*[m]\, W_N^{(N-m)k} = \sum_{m=0}^{N-1} x^*[m]\, W_N^{-mk} W_N^{Nk} \tag{3.69}$$

$$\mathcal{D}[x^*[-n]] \quad \left[\sum_{m=0}^{N-1} x[m]\, W_N^{mk} \right]^* = X^*[k]$$

3.7.2.4 DFT of the Real and the Imaginary Parts of the Sequence x[n]

To get these transforms, we will use the fact that the conjugate symmetric part of the DFT is the DFT of the real part of the sequence and the conjugate antisymmetric part of the DFT is the DFT of the imaginary part of the sequence. This is shown in Eq. (3.70).

$$\mathcal{D}[x_{\mathrm{Re}}[n]] = X_{\mathrm{cs}}[k] = \frac{1}{2}\big[X[k] + X^*[((-k))_N]\big]$$

$$\mathcal{D}[x_{\mathrm{Im}}[n]] = X_{\mathrm{ca}}[k] = \frac{1}{2}\big[X[k] - X^*[((-k))_N]\big] \tag{3.70}$$

3.7.2.5 DFT of the Conjugate Symmetric and the Conjugate Antisymmetric Part of the Sequence x[n]

We will get these transforms from the definitions of the conjugate symmetric sequence as shown in Eq. (3.71), where we have made use of Eq. (3.69),

$$x_{\mathrm{cs}}[n] = \frac{1}{2}\big[x[n] + x^*[((-n))_N]\big]$$

$$x_{\mathrm{cs}}[n] = \frac{1}{2}\big[X[k] + X^*[k]\big] \tag{3.71}$$

$$x_{\mathrm{cs}}[n] = X_{\mathrm{Re}}[k]$$

and the conjugate antisymmetric sequence as shown in Eq. (3.72), where we have made use of Eq. (3.69)

$$x_{ca}[n] = \frac{1}{2}\left[x[n]-x^*\left[((-n))_N\right]\right]$$

$$x_{ca}[n] = \frac{1}{2}[X[k]-X^*[k]] \tag{3.72}$$

$$x_{ca}[n] = X_{Im}[k]$$

3.8
Theorems of the DFT

Just like the other transforms that we have studied, the DFT also has its own theorems. Here, we discuss the more pertinent theorems.

3.8.1
The DFT is a Linear Transform

When you combine two time domain sequences linearly, the DFT of the resulting sequence is the linear combination of the DFT of the two sequences. When doing this extra care must be taken since the DFT of a sequence is the same length as the length of the sequence itself. So before we can combine the two sequences linearly, the length of the two sequences must be made equal. This we usually do by padding the short sequence with zeros. The linearity theorem is shown in Eq. (3.73). $x_1[n]$ and $x_2[n]$ are both length N sequences.

$$\mathcal{D}[x_3[n]] = \mathcal{D}[a_1x_1[n]+a_2x_2[n]] = X_3[k]$$

$$\text{So } X_3[k] = a_1X_1[k]+a_2X_2[K] \tag{3.73}$$

Example 3.13

Determine the DFT of the linear combination of the two sequences $x_1[n]=[1\ 3\ 3\ 1]$ and $x_2[n]=[1\ 2\ 3\ 4\ 3\ 2\ 1]$.

Solution 3.13: We first notice that the sequence $x_1[n]$ is a length 4 sequence, while the sequence $x_2[n]$ is a length 7 sequence. So we must first make the two sequences of equal length. We do this by padding three zeros to the end of sequence $x_1[n]$. The modified sequence $x_1[n]$ becomes $x_{1m}[n]=[1\ 3\ 3\ 1\ 0\ 0\ 0]$. Now, we can determine the DFT and combine the DFT to get the DFT of the combined sequence. The result is shown in the table below. Notice that the sum of the first two columns equals the third column.

The following table shows the relation of the DFT of the three sequences.

DFT of $x_1[n]$	DFT of $x_2[n]$	DFT of the linear combination
8.0000	16.0000	24.0000
$1.3019 + 5.7042i$	$-4.5489 + 2.1906i$	$-3.2470 + 7.8948i$
$-1.7470 + 0.8413i$	$0.1920 - 0.2408i$	$-1.5550 + 0.6005i$
$-0.0550 - 0.0689i$	$-0.1431 + 0.6270i$	$-0.1981 + 0.5581i$
$-0.0550 + 0.0689i$	$-0.1431 - 0.6270i$	$-0.1981 - 0.5581i$
$-1.7470 - 0.8413i$	$0.1920 + 0.2408i$	$-1.5550 - 0.6005i$
$1.3019 - 5.7042i$	$-4.5489 - 2.1906i$	$-3.2470 - 7.8948i$

3.8.2
The DFT of a Circular Shift in Time

When the time domain sequence is shifted in time, the sequence suffers a circular shift. The DFT of this shifted sequence is obtained as shown in Eq. (3.74). Here, we once again see that a shift in the time domain results in a rotation in the frequency domain.

$$\mathcal{D}\left[x\left[((n-m))_N\right]\right] = \sum_{n=0}^{N-1} x[n-m] W_N^{nk}$$

$$\mathcal{D}\left[x\left[((n-m))_N\right]\right] = \sum_{l=m}^{N+m-1} x[l] W_N^{(l+m)k} = W_N^{mk} \sum_{l=0}^{N-1} x[l] W_N^{lk} \tag{3.74}$$

$$\mathcal{D}\left[x\left[((n-m))_N\right]\right] = W_N^{mk} X[k]$$

3.8.3
The IDFT of a Circular Shift in Frequency

When the frequency domain sequence is shifted in frequency, the sequence undergoes a circular shift. The IDFT of this shifted sequence is obtained as shown in Eq. (3.75). Here, we once again see that a shift in the frequency domain results in a rotation in the time domain.

$$\mathcal{D}^{-1}\left[X\left[((k-m))_N\right]\right] = \frac{1}{N} \sum_{k=0}^{N-1} X[k-m] W_N^{-nk}$$

$$\mathcal{D}^{-1}\left[X\left[((k-m))_N\right]\right] = \frac{1}{N} \sum_{l=m}^{N+m-1} X[l] W_N^{-(l+m)n} = W_N^{-mn} \frac{1}{N} \sum_{l=0}^{N-1} X[l] W_N^{-nl} \tag{3.75}$$

$$\mathcal{D}^{-1}\left[X\left[((k-m))_N\right]\right] = W_N^{-mn} x[n]$$

3.8.4
Circular Convolution

Convolution of the DFT has to be treated as a special operation. To evaluate the DFT of the convolution of two finite domain sequences, we proceed from the definition of DFT as shown in Eq. (3.76)

$$\mathcal{D}\left[\sum_{r=0}^{N-1} x[r]h[n-r]\right] = \sum_{n=0}^{N-1}\sum_{r=0}^{N-1} x[r]h[n-r]W_N^{nk} \tag{3.76}$$

In Eq. (3.76), if we interchange the order of the two summations and rearrange, we get Eq. (3.77). So the DFT of the convolution is the product of the two individual DFTs. This is shown in Figure 3.18.

$$\sum_{n=0}^{N-1}\sum_{r=0}^{N-1} x[r]h[n-r]W_N^{nk} = \sum_{r=0}^{N-1} x[r]\sum_{m=0}^{N-1} h[m]W_N^{(m+r)k}$$

$$= \underbrace{\sum_{r=0}^{N-1} x[r]W_N^{rk}}_{\text{DFT of } x[n]} \underbrace{\sum_{m=0}^{N-1} h[m]W_N^{mk}}_{\text{DFT of } h[n]} \tag{3.77}$$

In Eq. (3.77), we are multiplying two length N sequences to get a length N sequence. So the convolution using the DFT is different from the linear convolution we have obtained in Chapter 1. Earlier, we have seen that the result of linear convolution is a sequence that is longer than either of the two sequences; in fact, it is equal to the sum of the two lengths minus 1. This is not so for the convolution seen here; the result of this convolution is the same length as either one of the two sequences. So, the convolution of finite length sequence in this sense must be special and different. We identify this convolution as circular convolution. In a circular convolution, the result of convolution is the same length as the two original sequences and the two sequences are of the same length. To understand circular convolution, look at Figure 3.18.

We know that convolution in the time domain translates to product in the frequency domain. So we can use this fact to determine the convolution of the two sequences. In Figure 3.18, we have a sequence h that is length N_1 and we have another sequence x of length

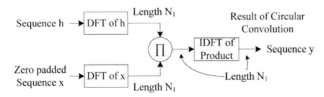

Figure 3.18 Performing convolution by multiplying the DFTs of two time domain sequences to get the resulting length N sequence.

N_2 that we have extended to make it length N_1 by padding N_1-N_2 zeros. We determine their DFTs that are also of length N_1 and multiply the two DFTs together. The resulting DFT is of length N_1. Finally, when we determine the IDFT we get the result of convolution but this result is of length N_1. This is definitely different from the linear convolution. We call this the circular convolution.

To understand the circular convolution, examine Figure 3.19. In Figure 3.19, we demonstrate how the circular convolution is accomplished. We begin with the two sequences, which are of equal length, lined up as shown in Figure 3.19a. The first step is to reverse one of the sequences; here, we have reversed the sequence h that is attached to the inner circle. This is shown in Figure 3.19b. In this position, when we perform the product of the corresponding terms and add the individual products together, we get the first output of the circular convolution. Next, we shift the sequence by one position; shifting in this case involves the rotation of the inner sequence. This is shown in Figure 3.19c. After the sequence is rotated, we perform the product and add the products to get the second output. This way we continue to rotate and compute the product. After $N-1$ rotations, the circular convolution is complete.

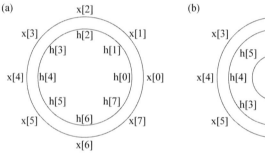

(a) Two sequences that have to be convolved

(b) The inner sequence has been reversed. Now we can compute the first output.

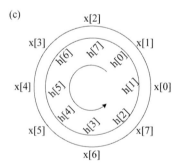

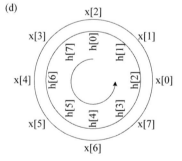

(c) Shift by one position. Now we can compute the second output.

(d) Shift by one more position. Now we can compute the third output.

Figure 3.19 Demonstrating the circular convolution using a pictorial representation of the two sequences and the shifting process that is circular.

Result of Circular
Convolution

Sequence h	1	2	4	3	3	1	Result
Reversed Sequence x	1	0	0	1	3	2	15
	2	1	0	0	1	3	10
	3	2	1	0	0	1	12
	1	3	2	1	0	0	**18**
	0	1	3	2	1	0	**23**
	0	0	1	3	2	1	**20**

Figure 3.20 Performing the circular convolution.

Example 3.14

Determine the circular convolution of the two sequences $h = [1\ 2\ 4\ 3\ 3\ 1]$ and $x = [1\ 2\ 3\ 1]$.

Solution 3.14: In order to determine the circular convolution, we must have both the sequences of the same size. So we will first extend sequence x to be of size 6 by padding it with two zeros. This makes the sequence $x = [1\ 2\ 3\ 1\ 0\ 0]$. With the new sequence formed, we can compute the convolution as shown in Figure 3.20. In Figure 3.20 in the topmost row, we have the sequence h listed. Then, in the next row we have the modified sequence x $[-n]$ listed; this is the reversed sequence. When we perform the individual products and then add them across, we get the first result of the convolution. In the next row, we have rotated the sequence x and by performing the same arithmetic we get the next result. Similarly, we rotate the sequence x and perform the required arithmetic that gives us the result of circular convolution.

It is interesting to note that the linear convolution of the two sequences in Example 3.14 would be as shown in Eq. (3.78). When we compare the result of the circular convolution with the result of the linear convolution, we find that part of the circular convolution is the same as the linear convolution.

$$x[n] * h[n] = \sum_{k=-\infty}^{\infty} x[n-k]h[k]$$

$$x[n] * h[n] = [1\ 2\ 4\ 3\ 3\ 1] * [1\ 2\ 3\ 1]$$

$$x[n] * h[n] = [14\ 11\ \mathbf{18}\ \mathbf{23}\ \mathbf{20}\ 14\ 6\ 1]$$

(3.78)

The items in bold are the same in both the linear and the circular convolution. So it seems possible to get linear convolution using circular convolution. We investigate this in the next section.

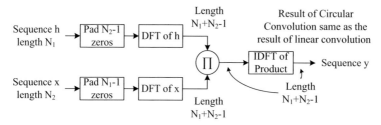

Figure 3.21 Using circular convolution to get the result of linear convolution.

3.8.5
Linear Convolution Using the Circular Convolution

In Example 3.14, we have seen that there are some outputs of the circular convolution that are exactly equal to the corresponding outputs from the linear convolution. On careful examination, we can say that if the sequence $h[n]$ is of length N_1 and the sequence $x[n]$ is of length N_2, then there are exactly (assuming $N_1 > N_2$) $N_1 - (N_2 - 1)$ outputs that are identical in both the linear and the circular convolution. This suggests to us that if we increase the length of the two sequences to be equal to $N_1 + N_2 - 1$, then the result of the circular convolution will be the same as the linear convolution. To increase the length of the two sequences, we pad the sequence with length N_1 with $N_2 - 1$ zeros, and we pad the sequence with length N_2 with $N_1 - 1$ zeros. Now, if we perform the circular convolution, the result will be the same as the linear convolution. This is shown in Figure 3.21.

By using MATLAB, we can evaluate convolution as shown in MATLAB 3.5. In the script, we have introduced two new commands; they are the "conv" and the "cconv" commands. The *conv* command determines the linear convolution. The *cconv* command determines the circular convolution. In the default condition, circular convolution is performed with $N_1 + N_2 - 1$ samples in each sequence. You can specify the length of the sequences to be used in the circular convolution as the third parameter in the *cconv* command. In the script of MATLAB 3.5, we determine the linear convolution of two sequences, then we determine the circular convolution of size N_1, and finally we determine the linear convolution using circular convolution by first extending both the sequences by padding zeros at the end of the sequences.

MATLAB 3.5

```
%% Evaluate the linear convolution of two un-equal length sequences
x=[ 1 2 4 3];
h=[ 1 2 2 2 2 1 1];
x_conv_h=conv(x,h),
%% Next determine the circular convolution of length equal
%% to the length of the longer sequence.
x_cconv_h=cconv(x,h,max(length(h),length(x))),
```

```
%% Now pad the two sequences so that they are length of the result
%% of the linear convolution and then perform the circular
%% convolution.
xlong=[ x zeros (1, (length (h) -1))];
hlong=[ h zeros (1, (length (x) -1))];
xl_cconv_hl=cconv (xlong, hlong, length (xlong)),
The result of running this program are as follows:
x_conv_h =1 4 10 17 20 19 17 12 7 3
x_cconv_h =
13.0000 11.0000 13.0000 17.0000 20.0000 19.0000 17.0000
xl_cconv_hl =
1.0000  4.0000  10.0000  17.0000  20.0000  19.0000  17.0000  12.0000
7.0000 3.0000
```

From the result you can clearly see that the circular convolution of length equal to the longer sequence (after making the shorter sequence equal to the longer sequence) gives you the result that is the same as the linear convolution only part of the time, the number of results that are the same as the linear convolution equal to $(N_1 - N_2 + 1)$ samples. When we obtained the circular convolution of the extended sequences (both sequences had length equal to $(N_1 + N_2 - 1)$, which is the length of the linear convolution), we found that it is the same as the linear convolution of the two sequences. We will use this fact in the next section where we evaluate the convolution of a very long sequence with a shorter sequence.

Example 3.15

We have two sequences: $x = [1\ 2\ 3\ 0\ 1]$ and $h = [1\ 2\ 3\ 4\ 4\ 3\ 2\ 1]$. We want to perform convolution of the two sequences in the following manner.

a) Pad sequence x with three zeros at the end, so $x = [1\ 2\ 3\ 0\ 1\ 0\ 0\ 0]$. Perform the circular convolution of the two sequences. Identify which samples from the circular convolution are the same as the linear convolution.

b) Pad sequence x with three zeros at the beginning, so $x = [0\ 0\ 0\ 1\ 2\ 3\ 0\ 1]$. Perform the circular convolution of the two sequences. Identify which samples from the circular convolution are the same as the linear convolution.

c) Pad sequence x with three zeros as shown here, so $x = [0\ 0\ 1\ 2\ 3\ 0\ 1\ 0]$. Perform the circular convolution of the two sequences. Identify which samples from the circular convolution are the same as the linear convolution.

Solution 3.15: The result of linear convolution is

$$\text{conv}(x, h) = 1\ 4\ 10\ 16\ 22\ 25\ 23\ 18\ 12\ 6\ 2\ 1$$

a) The result of circular convolution in part a is $\text{cconv}(xa, h, 8)$ = 13 10 12 17 **22 25 23 18**. The portion of the circular convolution highlighted in bold is the same as the linear convolution.

b) The result of circular convolution in part a is `cconv(xa,h,8)` = 25 23 18 13 10 12 17 **22**. The portion of the circular convolution highlighted in bold is the same as the linear convolution. We can think of the sequence in "b" as circularly shifted by three places from the sequence in "a." So the convolution result is also circularly shifted by three places.

c) The result of circular convolution in part a is `cconv(xa,h,8)` = 23 18 13 10 12 17 **22 25**. The highlighted portion of the circular convolution is the same as the linear convolution. We can think of the sequence in "c" as circularly shifted by two places from the sequence in "a." So the convolution result is also circularly shifted by two places.

3.9
DFT of Real Sequences

Earlier, we have seen that the DFT of a complex sequence is itself a complex sequence. We have also seen in the symmetry properties that we can determine the DFT of only the real part of the complex sequence and we can separate the imaginary part and the real part of the complex sequence. In most applications, the sequences that we have are only real sequences. So a question arises: is it possible to say take the first half of the sequence and treat it as the real part of the sequence and the second half of the sequence as the imaginary part of the sequence and determine the DFT of this complex sequence and from this separate out the DFT of the real part and the DFT of the imaginary part to get the DFT of the entire sequence? The answer of course is yes and this is how we do this.

3.9.1
DFT of a 2N Point Real Sequence Using N Point DFT

Consider a $2N$ point real sequence $x[n]$ with its DFT $X[k]$ also being $2N$ points. We divide this sequence so that the first half of the sequence $x[n]$ is the real part, while the second half of the sequence $x[n]$ is the imaginary part of the new sequence $w[n]$ as shown in Eq. (3.79).

$$w[n] = x[n] + ix[n+N] \quad 0 \le n \le \frac{2N-1}{2} \tag{3.79}$$

So we can write the DFT of the sequence $x[n]$ as shown in Eq. (3.80).

$$X[k] = \sum_{n=0}^{2N-1} x[n]\, W_{2N}^{nk} = \underbrace{\sum_{n=0}^{N-1} x[n]\, W_{2N}^{nk}}_{\substack{\text{Transform of real} \\ \text{part of } w[n]}} + \underbrace{\sum_{n=N}^{2N-1} x[n]\, W_{2N}^{nk}}_{\substack{\text{Transform of} \\ \text{imaginary part of } w[n]}} \tag{3.80}$$

We can write Eq. (3.80) as shown in Eq. (3.81).

$$X[k] = \sum_{n=0}^{N-1} x[n]\, W_{2N}^{nk} + \sum_{n=N}^{2N-1} x[n]\, W_{2N}^{nk}$$

$$X[k] = \sum_{n=0}^{N-1} x[n]\, W_{2N}^{nk} + \sum_{n=0}^{N-1} x[n+N]\, W_{2N}^{(n+N)k} \tag{3.81}$$

If we evaluate all the even points of $X[k]$ first and then all the odd points of $X[k]$, we get Eq. (3.82).

$$X[2k] = \sum_{n=0}^{N-1} x[n]\, W_{2N}^{n(2k)} + \sum_{n=0}^{N-1} x[n+N]\, W_{2N}^{(n+N)(2k)}$$

$$X[2k] = \sum_{n=0}^{N-1} x[n]\, W_{N}^{n(k)} + \sum_{n=0}^{N-1} x[n+N]\, W_{N}^{n(k)}\, W^k \tag{3.82}$$

When we examine Eq. (3.82) carefully, we see that all the even points of $X[k]$ are obtained by the transform of $w[n]$ since W^k is just 1 for all values of k. Similarly, when we examine all the odd values of $X[k]$, we get Eq. (3.83).

$$X[2k+1] = \sum_{n=0}^{N-1} x[n]\, W_{2N}^{n(2k+1)} + \sum_{n=0}^{N-1} x[n+N]\, W_{2N}^{(n+N)(2k+1)}$$

$$X[2k+1] = \sum_{n=0}^{N-1} W_{2N}^{n} x[n]\, W_{N}^{n(k)} - \sum_{n=0}^{N-1} W_{2N}^{n} x[n+N]\, W_{N}^{n(k)} \tag{3.83}$$

When we examine Eq. (3.83) carefully, we find that the transform of all the odd values of $X[k]$ is obtained by multiplying the conjugate of $w[n]$ by W_{2N}^n and then determining its DFT. This way we have obtained a DFT of a sequence of length $2N$ by determining two DFT of length N. This is a saving in the computation effort required as we will see later in this book when we study an efficient algorithm that computes the DFT known as the fast Fourier transform.

3.9.2
Two N Point DFT of Real Sequences Using a Single N Point DFT

Sometimes, you have two N point sequences; here they are $g[n]$ and $h[n]$ and you want the DFT of both the sequences individually. This can also be done by combining the two sequences into one sequence, where one forms the real part while the other forms the imaginary part of the new sequence as shown in Eq. (3.84)

$$x[n] = g[n] + jh[n] \tag{3.84}$$

Then, the DFT of $g[n]$ and the DFT of $h[n]$ are obtained directly from the symmetry relations as shown in Eq. (3.85).

$$G[k] = \frac{1}{2}\left[X[k] + X^*\left[\langle -k\rangle_N\right]\right]$$

$$H[k] = \frac{1}{2j}\left[X[k] - X^*\left[\langle -k\rangle_N\right]\right] \tag{3.85}$$

3.10
Convolution of Very Long Sequences

In many applications, in which we are interested in performing the linear convolution of two sequences, one of the two sequences is considered very long compared to the other sequence. This is certainly true in, say, a speech signal or a song performance. To actually perform the convolution in the time domain would often be prohibitively expensive as far as computing facilities are concerned. Such convolutions are performed as we have seen in Figure 3.21 with a slight modification. We saw in MATLAB 3.5 that it is possible to perform linear convolution using circular convolution. Then in Example 3.15, we saw that when we perform a circular convolution, part of the result is the same as the result that we get from linear convolution. These two facts are the key to the method that we will employ to get linear convolution while we use circular convolution.

In general, when we want to convolve two sequences $x_1[n]$ of length N_1 and another sequence $x_2[n]$ of length N_2, there linear convolution will result in a sequence of length $(N_1 + N_2 - 1)$. Thus, we would first obtain two DFT of length $(N_1 + N_2 - 1)$, multiply them together, and get an IDFT of length $(N_1 + N_2 - 1)$ to get the linear convolution. This would work if the two sequences were of finite length and we could store the sequences. But consider a sequence such as speech or a radar echo. These signals are not generally of finite length. Also, we are not usually at liberty to store the signal, as storing the signal would introduce a very large delay in the processing of the signal. Generally, we would like to avoid such a delay in the processing of signals.

So to filter the long signal while still avoiding the large delay and at the same time using DFT, we will divide the signal into segments of length L. Each segment will then be convolved with the filter sequence by determining the DFT of the smaller segment and multiplying them together. Finally, we will keep only those samples that are the same as the linear convolution and fit them in an appropriate manner so that after the complete operation, we will be left with the linear convolution of the two sequences that is computed in segments.

3.10.1
Overlap and Add Method

To show how we fit the filtered sections together, consider the unit sample response of the filter $h[n]$ to be length M and the signal to be filtered to be of infinite length divided into segments of length L. The two sequences are shown in Figure 3.22. Let us first divide the infinite sequence $x[n]$ into many sections of length L. The kth section of the infinite sequence is then denoted by $x_k[n]$ and it consists of

$$x_k[n] = \begin{cases} x[n] & kL \leq n \leq (k+1)L-1 \\ 0 & \text{otherwise} \end{cases} \quad k = 0, 1, 2, \ldots \tag{3.86}$$

With this definition of $x_k[n]$, we can write the sequence $x[n] = \sum_{k=0}^{\infty} x_k[n]$ and when we convolve $x[n]$ with $h[n]$, we get the convolution result as shown in Eq. (3.87).

$$x[n] * h[n] = \sum_{k=0}^{\infty} x_k[n] * h[n] \tag{3.87}$$

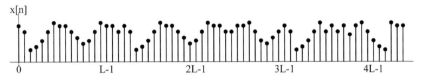

Figure 3.22 Finite duration sequence *h*[*n*] to be convolved with an infinite duration sequence *x*[*n*].

We know that each of the $x_k[n]$ has only L points and $h[n]$ has only M points, so each of the convolution $x_k[n]^*h[n]$ will be $(L + M - 1)$ points long. To get this convolution using circular convolution, we will pad the sequence $x_k[n]$ with zeros to length $L + M - 1$, as shown in Figure 3.23a.

When we convolve each of the sections with the filter transfer function that is also zero padded to make it length $L + M - 1$, we get the individual convolutions as shown in Figure 3.23b. Consider the first convolution for now, it has $(L + M - 1)$ points from these $(L + M - 1)$ points; first L points of the circular convolution are the same as the linear convolution and the last $M - 1$ points are not the same as those that we would get from a linear convolution.

We call these $M - 1$ points as the tail of the convolution; the tail is not the correct result because it is incomplete. Now, consider the second convolution; in the second convolution, the first $M - 1$ points also represent an incomplete convolution, but when these $M - 1$ points are added to the tail of the previous convolution, they together represent the result that corresponds to the result of linear convolution. This is where the method gets its name; it adds the overlapped $M - 1$ points from two successive convolutions to get the linear convolution. From the remaining points, the last $M - 1$ points once again form the tail and the remaining points are the same as the result of the linear convolution. Continuing this way, we get the complete linear convolution of an infinite sequence using smaller sequence.

There is one more modification to the last convolution. In the last convolution, the long sequence may not have L points. This can be easily corrected by extending the shorter sequence to length $(L + M - 1)$ by padding it with the appropriate number of zeros. Then, after adding the leading $M - 1$ terms to the previous tail, we store all the remaining points and the convolution is complete.

3.10.2
Overlap and Save Method

Another method that we can use to convolve a very long sequence with a short sequence is the overlap and save method. In this method also, we divide the long sequence into shorter

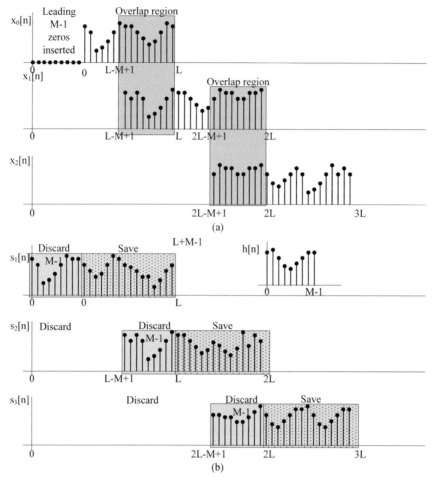

Figure 3.23 Dividing the long sequence for convolution. Performing the convolution using the overlap and save method. (a) How the sequence x[n] is divided to perform overlap and add method. (b) How the linear convolution is obtained using overlap and save method.

sequences, but unlike the overlap and add method where we divided the long sequence in consecutive blocks of length L, in this method we overlap several sample points of one segment with the next segment. The division of sequences is explained in Eq. (3.88) and Figure 3.23.

In Figure 3.23a, we see that each segment of the long sequence gets L new sample points and it repeats $M - 1$ sample points from the previous segment. This is true for all segments except the first segment. Since this segment does not have a previous segment, we pad it in the front with $M - 1$ zeros. This happens only in the first segment. From then on, all the other segments choose the last $M - 1$ samples from the previous segment. Thus, the length of the segment is $L + M - 1$ and each segment gets L new sample points. So we can write the segments as shown in Eq. (3.88).

$$
\left.
\begin{array}{rcl}
x_1[n] &=& [\text{zeros}(1,(M{-}1))\, x[1:L]] \\
x_2[n] &=& [x[(L{-}M{+}1):2L]] \\
x_3[n] &=& [x[(2L{-}M{+}1):3L]] \\
\vdots & & \vdots \\
x_k[n] &=& [x[((k{-}1)L{-}M{+}1):kL]]
\end{array}
\right\} \; 0 \le n \le L+M-1
\qquad (3.88)
$$

Now, if we let each convolution of the small segment $x_k[n]$ with the filter sequence $h[n]$ (which is padded by zeros to make it also of length $(L + M - 1)$) be denoted by $s_k[n]$, then these individual convolutions give us the correct result that we are looking for in the last L terms after we discard the first $M - 1$ terms from the result of the convolution. This is shown in Figure 3.23b. By concatenating the last L samples from each of the individual s_k convolutions, we get the result that we are looking for. One more point that you have to remember is that the very last segment also needs to have its last $M - 1$ points overlapped, or else the convolution is not complete. So if the division of the sequences brings you to the end of the long sequence, then you need to add one more segment with only the last $M - 1$ terms that have not overlapped with any other sequence so far. Doing this will complete the convolution. This is demonstrated in Example 3.16

Example 3.16

Consider the following two sequences. Sequence $h = [1\ 2\ 3\ 0\ 1]$ is the short sequence while $x = [1\ 2\ 1\ 3\ 2\ 4\ 3\ 5\ 4\ 3\ 4\ 2\ 3\ 1\ 2\ 1\ 1\ 2\ 4\ 3\ 2\ 1\ 2\ 3]$ is the long sequence. Divide the long sequence into four equal segments and then perform the linear convolution using both the overlap and add method and the overlap and save method.

Solution 3.16: First, consider the overlap and add method. Since the long sequence is divided into four segments of length 6 each and the short sequence is of length 5, we first need to make both the sequences of length $6 + 5 - 1 = 10$. The length 10 sequences are listed as follows:

```
x1[ n] =[ 1 2 1 3 2 4 0 0 0 0]  x2[ n] =[ 3 5 4 3 4 2 0 0 0 0]
x3[ n] =[ 3 1 2 1 1 2 0 0 0 0]  x4[ n] =[ 4 3 2 1 2 3 0 0 0]  h[ n] =[ 1 2 3 0 1 0 0 0 0 0]]]
```

The result of convolving these sequences using MATLAB is given below.

```
xa = cconv (x1,h,10) =[ 1 4 8 11 12 19 15 15 2 4]
xb = cconv (x2,h,10) =[ 3 11 23 26 25 24 20 9 4 2]
xc = cconv (x3,h,10) =[ 3 7 13 8 12 8 9 7 1 2]
xd = cconv (x4,h,10) =[ 4 11 20 14 14 13 14 10 2 3]
```

Now, when we add the tail of x_a to the leading four values of x_b, then add the tail of x_b to the leading four values of x_c, and finally add the tail of x_c to the leading four values of x_d, we have the same result as we would have obtained if we had performed a linear convolution; this is done for you in Figure 3.24.

$x_a = cconv(x_1,h,10)$ [1 4 8 11 12 19 15 15 2 4]

$x_b = cconv(x_2,h,10)$ [3 11 23 26 25 24 20 9 4 2]

$x_c = cconv(x_3,h,10)$ [3 7 13 8 12 8 9 7 1 2]

$x_d = cconv(x_4,h,10)$ [4 11 20 14 14 13 14 10 2 3]

Resultant Linear Convolution   1 4 8 11 12 19 18 26 25 30 25 24 23 16 17 10 12 8 13 18 21 16 14 13 14 10 2 3

Figure 3.24 Using overlap and add method to perform linear convolution.

Next, we consider the overlap and save method. Again, since the long sequence is divided into four segments of length 6 each and the short sequence is of length 5, the overlap portion will be of length 4 and the first sequence will begin with four zeros. The length 10 sequences are listed below. Notice we had to add the sequence $x_5[n]$ to make sure that the last four samples were properly overlapped.

x1[n] =[0 0 0 0 1 2 1 3 2 4] x2[n] =[1 3 2 4 3 5 4 3 4 2]
x3[n] =[4 3 4 2 3 1 2 1 1 2] x4[n] =[2 1 1 2 4 3 2 1 2 3]
x5[n] =[2 1 2 3 0 0 0 0 0 0] h[n] =[1 2 3 0 1 0 0 0 0 0]

The result of convolving these sequences using MATLAB is given below.

```
xa = cconv(x1,h,10) = [ 15 15 2 4 1 4 8 11 12 19]
xb = cconv(x2,h,10) = [ 21 14 15 19 18 26 25 30 25 24]
xc = cconv(x3,h,10) = [ 13 18 23 21 23 16 17 10 12 8]
xd = cconv(x4,h,10) = [ 16 15 11 10 13 18 21 16 14 13]
xe = cconv(x5,h,10) = [ 2 5 10 10 14 10 2 3 0 0
```

Now from these individual convolutions, when we strip away the part of the result that is not linear convolution (the first four or $M-1$ points from each of the smaller convolutions), we will get the result that is the same as the linear convolution as shown in Figure 3.25.

$x_d = cconv(x_4,h,10) = [16 15 11 10 13 18 21 16 14 13]$
$x_e = cconv(x_5,h,10) = [2 5 10 10 14 10 2 3 0 0]$

$x_a = cconv(x_1,h,10)$ [15 15 2 4 1 4 8 11 12 19]
$x_b = cconv(x_2,h,10)$ [21 14 15 19 18 28 25 30 25 24]
$x_c = cconv(x_3,h,10)$ [13 18 23 21 23 16 17 10 12 8]
$x_d = cconv(x_4,h,10)$ [6 15 11 10 13 18 21 16 14 13]
$x_e = cconv(x_5,h,10)$ [2 5 10 10 14 10 2 3 0 0]

Discard all the shaded points.

Resultant Linear Convolution   1 4 8 11 12 19 18 26 25 30 25 24 23 16 17 10 12 8 13 18 21 16 14 13 14 10 2 3

Figure 3.25 Using overlap and save method to perform linear convolution.

3.11
Chapter Summary

We began this chapter with a review of the Fourier transform. From the definition, we saw that the Fourier transform of a time signal that was continuous was a continuous function in the frequency domain. We also saw that the Fourier transform of the signal exists if the time domain signal satisfies the Dirichlet conditions. A signal that satisfies the Dirichlet conditions should have finite maxima or minima and finite number of discontinuities in a finite duration of time.

We saw the block diagram of the sampling process. This showed how we convert a continuous time signal into a sampled signal.

Next, we studied the discrete time Fourier transform. Here, the time domain signal is a sampled infinite duration signal, while the Fourier domain signal is a continuous periodic signal with period 2π. We studied several properties of the DTFT, and among the properties we studied were the linearity property, the time reversal property, the shift in time, the shift in frequency, the convolution in time, and the Parseval's relation. All these properties are listed in Table 3.1 for you to read and review.

Sampling a continuous time signal was studied next. Here, we saw that the frequency content of the sampled signal is preserved in the sampled signal as long as we sample the signal at least at the Nyquist rate. This is a frequency that satisfies the Nyquist theorem and for a band-limited signal, the Nyquist rate is twice the highest frequency present in the signal. We also saw that when we sample the signal we create images of the frequency content of the signal at intervals of 2π. The original signal can be recovered from the sampled signal by using the interpolation formula. If the signal is sampled at a rate that is slower than the Nyquist rate, then we say that the sampled signal is aliased and the original signal is lost.

The discrete Fourier transform is the transform of a finite duration, the discrete time domain signal, which gives us a finite duration discrete frequency domain sequence. To understand this transform, we interpreted the finite duration signal in one of the two different ways. One way to visualize the time domain signal was to think of the signal as a finite duration signal; another interpretation that we used was that this finite duration signal was one cycle of a periodic signal. Both the viewpoints were essential to understand the properties of the transform. The properties of the transform were very similar to those of the DTFT.

There was an interesting twist when we examined the convolution property of this transform. Since this signal is finite in length and duration, the result of convolution had to be of finite length and duration that was the same as the signal. This led us to the concept of circular convolution. When we were studying the circular convolution, we found that it was possible to get the same result from circular convolution as we got from linear convolution if we extended the sequence by padding it with zeros and making its length $L + M - 1$ that would also be the length of the linear convolution.

In the study of the DFT, we saw that one of the symmetry properties is that we can determine the DFT of the real part of the sequence and the DFT of the imaginary part of the sequence and separate them. We then used this property in various ways to reduce the computation load that the DFT has. One such way was combine two real sequences of length N to form one complex sequence of length N and then evaluate the N point DFT. This DFT can be separated so that we have the transform of just the real part and also the transform of just the imaginary part.

Finally, we closed the chapter with the study of how we perform a linear convolution of a very long sequence with a shorter sequence by taking smaller chunks of the long sequence. We found that there were two different methods we could use: one was the overlap and add method in which the tail from one section was added to the front of the next section to get the result that is the linear convolution of the two sequences. In the other method – the overlap and save method – we overlap the last $M - 1$ points of the previous sequence with the first $M - 1$ points of the present sequence. In this method, we discard the first $M - 1$ points of any convolution and save the last L points of the convolution. Concatenating these L points from the various convolutions gives us the result that is identical to the linear convolution.

MATLAB commands introduced in this chapter:

CONV: Perform the convolution of two sequences

CCONV: Perform the circular convolution of the two sequences. If the two sequences are of unequal length, then the shorter sequence is zero padded to the length of the longer sequence. If a third parameter is present, then both the sequences are zero padded at the end to the length specified in the third parameter.

FFT: Perform the DFT of the sequence specified. The second parameter can be specified. If the second parameter is specified, then the DFT is computed assuming that the sequence is of length specified. The sequence is extended to the specified length by zero padding.

IFFT: Determine the inverse DFT of the given sequence.

4
The Z-Transform

4.1
Introduction

In systems that operate on continuous time signals, we used the Laplace transform as a general tool to analyze the system. The Laplace transform tool was more powerful than the Fourier transform tool as the Laplace transform converges for many more signals and systems than the Fourier transform does. This led us to say that the Laplace transform is generalizing the Fourier transform. In a similar manner for discrete time systems, we have seen the use of the DTFT as an analysis tool. Just as we used the Laplace transform for systems that did not converge for the Fourier transform in the continuous domain, we will use the Z-transform for systems that do not converge for the DTFT in the discrete domain. So the Z-transform is generalizing the DTFT for discrete time systems. In this chapter, we will study the Z-transform and use it to represent a system, to analyze a system, and finally to design a system.

In studying the Z-transform, we will need to use many results from the theory of complex numbers. The results themselves are simple to use once you know where and how to apply them. These results have been proved rigorously in many books on complex numbers. We will not employ so much mathematical rigor in deriving the results, and most of the time we will state the result and the conditions where the result applies and then proceed to show how to use the result. This does not mean that we will not be precise and accurate in defining and using the result. We will just take a few liberties with regard to the mathematical rigor.

4.2
Definition of the Z-Transform

The Z-transform, which is denoted as $X(z)$ of a discrete sequence $x[n]$, is defined by an infinite two-sided summation as shown in Eq. (4.1).

$$X(z) = \sum_{n=-\infty}^{\infty} x[n]z^{-n} \tag{4.1}$$

In Eq. (4.1), the variable z is a complex number and the z-transform itself is known as the two-sided Z-transform. We can express the complex variable z in its polar form to get a better

Digital Filters: Theory, Application and Design of Modern Filters, First Edition. Rajiv J. Kapadia.
© 2012 Wiley-VCH Verlag GmbH & Co. KGaA. Published 2012 by Wiley-VCH Verlag GmbH & Co. KGaA.

idea of what the z-transform is and does. In its polar form, the complex variable has a radius or the length and a phase of rotation so we can write $z = re^{j\omega}$. When we replace z in Eq. (4.1) by its polar representation, we get a very interesting interpretation of the definition of the Z-transform. This is shown in Eq. (4.2).

$$X(z)|_{z=re^{j\omega}} = \sum_{n=-\infty}^{\infty} x[n]\left(re^{j\omega}\right)^{-n} = \sum_{n=-\infty}^{\infty} (x[n]r^{-n})e^{j\omega n} \tag{4.2}$$

Equation (4.2) is the definition of the Z-transform and can also be interpreted as the DTFT of the sequence $x[n]r^{-n}$ that is the sequence $x[n]$ multiplied by an exponential sequence. This is where the Z-transform gets most of its usefulness and generality. Now, if the sequence $x[n]$ does not converge, then we cannot determine its DTFT, but when the same sequence $x[n]$ is multiplied by the exponential r^{-n}, it can be made to converge[1] for some value of r. The value of r for which the summation in Eq. (4.2) converges is known as the region of convergence (RoC). It is this ability to define the region of convergence that makes the Z-transform a more powerful tool in analyzing and studying the discrete time systems.

In Eq. (4.2), if we make r exactly equal to 1, then we see that the Z-transform and the DTFT both are exactly equal to each other. So sometimes, we say that the DTFT is the Z-transform evaluated on the unit circle in the Z-plane. The unit circle and the concept of what happens inside and outside the unit circle is so important that most of the time when we draw any representation of a system using the z-transform we will also include the unit circle. In layman's terms, the Z-transform and the unit circle take on a special meaning for the following reason.

We know that we cannot have any poles in the region of convergence because by definition of a pole, the function would not converge at the location of the pole, hence the summation would become infinite, and so all the poles of the Z-transform must lie outside the region of convergence. Next, when we make $r = 1$, the Z-transform is evaluated on the unit circle and if the Z-transform exists, then it has all its poles within (inside) the unit circle since the region of convergence extends from $r = 1$ to $r = \infty$. We have also seen that when $r = 1$, the DTFT and the Z-transform are identical and for a stable system the summation in Eq. (4.2) with $r = 1$ must also be finite. So we can conclude that for a stable system, the region of convergence must not contain any poles of the system, that the region of convergence must include the unit circle (circle in the z-plane with $r = 1$), and that all the poles must lie inside the unit circle. For this reason, whenever we represent any discrete system using the Z-transform, we also include the unit circle.

4.2.1
The Importance of the Region of Convergence

We have just seen that the Z-transform is the summation as shown in Eq. (4.1). Let us see how we use this summation to determine the Z-transform of different systems. Toward this end, pay particular attention to how the sum is evaluated and examine Examples 4.1 and 4.2.

---

[1] For those who are sticklers for math rigor, it is definitely possible to come up with a sequence that will not converge for any value of r. But here, we are referring to the most common signals that we will come across in dealing with everyday signals and systems.

Example 4.1

Determine the Z-transform and the region of convergence of $x[n] = a^n u[n]$.
   Solution 4.1: Using the definition of the Z-transform, we can write Eq. (4.3).

$$X(z) = \sum_{n=-\infty}^{\infty} x[n] z^{-n} = \sum_{n=0}^{\infty} a^n z^{-n}$$

$$(4.3)$$

$$X(z) = \sum_{n=0}^{\infty} \left(az^{-1}\right)^n = \frac{1}{1-az^{-1}} \quad az^{-1} < 1$$

   In evaluating the summation, which is a geometric progression, we had to make sure that the geometric progression is less than 1. This condition, which guarantees the summation will not become infinite, gives us the region of convergence, which in this case $az^{-1} < 1$ or $z > a$. So, we say that the Z-transform of $x[n] = a^n u[n]$ is $X(z) = 1/(1-az^{-1})$ and the RoC is $z > a$. This is shown in Figure 4.1.

Example 4.2

Determine the Z-transform and the region of convergence of $x[n] = -a^n u[-n-1]$.
   Solution 4.2: Using the definition of the Z-transform, we can write Eq. (4.4) as follows:

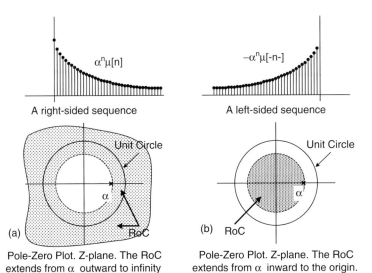

A right-sided sequence   A left-sided sequence

(a) Pole-Zero Plot. Z-plane. The RoC extends from α outward to infinity

(b) Pole-Zero Plot. Z-plane. The RoC extends from α inward to the origin.

Figure 4.1   Two different time domain sequences with the same Z-domain function but each with a different RoC.

$$X(z) = \sum_{n=-\infty}^{\infty} x[n]z^{-n} = \sum_{n=-\infty}^{-1} -a^n z^{-n} = \left(\sum_{n=0}^{\infty} -a^{-n} z^n + 1 \right)$$

$$X(z) = \left(1 - \frac{1}{1-a^{-1}z} \right) = \left(\frac{1-a^{-1}z-1}{1-a^{-1}z} \right) \quad a^{-1}z < 1 \tag{4.4}$$

$$X(z) = \frac{-a^{-1}z}{1-a^{-1}z} = \left(\frac{1}{1-az^{-1}} \right) \quad a^{-1}z < 1$$

In evaluating the Z-transform in Example 4.2, we used the summation of the geometric series and the condition for the summation to converge in this case was $a^{-1}z < 1$.

Examine the results for Examples 4.1 and 4.2. We see that the two time domain sequences are different, but the function that represents $X(z)$, which is $1/(1 - az^{-1})$, is the same for both the sequences. The difference between the two Z-transforms is the region of convergence. So we see that it is very important to include the RoC when we speak of a Z-transform. Without the RoC, the Z-transform is incomplete. Also, the RoC is the condition or the criterion that must be satisfied so that the summation in Eq. (4.1) converges.

The two examples above show us such a dramatic effect; even though the two sequences are different, their Z-transform expressions are exactly identical. What separates the two is the RoC and for each sequence the RoC is a different RoC. When we draw the pole–zero plot of the two Z-transforms, this fact becomes more clear. This plot of the two Z-transforms and the RoC of the two sequences is drawn in the bottom of Figure 4.1. The pole–zero plot is drawn in Figure 4.1. In Figure 4.1a, we have drawn the sequence from Example 4.1 and its Z-transform showing that the pole is located at $z = a$ and the RoC extends from a outward to infinity. In Figure 4.1b, we have drawn the sequence from Example 4.2 and its Z-transform showing that the pole is located at $z = a$ and the RoC extends from a inward to zero.

4.2.2
Region of Convergence of Left- and Right-sided Sequences

In Example 4.1, we saw that the RoC extends outward toward the infinity, while in Example 4.2 we saw that the RoC shrank inward toward the origin. There is a special relation between the RoC and the sequences.

a) **Right-sided sequences**: When a sequence is zero for $n < 0$ and nonzero for $n \geq 0$, we say that the sequence is a right-sided sequence. Such a sequence is given by Eq. (4.5).

$$X_R(z) = \sum_{n=0}^{N} x_R[n]z^{-n} \quad N > 0 \tag{4.5}$$

This sequence will converge from its pole furthest from the origin toward the infinity. If the upper limit N is less than ∞, then there will also be poles present at $z = \infty$ and the RoC will not include the point $z = \infty$, if the sequence is infinite in length, then the RoC is the entire z-plane outside the circle of radius of the largest pole in the sequence.

b) **Left-sided sequence**: When a sequence is zero for $n < 0$ and nonzero for $n \geq 0$, we say that the sequence is a left-sided sequence. Such a sequence is given by Eq. (4.6).

$$X_L(z) = \sum_{n=N}^{-1} x_L[n]z^{-n} \quad N < 0 \tag{4.6}$$

This sequence will converge from its pole closest to the origin inward toward the origin. If the value of N is less than $-\infty$, then there will also be poles present at $z = 0$ and the RoC will not include the point $z = 0$, if the sequence is infinite in length, then the RoC is the entire z-plane inside the circle of radius of the smallest pole in the sequence.

c) **Two-sided sequence**: When the sequence has nonzero points on both sides of zero, we say that the sequence is a two-sided sequence. Such a sequence is given by Eq. (4.7).

$$X_T(z) = \sum_{n=N_1}^{N_2} x_T[n]z^{-n} \quad N_1 < 0 < N_2 \tag{4.7}$$

We can always rewrite the two-sided sequence as a left- and a right-sided sequence shown in Eq. (4.8) as follows:

$$X_T(z) = \underbrace{\sum_{n=N_1}^{-1} x_T[n]z^{-n}}_{\text{Left-sided sequence}} + \underbrace{\sum_{n=0}^{N_2} x_T[n]z^{-n}}_{\text{Right-sided sequence}} \quad N_1 < 0 < N_2 \tag{4.8}$$

In the two-sided sequence, the right-sided sequence will behave like the sequence in Eq. (4.5) and its RoC will be similar to the RoC of the sequence shown in Figure 4.2, while the left-sided sequence will behave like the sequence in Eq. (4.6) and its RoC will be similar to the RoC of the sequence shown in Figure 4.3. But the sequence that we have is a two-sided sequence, so the RoC of the sequence must be such that it satisfies both the types of RoCs as shown in Figures 4.2 and 4.3. This gives us two different possibilities as shown in Figure 4.4.

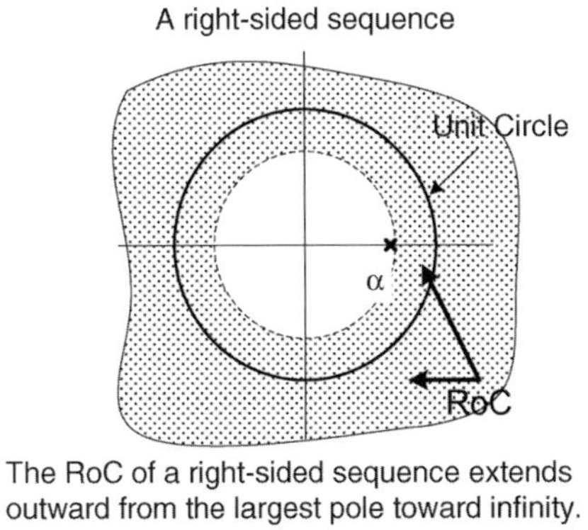

A right-sided sequence

The RoC of a right-sided sequence extends outward from the largest pole toward infinity.

Figure 4.2 The RoC of a right-sided sequence.

A left-sided sequence

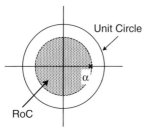

The RoC of a left-sided sequence extends inward
from the smallest pole toward the origin.

Figure 4.3 The RoC of a left-sided sequence.

The first case that we look at is when the largest pole of the right-sided sequence is smaller than the smallest pole of the left-sided sequence. In this case, the RoC for the right-sided sequence and the RoC of the left-sided sequence will have an overlap region as shown on the left-hand side of Figure 4.4. This annular ring-like overlap region is the common RoC and the Z-transform exists only in this region.

The second case that we look at is when the largest pole of the right-sided sequence is greater than the smallest pole of the left-sided sequence. In this case, the RoC for the right-sided sequence and the RoC of the left-sided sequence will have no overlap region as shown on the right-hand side of Figure 4.4. Since there is no common RoC for both the sequences, there is no RoC of the entire sequence and the Z-transform of the sequence does not exist.

So in general, there are four different possible cases for the RoC of a sequence.

a) The sequence is right sided and the RoC is like the one shown in Figure 4.2.
b) The sequence is left sided and the RoC is like the one shown in Figure 4.3.
c) The sequence is two sided and the RoC is like the one shown in Figure 4.4.

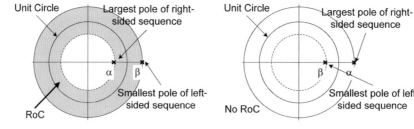

When the largest pole of the right-sided sequence is smaller than the smallest pole of the left-sided sequence. The RoC is the shaded region.

When the largest pole of the right-sided sequence is larger than the smallest pole of the left-sided sequence. There is no common RoC. So this Z-Transform does not exist.

Figure 4.4 Two possible RoCs of a two-sided sequence.

4.3
The Inverse Z-Transform

The Z-transform we have seen takes us from the discrete time domain to the frequency domain. The inverse Z-transform will take us from the frequency domain back to the discrete time domain. There are in general three different methods that we can use to determine the inverse transform of the Z-domain transfer function. They are the *Cauchy integral theorem* that uses the results from the complex number theory to evaluate the inverse of the z-domain function, then there is the *partial fraction expansion* that is similar to the method that we used to determine the inverse Laplace transform and uses the results directly from the Cauchy integral theorem, and finally there is the *long division* method that expresses the Z-domain function in terms of its Laurent series expansion. In this expansion, the coefficient of the z-term indicates the position of the discrete value in the discrete time sequence.

4.3.1
The Cauchy Integral Theorem

We have seen that the Z-transform function is a function of complex variable, so to work with this function we must use the results from the theory of complex numbers. The inverse Z-transform can be derived by using the Cauchy integral theorem. This theorem states that the closed contour integral around any first-order pole in the complex plane is equal to $2\pi j$. This can be stated in the form of Eq. (4.9) as follows:

$$\frac{1}{2\pi j} \int_c z^{k-1} dz = \begin{cases} 1 & k = 0 \\ 0 & k \neq 0 \end{cases} \tag{4.9}$$

In Eq. (4.9), the integral is performed around a closed region on the contour c and the integral is performed in the counterclock-wise direction and the contour c encircles the origin. To use the Cauchy integral, we begin with the definition of Z-transform as given in Eq. (4.1). Next, we multiply both sides of Eq. (4.1) by z^{k-1} and integrate the equation around the closed contour c that encircles the origin and also lies in the RoC of the Z-transform function. By doing this, we get Eq. (4.10).

$$\frac{1}{2\pi j} \int_c X(z) z^{k-1} dz = \frac{1}{2\pi j} \int_c \sum_{n=-\infty}^{\infty} x[n] z^{k-n-1} dz \tag{4.10}$$

On interchanging the order of the summation and the integration in Eq. (4.10), we get the first equality in Eq. (4.11). Then using Eq. (4.9) on the center term, we get the last equality in Eq. (4.11).

$$\frac{1}{2\pi j} \int_c X(z) z^{k-1} dz = \sum_{n=-\infty}^{\infty} x[n] \frac{1}{2\pi j} \int_c z^{k-n-1} dz = x[k] \tag{4.11}$$

Therefore, the inverse of the of the Z-domain function can be written as

$$x[n] = \frac{1}{2\pi j} \int_c X(z) z^{n-1} dz \tag{4.12}$$

In Eq. (4.12), the closed contour is traversed in the counterclock-wise direction and it also includes the origin. The method of evaluation of the inversion integral is by the method of the residues. The method of residues says that the inverse z-transform as given by Eq. (4.12) is evaluated by adding together the residues of the function $X(z)z^{n-1}$ at the location of each pole of the function inside the closed counterclock-wise contour that encircles the origin.

For $X(z)z^{n-1}$ that are rational functions of z, the function can be expressed as follows (by isolating all the poles at any particular location, for example, at $z = z_0$ in the following expression). We assume that there are r identical poles at location $z = z_0$ in the function $X(z)$ z^{n-1}. This is shown in Eq. (4.13).

$$X(z)z^{n-1} = \frac{\Phi(z)}{(z-z_0)^r} \tag{4.13}$$

In Eq. (4.13), $X(z)z^{n-1}$ contains r identical poles at the location $z = z_0$ that we have isolated on the right-hand side of the equation. The function $\Phi(z)$ has no poles at $z = z_0$. Now, we can determine the residue at the pole location $z = z_0$ as shown in Eq. (4.14)

$$\text{Residue}\left[X(z)z^{n-1}\ z = z_0\right] = \frac{1}{(r-1)!}\left[\frac{d^{r-1}\Phi(z)}{dz^{r-1}}\right]_{z=z_0} \tag{4.14}$$

It is very interesting to see what we get when we have just a simple pole at $z = z_0$; in this case $r = 1$, so the residue at the simple pole $z = z_0$ is given by Eq. (4.15).

$$\text{Residue}\left[X(z)z^{n-1}\ \text{at } z = z_0\right] = \Phi(z_0) \tag{4.15}$$

I am sure some of you recognized that you have used the functions on the right-hand side of Eqs. (4.14) and (4.15) in evaluating the residues at the poles when you were trying to determine the inverse Laplace transform. These are the same expressions. We will see that the similarity is repeated when we examine the partial fraction expansion method. The following example will not only show us how we use the inversion integral but also show us the difficulty involved in using this method and why we prefer to use one of the other two methods.

Example 4.3

Determine the inverse Z-transform of the rational function $X(z) = 5z/(z-0.5)(z-0.4)$ using the inversion integral.

Solution 4.3: In the function given we have two poles, a single pole at $z = 0.5$ and a single pole at $z = 0.4$, we also have a zero at the origin. We will first determine the residue at the single pole at $z = 0.5$ as shown in Eq. (4.16).

$$\left[\text{Residue } X(z)z^{n-1} \text{ at } z = 0.5\right] = \left[\frac{5 \cdot z \cdot z^{n-1}}{(z-0.4)}\right]_{z=0.5} \tag{4.16}$$

To evaluate the function in Eq. (4.16), we will first assume that $n \geq 0$, or else there is also a pole at $z = 0$ when $n < 0$; more about this later. To evaluate this residue, substitute $z = 0.5$ in Eq. (4.16) and we get

$$\left[\frac{5 \cdot z \cdot z^{n-1}}{(z-0.4)}\right]_{z=0.5} = \frac{5(0.5)(0.5)^{n-1}}{(0.5-0.4)} = \frac{2.5}{(0.1)}(0.5)^{n-1} = 25(0.5)^{n-1} \quad n > 0$$

To determine the residue at the other pole, we use Eq. (4.17)

$$\left[\text{Residue } X(z)z^{n-1} \text{ at } z = 0.4\right] = \left(\frac{5 \cdot z \cdot z^{n-1}}{(z-0.5)}\right)_{z=0.4} = \frac{5(0.4)(0.4)^{n-1}}{(0.4-0.5)} = -20(0.4)^{n-1}$$

$$(4.17)$$

In Eq. (4.17) also, we have assumed that $n > 0$, or else we have additional poles at the origin. It is this assumption that is at fault and causes us difficulty in using this method. Assuming that $n \geq 0$, the sequence is a right-sided sequence. If this is not known *a priori*, we have to evaluate the residue for each successive value of $n < 0$ as each additional value of n will add a new pole of the function $X(z)z^{n-1}$ when $n < 0$. This evaluation becomes very long and tedious and we will leave it to the mathematicians to work out the details of what happens in these cases. So the inverse Z-transform for the given function is given by $x[n] = (25(0.5)^n - 20(0.4)^n)\mu[n]$. For ourselves, we will almost always use one of the next two methods.

4.3.2
Partial Fraction Expansion

This is a technique that uses the results of the previous method to determine the inverse z-transform of a rational Z-domain function. We are very familiar with this method already as this method is very similar to the method that we employed to determine the inverse Laplace transform. In principle, to use this method we break down the z-domain function into simpler terms that contain one or two poles. Then, we compute the residue at that pole and determine the inverse z-transform by looking up the Z-transform table to determine the inverse transform. If $X(z)$ is a z-domain function, then we can express $X(z)$ as a ratio of polynomials or in its partial fraction expansion as shown in Eq. (4.18).

$$X(z) = \frac{P(z)}{Q(z)} = \sum_{r=1}^{N} \frac{A_r}{(z-z_r)} \qquad (4.18)$$

In Eq. (4.18), the A_r terms represent the residue at the pole $z = z_r$. This residue is evaluated as shown in Eq. (4.19).

$$A_r = (z-z_r)X(z)\big|_{z=z_r} \qquad (4.19)$$

This is exactly how we evaluated the residue to determine the inverse Laplace transform. The method of evaluating the residue works fine when the function $X(z)$ is a rational function. A rational function is the function that has the order of its numerator less than the order of the denominator. When the order of the numerator is more than the order of the denominator, we first perform long division (as shown in Section 4.3.3) till we get a function that is rational. Then, we add the result of this long division to the rational function and proceed as we did in Eq. (4.19). So if the order of the polynomial $P(z)$ in Eq. (4.18) is greater than the order of the polynomial $Q(z)$, then we divide the polynomial $P(z)$ by the polynomial

$Q(z)$ to get a result polynomial $R(z)$ whose order is the difference between order of $P(z)$ – order of $Q(z)$. A typical result with the order of $P(z) = M$ and the order of $Q(z) = N$ is shown in Eq. (4.20).

$$X(z) = \underbrace{B_{M-N}z^{M-N} + B_{M-N-1}z^{M-N-1} + \cdots + B_1z + B_0}_{R(z)} + \sum_{r=1}^{N}\frac{A_r}{(z-z_r)} \qquad (4.20)$$

In equations (4.18) to (4.20), we have assumed that all the poles are single poles. If some of the poles in $X(z)$ are repeated poles, the expression for partial fraction expansion needs to be modified to include this fact. This expansion is easy and it is similar to the partial fraction expansion we used for repeated poles for the Laplace transform. This expression is shown in Eq. (4.21). In Eq. (4.21), we have included a repeated pole at $z = z_i$ and this pole is of order s.

$$X(z) = \underbrace{B_{M-N}z^{M-N} + B_{M-N-1}z^{M-N-1} + \cdots + B_1z + B_0}_{R(z)} + \sum_{r=1}^{N-s}\frac{A_r}{(z-z_r)} + \sum_{k=1}^{s}\frac{C_k}{(z-z_i)^k}$$

$$(4.21)$$

The coefficients C_k are computed as shown in Eq. (4.22).

$$C_k = \frac{1}{(s-k)!}\left[\frac{d^{s-k}[z-z_i]^sX(z)}{dz^{s-k}}\right]\Bigg|_{z=z_i} \qquad (4.22)$$

In all the above equations, we have considered polynomials in z. The Z-domain functions we will find are generally expressed as polynomials or rational fractions in z^{-1}. To use the partial fraction expansion method, we can use the Z-domain functions in either z or in z^{-1}. If we use the same example as we did for the method of inversion integral, we can show that both the methods lead to the same time domain sequence.

Example 4.4

Determine the inverse Z-transform of the rational function $X(z) = 5z/(z-0.5)(z-0.4)$ using the partial fraction expansion method.

Solution 4.4: In the function given, we have two poles, a single pole at $z = 0.5$ and a single pole at $z = 0.4$, and we also have a zero at the origin. We will first write the given function as partial fractions and then determine the coefficients of the two partial fractions. This is done as follows:

$$X(z) = \frac{5z}{(z-0.5)(z-0.4)} = \frac{A}{(z-0.5)} + \frac{B}{(z-0.4)}$$

In this partial fraction expansion, we have to determine the coefficients A and B. These coefficients are evaluated using Eq. (4.19). Using this equation, we can evaluate the coefficients as shown in Eq. (4.23).

$$A = (z-0.5)X(z)|_{z=0.5} = \frac{5(0.5)}{(0.5-0.4)} = 25$$

$$B = (z-0.4)X(z)|_{z=0.4} = \frac{5(0.4)}{(0.4-0.5)} = -20$$

(4.23)

Now that we have evaluated the coefficients, we can write the inverse directly as $x[n] = (25(0.5)^n - 20(0.4)^n)$, which is the same result as we got in Example 4.3.

Example 4.5

Determine the inverse Z-transform of the rational function using the partial fraction expansion method. $X(z) = (2z^2-2.05z)/(z-0.8)(z-1.25)$ and $(2-2.05z^{-1})/(1-0.8z^{-1})(1-1.25z^{-1})$ For this function, there are three different regions of convergences; determine the corresponding time domain sequence for each RoC.

Solution 4.5: We will first show that the z-domain function can be written in powers of either z or z^{-1} and the result of the inverse transform is the same time domain sequence. First, notice that the two functions are exactly the same. If you multiply and divide the second function by z^2, then you do not change the function but you end up with the function written in powers of z, which is the first function (Figure 4.5a–c).

We will first work with the first function. This function is not a proper rational fraction, so we must first convert it to a proper fraction. We will do this by changing the function $X(z)$ to the function $X(z)/z$. Now, we have a proper fraction and it has three poles as shown here

$$\frac{X(z)}{z} = \frac{z^2-2.05z}{z(z^2-2.05z+1)} = \frac{A_0}{z} + \frac{A_1}{(z-0.8)} + \frac{A_2}{(z-1.25)}$$

By evaluating the three residues, we get

$$A_0 = zX(z)|_{z=0} = 0$$

$$A_1 = (z-0.8)X(z)|_{z=0.8} = 1$$

$$A_2 = (z-1.25)X(z)|_{z=1.25} = 1$$

Now that we have these residues, we can write the time domain function since we know that $x[n] = Z^{-1}(z/(z-a)) = (a)^n$. The inverse Z-transform of the function given is, depending on the RoC,

$$x[n] = \{(0.8)^n + (1.25)^n\}\mu[n] \qquad |z| > 1.25$$

$$x[n] = (0.8)^n\mu[n] - (1.25)^n\mu[-n-1] \qquad 0.8 \le |z| \le 1.25 \qquad (4.24)$$

$$x[n] = \{-(0.8)^n - (1.25)^n\}\mu[-n-1] \qquad |z| < 0.8$$

We next repeat the same example with the function given second in the problem statement. This time the function is a polynomial in z^{-1} rather than in terms of z. This time the function is a proper rational function, so we can immediately write its partial fraction expansion as shown here.

(a)

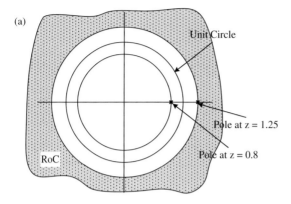

(b)

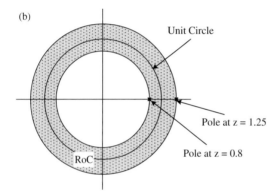

(c)

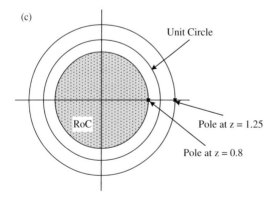

Figure 4.5 Region of convergence for (a) $|z| > 1.25$; (b) $0.8 \leq |z| \leq 1.25$; and (c) $|z| < 0.8$.

$$X(z) = \frac{2-2.05z^{-1}}{1-2.05z^{-1}+z^{-2}} = \frac{B_0}{(1-0.8z^{-1})} + \frac{B_1}{(1-1.25z^{-1})}$$

By evaluating the two residues, we get

$$B_0 = (1-0.8z^{-1})X(z)|_{z=0.8} = 1$$

$$B_1 = (1-1.25z^{-1})X(z)|_{z=1.25} = 1$$

Now that we have these two residues, we can write the time domain function since we know that $x[n] = Z^{-1}(1/(1-az^{-1})) = (a)^n$. The inverse Z-transform of the function given is, depending on the RoC,

$$x[n] = \{(0.8)^n + (1.25)^n\}u[n] \qquad\qquad |z| > 1.25$$

$$x[n] = (0.8)^n u[n] - (1.25)^n u[-n-1] \qquad 0.8 \le |z| \le 1.25$$

$$x[n] = \{-(0.8)^n - (1.25)^n\}u[-n-1] \qquad\qquad |z| < 0.8$$

Example 4.6

Determine the inverse Z-transform of the rational function $X(z) = z/(z^2-0.3z+0.3)$ using the partial fraction expansion method.

Solution 4.6: The Z-domain function is a proper rational function. It has two poles that are complex conjugate poles located at $0.15 \pm j0.4058$. So, we can write the function as

$$\frac{X(z)}{z} = \frac{1}{(z-0.15+j0.4058)(z-0.15-j0.4058)}$$

$$\frac{X(z)}{z} = \frac{A_0}{(z-0.15+j0.4058)} + \frac{A_1}{(z-0.15-j0.4058)}$$

By evaluating the two residues, we get

$$A_0 = (z-0.15+j0.4058)X(z)|_{z=0.15-j0.4058} = \frac{1}{j0.8116}$$

$$A_1 = (z-0.15-j0.4058)X(z)|_{z=0.15+j0.4058} = \frac{-1}{j0.8116}$$

The inverse transform can then be written as

$$x[n] = \frac{1}{j0.8116}[(0.15+j0.4058)^n - (0.15-j0.4058)^n]u[n]$$

4.3.2.1 Computing Residues Using MATLAB

For partial fraction expansion, the residues of the function can also be obtained using the MATLAB command *residue* as shown in script MATLAB 4.1. The residue function in MATLAB can be used in one of the two ways. If you already have the residues and want the polynomial, then the residue function will determine the polynomial. The other way to use this function is to begin with the numerator and the denominator polynomials and determine the residue at the location of the poles. This is the use that we will employ most often. In MATLAB, this command looks like *[r p k] = residue(b,a)*. In this command, the column vector *r* is the vector of residues, the column vector *p* is the vector of pole locations where the residues are, and the vector *k* is the vector of direct terms if the order of *b* is greater than the order of *a*, or else it is a null vector. The script below will compute the residues of Z-domain function given in Example 4.7.

Example 4.7

Determine the residues of the rational Z-domain function $X(z) = (2z^2 - 2.05z)/(z^2 - 2.05z + 1)$ and $(2 - 2.05z^{-1})/(1 - 2.05z^{-1} + z^{-1})$.

Solution 4.7: Here, we will use the MATLAB script 4.2 to compute the residues of the two functions. First, note that both the functions are identical except one that is written in terms of the variable z, while the other is written in terms of the variable z^{-1}.

MATLAB 4.1

```
b=[ 2 -2.05 0] ;    %%Define the polynomials as coefficients of
a=[ 1 -2.05 1] ;    %%decreasing powers of the variable. a is
b2=[ 2 -2.05] ;     %%denominator polynomial, b is the numerator
[ r1 p1 k1] =residue(b,a) %%Residues of function in powers of z
[ r2 p2 k2] =residue(b2,a) %%Residues of function in powers of z⁻¹
```

The result of the above MATLAB script agrees with the results that we obtained in Example 4.5 earlier. Once we have the residues, we can write the inverse of the Z-transform by noting if the RoC represents a right-handed or a left-handed sequence.

4.3.3
Long Division

The third method of determining the inverse Z-transform is the method of long division. In this method, we write the z-domain function as shown in the middle of Eq. (4.18). Then, after we have the numerator and the denominator polynomials, we divide the numerator polynomial with the denominator polynomial. If the RoC of the Z-domain function extends from the largest pole outward, then we perform the long division so that the result polynomial is in terms of z^{-1} and if the RoC extends from the smallest pole inward to the origin, then we perform the long division so that the result polynomial is in terms of z. If the RoC of the Z-domain is an annular ring like the one shown in Figure 4.3, then you first divide the function into two parts, one represents the right-sided function and the other represents the left-sided function. Then, perform two separate long divisions, one to represent the right-sided sequence and the other to represent the left-sided sequence. Example 4.8 will clear this concept.

When we use long division, we do not get back a closed form expression of the time domain function as we saw with the inversion integral and the partial fraction expansion. We get a time domain sequence. The various powers of z indicate the position of the sample value in the sequence. So, for example, the coefficient of z^{-4} represents the sample value of the fourth sample after the time origin and the coefficient of z^6 represents the sixth sample before the time origin. We will repeat Example 4.5 here again.

Example 4.8

Determine the inverse Z-transform of the rational function $X(z) = (2-2.05z^{-1})/(1-2.05z^{-1}+z^{-2})$ using the long division method. For this function, there are three different regions of convergences; for each RoC, determine the corresponding time domain sequence.

   Solution 3.8: Since this time we are going to use the method of long division, we will leave the z-domain expression as it is. Also, since we have to develop the z-domain expression for each possible RoC, we will break the example down in terms of the RoC as follows.

a)   RoC extends outward toward the infinity from the largest pole, $|z| > 1.25$ in this case. This time we want to complete the division and end up with a result polynomial in terms of z^{-1}. This long division is shown in Figure 4.6a. So, the first sample at time zero is 2, the next sample is 2.05, the third sample is 6.2025, and the rest of the samples can be computed this way. Comparing the result from the earlier example using Eq. (4.24) for RoC $|z| > 1.25$, we can evaluate the first three samples by substituting in Eq. (4.24)$n = 0$, then 1, and finally 2. We find that they are identical to the result we have just obtained.

b)   RoC extends inward toward the origin from the smallest pole, $|z| < 0.8$ in this case. This time we want to complete the division and end up with a result polynomial in terms of z. This long division is shown in Figure 4.6b. This time the first sample is at time -1 and it is -205, the previous sample is 2.2025, the third sample is 6.2525, and the rest of the samples can be computed this way. Comparing the result from the earlier example using Eq. (4.24) for RoC $|z| < 0.8$, we can evaluate the first three samples by substituting in Eq. (4.24)$n = -1$, then -2, and finally -3. We find that they are identical to the result we have just obtained.

c)   In the third case when the RoC is an annular ring, we have one pole that is right sided and one pole that is left sided. So first, we will separate the two poles. Then, for the pole that has the RoC $|z| > 0.8$, we will perform the long division and get the result polynomial in negative powers of z. For the pole that has the RoC $|z| < 1.25$, we will perform the long division and get the result polynomial in positive powers of z. Then, when we put the two results together, we get the inverse Z-transform that we are looking for. This long division is shown in Figure 4.6c. From Eq. (4.24) for RoC $0.8 < |z| < 1.25$, we can evaluate the first three samples by substituting in Eq. (4.24)$n = 0$, then 1, and finally 2 and $n = -1$, -2, and -3. We find that they are identical to theresult we have just obtained. Remember that one sequence is left sided, so it will have zero values for $n \geq 0$, while the other is right sided, so it will have zero values for $n < 0$.

4.4
Theorems and Properties of the Z-Transform

Just as we discussed for the DTFT and the DFT, the Z-transform also has specific properties and theorems that we will discuss here. These properties and theorems are important in

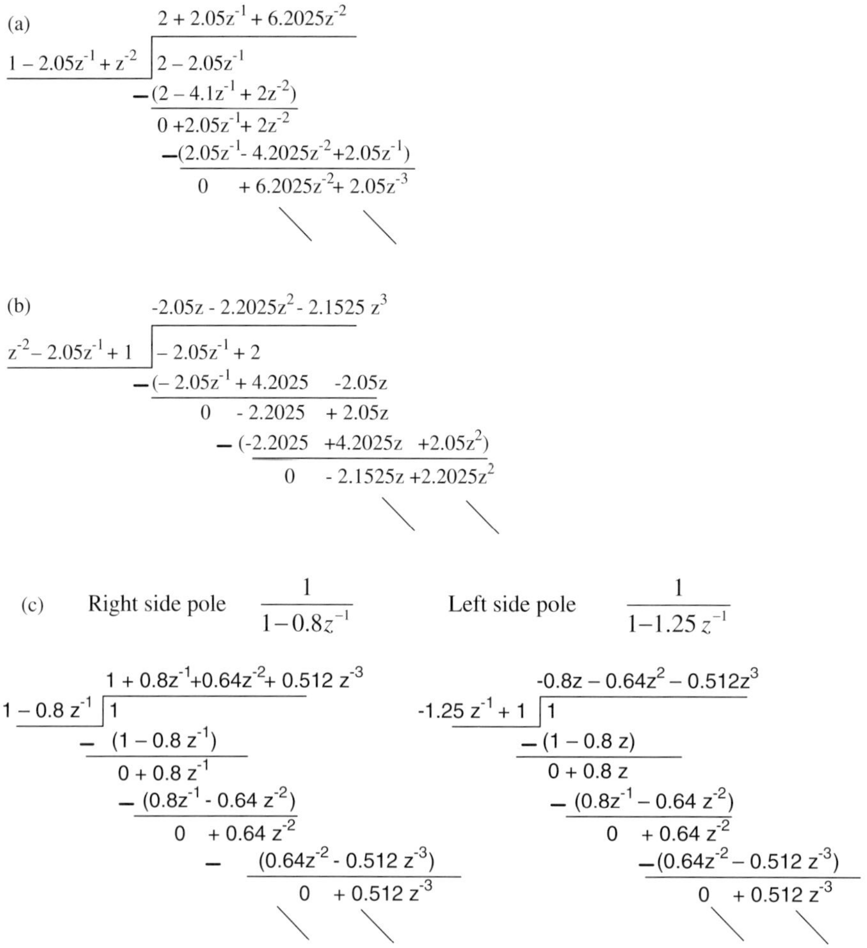

(a)

$$\begin{array}{r} 2 + 2.05z^{-1} + 6.2025z^{-2} \\ 1 - 2.05z^{-1} + z^{-2} \overline{\smash{\big)}\ 2 - 2.05z^{-1}} \\ \underline{-(2 - 4.1z^{-1} + 2z^{-2})} \\ 0 + 2.05z^{-1} + 2z^{-2} \\ \underline{-(2.05z^{-1} - 4.2025z^{-2} + 2.05z^{-1})} \\ 0 \quad + 6.2025z^{-2} + 2.05z^{-3} \end{array}$$

(b)

$$\begin{array}{r} -2.05z - 2.2025z^2 - 2.1525\,z^3 \\ z^{-2} - 2.05z^{-1} + 1 \overline{\smash{\big)}\ -2.05z^{-1} + 2} \\ \underline{-(-2.05z^{-1} + 4.2025) \quad -2.05z} \\ 0 \quad -2.2025 \quad +2.05z \\ \underline{-(-2.2025 \quad +4.2025z \quad +2.05z^2)} \\ 0 \quad -2.1525z +2.2025z^2 \end{array}$$

(c) Right side pole $\dfrac{1}{1-0.8z^{-1}}$ Left side pole $\dfrac{1}{1-1.25\,z^{-1}}$

$$\begin{array}{r} 1 + 0.8z^{-1} + 0.64z^{-2} + 0.512\,z^{-3} \\ 1 - 0.8\,z^{-1} \overline{\smash{\big)}\ 1} \\ \underline{-\ (1 - 0.8\,z^{-1})} \\ 0 + 0.8\,z^{-1} \\ \underline{-\ (0.8z^{-1} - 0.64\,z^{-2})} \\ 0 \quad + 0.64\,z^{-2} \\ \underline{-\quad (0.64z^{-2} - 0.512\,z^{-3})} \\ 0 \quad + 0.512\,z^{-3} \end{array}$$

$$\begin{array}{r} -0.8z - 0.64z^2 - 0.512z^3 \\ -1.25\,z^{-1} + 1 \overline{\smash{\big)}\ 1} \\ \underline{-\ (1 - 0.8\,z)} \\ 0 + 0.8\,z \\ \underline{-\ (0.8z^{-1} - 0.64\,z^{-2})} \\ 0 \quad + 0.64\,z^{-2} \\ \underline{-(0.64z^{-2} - 0.512\,z^{-3})} \\ 0 \quad + 0.512\,z^{-3} \end{array}$$

Figure 4.6 RoC extends outward toward (a) the infinity and (b) inward toward the origin. (c) RoC is an annular region from $|z| > 0.8$ to $|z| < 1.25$.

what the Z-transform represents to us in terms of system analysis and how the various systems with different Z-transform respond when applied the same input. In this section, we present only some of the most frequently used theorems and properties, other theorems and properties can be found in the references cited at the end of the book.

4.4.1
The Region of Convergence

Earlier, we saw that there are only three different types of RoCs. Each RoC had one common characteristic: the RoC cannot contain any pole within it; it is bounded by poles or by infinity or the origin. So, the RoC may extend from one pole to another pole that is adjacent in terms of the radius of the pole, or it may extend from a pole to infinity, or the third option is for the

RoC to start from a pole and extend to the origin. The RoC is always a continuous area. There cannot be breaks in the RoC or the RoC cannot be a disjointed area in the z-plane. Finally, the RoC is the area where the sequence $x[n]z^{-n}$ is absolutely summable. The diagrams showing the three possible RoCs are given in Figure 4.5. Figure 4.6a depicts the RoC that extends from the largest pole outward toward the infinity, Figure 4.6b depicts the RoC that extends from one pole to another pole, and Figure 4.6c depicts the RoC that extends from the smallest pole inward toward the origin. Similarly, the different possible RoCs are also shown in Figure 4.7.

4.4.2
Linearity

Two sequences $x[n]$ and $y[n]$ with their z-transforms $X(z)$ and $Y(z)$ having RoC, as shown in Eq. (4.25), can be combined as shown in Eq. (4.26).

$$\mathcal{Z}[x[n]] = X(z) \quad \mathbf{R}_{x^-} < |z| < R_{x^+}$$
$$\mathcal{Z}[y[n]] = Y(z) \quad \mathbf{R}_{y^-} < |z| < R_{y^+} \tag{4.25}$$

$$\mathcal{Z}[\alpha x[n] + \beta y[n]] = \alpha X(z) + \beta Y(z) \quad \mathbf{R}_- < |z| < R_+ \tag{4.26}$$

The region of convergence is at least the overlap region of the two individual RoCs. The RoC can be larger than the region of convergence of any one of the functions or both the functions if the pole that defines the boundary of the RoC of one of the sequences is canceled by a zero of the combined sequence. A simple example of this is when both the sequences $x[n]$ and $y[n]$ are of infinite duration. Their RoC extends from their largest pole to

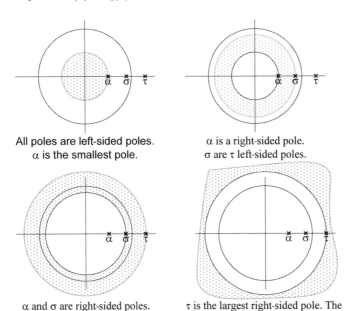

All poles are left-sided poles.
α is the smallest pole.

α is a right-sided pole.
σ are τ left-sided poles.

α and σ are right-sided poles.
τ is a left-sided pole.

τ is the largest right-sided pole. The other two also have to be right-sided.

Figure 4.7 Examples of different RoCs. This Z-transform has three poles, so there are three possible RoCs.

infinity. Now, if the sum of the two sequences is such that the resulting sequence is of finite duration, for example, when $x[n] = a^n \mu[n]$ and $y[n] = a^n \mu[n - n_0]$, the RoC is $|a| < |z|$. Now if we form a new sequence that is the difference of the two sequences $x[n] - y[n]$, then the resulting sequence is of finite length, it is a sequence of length n_0 and it is of finite duration. This finite duration sequence has a RoC that is the entire z-plane with $|z| > a$.

4.4.3
Shift in the Time Domain

When a time sequence gets shifted, its Z-transform undergoes a rotation. This is shown by Eq. (4.27).

$$\mathcal{Z}[x[n-n_0]] = \sum_{n=-\infty}^{\infty} x[n-n_0]z^{-n} \quad \text{let } n-n_0 = m$$

$$\mathcal{Z}[x[n-n_0]] - \sum_{m=-\infty}^{\infty} x[m]z^{-(m+n_0)} = z^{-n_0} \sum_{m=-\infty}^{\infty} x[m]z^{-m} = z^{-n_0}X(z)$$

(4.27)

In the above equation, the RoC of the shifted sequence is the same as that of the original sequence except for possibly excluding the poles at the origin if n_0 is positive and the origin was also part of the RoC, and excluding the poles at ∞ if n_0 is negative and infinity was also part of the RoC.

4.4.4
Scaling in the Frequency Domain

When the sequence $x[n]$ is multiplied by a sequence a^n where a may be complex, the Z-transform of the resulting sequence can be obtained as shown in Eq. (4.28) as follows:

$$\mathcal{Z}[a^n x[n]] = \sum_{n=-\infty}^{\infty} a^n x[n]z^{-n} = \sum_{n=-\infty}^{\infty} x[n](a^{-1}z)^{-n} = X(a^{-1}z)$$

(4.28)

In Eq. (4.28), the RoC of the new function is $|a|R_{x^-} < |z| < |a|R_{x^+}$, that is, the RoC gets scaled by the sequence parameter a. When the parameter a is a real quantity, the poles of the product sequence $(a^n x[n])$ shift along radial lines compared to the poles of $x[n]$. If the parameter a is complex with unit magnitude, the scaling corresponds to rotation in the z-plane. In this case, the poles and zeros change along the circular direction; this scaling keeps the RoC the same.

4.4.5
Conjugation of a Complex Sequence

When we know the Z-transform of a sequence, the Z-transform of its conjugate can be obtained from the same sequence by first conjugating the variable z and then conjugating the entire z-transform as shown in Eq. (4.29)

$$\mathcal{Z}[x^*[n]] = \sum_{n=-\infty}^{\infty} x^*[n]z^{-n} = \left[\sum_{n=-\infty}^{\infty} x[n](z^*)^{-n} \right]^* = X^*(z^*)$$

(4.29)

The RoC of the resulting Z-domain function is the same since the process of conjugation does not change any pole locations. This is true since the Z-domain functions that we will come across all have real coefficients implying that all the poles and zeros occur as complex conjugate pairs. When the poles and the zeros occur as complex conjugate pairs, the poles and zeros the poles and zeros exchange the location with their conjugates and as a result the poles and zeros are at the same radius from the origin as the original pole zero locations.

4.4.6
Differentiation in the Z-Domain

When the time domain sequence is multiplied by a linear ramp, its Z-transform is obtained by differentiating the Z-transform of the function before it was multiplied by the linear ramp. To see how the Z-transform is obtained, first look at what the derivative of the Z-transform is, this is shown in Eq. (4.30).

$$\frac{d[X[z]]}{dz} = \frac{d\left[\sum_{n=-\infty}^{\infty} x[n]z^{-n}\right]}{dz} = \sum_{n=-\infty}^{\infty} -nx[n]z^{-n-1} = -z^{-1} \sum_{n=-\infty}^{\infty} nx[n]z^{-n} \quad (4.30)$$

From Eq. (4.30), we can write Eq. (4.31) by rearranging some terms.

$$\sum_{n=-\infty}^{\infty} nx[n]z^{-n} = -z\frac{d[X[z]]}{dz} \quad (4.31)$$

4.4.7
Convolution of Two Time Domain Sequences

The Z-transform of the convolution of two time domain sequences is obtained by first writing the Z-transform definition of the convolution as shown in Eq. (4.32).

$$\mathcal{Z}\left[\sum_{r=-\infty}^{\infty} x[r]h[n-r]\right] = \sum_{n=-\infty}^{\infty}\left[\sum_{r=-\infty}^{\infty} x[r]h[n-r]\right]z^{-n} \quad (4.32)$$

If we interchange the order of the summation in Eq. (4.32), we get

$$\mathcal{Z}\left[\sum_{r=-\infty}^{\infty} x[r]h[n-r]\right] = \sum_{r=-\infty}^{\infty} x[r]\left[\sum_{n=-\infty}^{\infty} h[n-r]\right]z^{-n}$$

$$\mathcal{Z}\left[\sum_{r=-\infty}^{\infty} x[r]h[n-r]\right] = \sum_{r=-\infty}^{\infty} x[r]\left[\sum_{m=-\infty}^{\infty} h[m]\right]z^{-(m+r)} \quad (4.33)$$

$$\mathcal{Z}\left[\sum_{r=-\infty}^{\infty} x[r]h[n-r]\right] = \underbrace{\sum_{r=-\infty}^{\infty} x[r]z^{-r}}_{\text{Z-transform of } h[n]} \underbrace{\left[\sum_{m=-\infty}^{\infty} h[m]z^{-m}\right]}_{\text{Z-transform of } h[n]}$$

Equation (4.33) once again gives us the familiar result that convolution in the time domain is transformed to multiplication in the frequency domain.

Table 4.1 Z-transform theorems and properties.

Sequence x[n]		Z-transform
$ax[n] + by[n]$	$aX(z) + bY(z)$	$\text{Max}(R_{x-}; R_{y-}) < \backslash z \backslash < \text{Min}(R_{x+}; R_{y+})$
$x(n + n_0)$	$z^{n_0}X(z)$	$R_{x-} < \lvert z \rvert < R_{x+}$
$a^n x[n]$	$X(a^{-1}z)$	$\lvert a \rvert R_{x-} < \lvert z \rvert < \lvert a \rvert R_{x+}$
$nx[n]$	$-z(dX[z]/dz)$	$R_{x-} < \lvert z \rvert < R_{x+}$
$x^*[n]$	$X^*(z^*)$	$R_{x-} < \lvert z \rvert < R_{x+}$

4.4.8
Summary of Z-Transform Theorems and Properties

Table 4.1 lists some of the Z-transform theorems and properties. Table 4.2 lists some of the common Z-transform pairs. The Z-transform is used just as we used the Laplace transform. The Laplace transform was used for analog signals while the Z-transform is used for discrete signals. We also saw similarity between the two transforms when we were trying to determine the inverse Z-transform.

Table 4.2 Some Common Z-transform Pairs.

Signal x[n]	Transform X(z)	Region of Convergence
$\delta(n)$	1	Entire z plane
$\mu(n)$	$\dfrac{1}{1-z^{-1}}$	$\lvert z \rvert > 1$
$a^n\mu(n)$	$\dfrac{1}{1-az^{-1}}$	$\lvert z \rvert > \lvert a \rvert$
$na^n\mu(n)$	$\dfrac{az^{-1}}{(1-az^{-1})}$	$\lvert z \rvert > \lvert a \rvert$
$-a^n\mu(-n-1)$	$\dfrac{1}{1-az^{-1}}$	$\lvert z \rvert < \lvert a \rvert$
$-na^n\mu(-n-1)$	$\dfrac{az^{-1}}{(1-az^{-1})}$	$\lvert z \rvert < \lvert a \rvert$
$\cos(\omega_0 n)\mu(n)$	$\dfrac{1-z^{-1}\cos(\omega_0)}{1-2z^{-1}\cos(\omega_0)+z^{-2}}$	$\lvert z \rvert > 1$
$\sin(\omega_0 n)\mu(n)$	$\dfrac{z^{-1}\sin(\omega_0)}{1-2z^{-1}\cos(\omega_0)+z^{-2}}$	$\lvert z \rvert > 1$
$(a^n\cos(\omega_0 n))\mu(n)$	$\dfrac{1-az^{-1}\cos(\omega_0)}{1-2az^{-1}\cos(\omega_0)+a^2z^{-2}}$	$\lvert z \rvert > \lvert a \rvert$
$(a^n\sin(\omega_0 n))\mu(n)$	$\dfrac{az^{-1}\sin(\omega_0)}{1-2az^{-1}\cos(\omega_0)+a^2z^{-2}}$	$\lvert z \rvert > \lvert a \rvert$

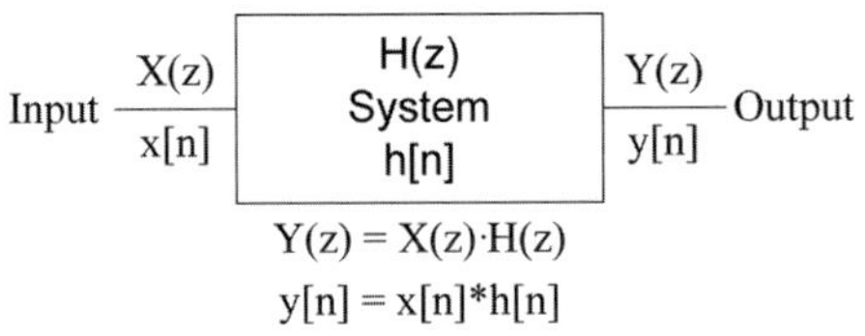

$$Y(z) = X(z) \cdot H(z)$$
$$y[n] = x[n]*h[n]$$

Figure 4.8 Representation of a system in the time domain and the frequency domain.

4.5
Application of Z-Transforms to Systems

Earlier in Chapter 1, we described a linear shift-invariant system; here, we will describe the linear shift-invariant system in a more general way compared to how we described the system in Chapter 1 by using the Z-transform of the unit sample response. The typical system is shown in Figure 4.8 and it is described by the input–output relations both in the time and in the frequency domains.

The system function of the system is identified as $H(z)$ that is also the Z-transform of the unit sample response of the system $h[n]$. Earlier in Chapter 1, we saw that if the unit sample response, which is the sequence $h[n]$, is absolutely summable, the system is stable. If we now extend this idea to the system function in the Z-domain, the RoC of the sequence $h[n]z^{-n}$ is the region where $h[n]z^{-n}$ is absolutely summable. If this RoC includes the unit circle where $|z| = 1$, then the system will be stable. So, the test for stability of the system in the Z-domain reduces to the inclusion of the unit circle in the RoC. This in turn implies that the poles of the system function will all lie inside the unit circle as the poles cannot lie in the RoC and the RoC has to include the entire Z-plane where $|z| > a$. So, for the system to be stable the system function must have all the poles inside the unit circle and the unit circle must be part of the RoC of the system function.

For a system that is described by linear constant coefficient, linear difference equation is generally written as shown in Eq. (4.34).

$$\sum_{k=0}^{K} a_k y[n-k] = \sum_{r=0}^{R} \beta_r x[n-r] \tag{4.34}$$

Next, if we were to determine the Z-transform of Eq. (4.34) using the shifting theorem and the linearity theorem, we get the two sides of Eq. (4.34) in the Z-domain as shown in Eq. (4.35).

$$Z\left[\sum_{k=0}^{K} a_k y[n-k]\right] = \sum_{k=0}^{K} a_k Z[y[n-k]] = \sum_{k=0}^{K} a_k z^{-k} Y(z)$$

$$Z\left[\sum_{r=0}^{R} \beta_r x[n-r]\right] = \sum_{r=0}^{R} \beta_k Z[x[n-k]] = \sum_{r=0}^{R} \beta_r z^{-r} X(z) \tag{4.35}$$

From Eq. (4.35), we can write the system function by dividing the term $Y(z)$ by the term $X(z)$ as shown in Eq. (4.36).

$$\frac{Y(z)}{X(z)} = \frac{\sum_{r=0}^{R} \beta_r z^{-r}}{\sum_{k=0}^{K} a_k z^{-k}} = H(z) \tag{4.36}$$

Equation (4.36) tells us that the system function is a ratio of polynomials in z^{-1}. Being a ratio of polynomials, the system function can also be written in terms of its poles and zeros. Writing the function in terms of its poles and zeros gives us another way of determining if the system is stable or not.

We remember from our earlier discussion that for the system function to represent a stable system in the Z-domain, the RoC must include the unit circle. We also remember that the RoC is bounded by the poles of the system and that the RoC cannot contain any poles of the system. Hence, the poles of the system must lie outside the RoC and the RoC must include the unit circle and it must extend outward toward infinity. So for a system to be stable, all the poles of the system function must lie inside the unit circle. For this reason, whenever you see a pole–zero plot of the system function, you also see the unit circle drawn, so it is immediately obvious if this particular system is stable or not.

4.6
Responses to Typical Pole–Zero Patterns

We have represented the Z-transform in terms of its poles and zeros. The location of the poles has a very major influence on how the system responds. Here, we see some typical pole patterns and relate the pole pattern to the expected response.

4.6.1
First-Order Poles

All the Z-domain functions that we have seen so far and the functions of real systems must have real coefficients. This implies that if we have single poles, then these poles must be real poles. All the first-order poles of the Z-transform must be real poles and hence must lie on the real axis in the Z-domain. Some of the first-order poles are shown in Figure 4.9a. Here, we have four possible choices. Two of the choices have the poles inside the unit circle (the two figures on the left-hand side) and the other two have the poles outside the unit circle (the two figures on the right-hand side).

The responses to these poles are shown in Figure 4.7b. Notice that the poles that lie inside the unit circle (the two figures on the left-hand side of Figure 4.7b) produce a typical response that decays and show that the system is a stable system. The poles that lie outside the unit circle produce a typical response that grows exponentially (the two figures on the right-hand side of Figure 4.7b). The poles that lie on the right half of the Z-plane have a response that does not change signs (the two figures on the top half of Figure 4.7b), while the poles that lie on the left-hand side of the z-plane exhibit a response that alternates between positive and negative (the two figures on the bottom half of Figure 4.7b). The decay rate of the poles that lie inside the unit circle is proportional to the location of the pole. If the pole is close to the origin, it decays very fast. If the pole lies on the real axis and the unit circle at the same time, then the pole is located at $z = \pm 1$. These poles are marginally stable in that their response not only does not grow without bound but also does not decay to zero either.

4.6.2
The Second-Order Poles

Poles in a second-order system either are all real poles or they occur as complex conjugate pairs of poles. The relative placement of the poles will affect the response from the system.

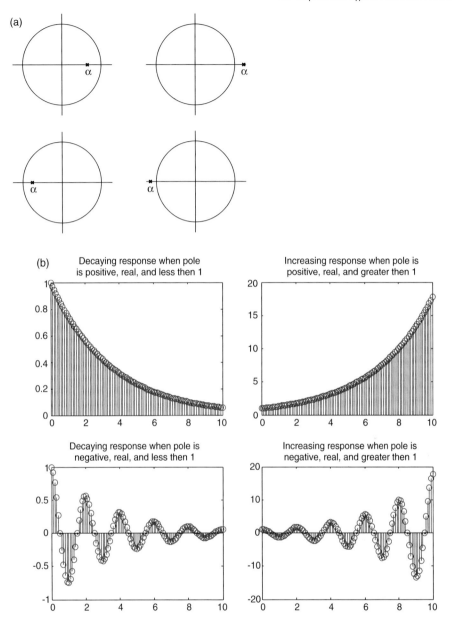

Figure 4.9 (a) Four possible locations for first-order single poles. (b) Some typical responses for simple single poles.

Since there are many possibilities with these types of poles, we will list some on the typical pole locations and their responses that you might expect. To see the rest of the responses, you may use MATLAB scripts such as the one shown here in MATLAB 4.2. Figure 4.10a shows four different pole locations that are of interest. In the left half, we see two poles that are

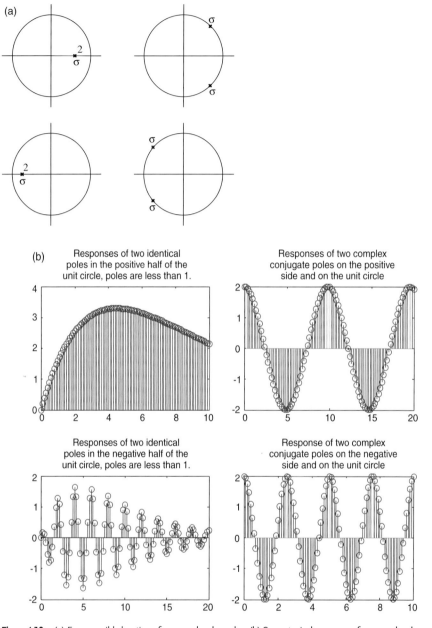

Figure 4.10 (a) Four possible locations for second-order poles. (b) Some typical responses for second-order poles.

identical and real, one pair of poles are in the left half of the Z-plane while the other is in the right half of the Z-plane. In the right half, of Figure 4.10a we see the poles that are complex conjugates. In the top, we see two poles in the right half and on the unit circle and in the bottom we see poles on the unit circle and in the left half.

The response that you would expect from these poles is shown in Figure 4.8b. The left half of Figure 4.8b shows the response due to the two real poles that are inside the unit circle. The top-left corner represents the poles in the right half of the unit circle and the bottom-left corner represents the poles in the left half of the unit circle. Notice that now the response first grows and then decays quickly to zero. The response that we have drawn is for two identical poles, but the response would be similar even if the two real poles were not identical. When both the poles lie in the right half, the response would be similar to the response shown in the top-left corner of Figure 4.8b, while if the two unequal real poles were inside the left half of the unit circle, the response would be similar to the response shown in the bottom-left of Figure 4.8b. If one or both the poles lie outside the unit circle, then the response would grow without bound and the system would be an unstable system.

When the poles are complex conjugates, the pole locations are as shown in the right half of Figure 4.7a and the response is as shown in the right half of Figure 4.8b. Notice that when the poles lie exactly on the unit circle, the response is exactly sinusoidal and it never grows or decays. The response in the top-right corner of Figure 4.8b represents the pole locations on the unit circle and in the right half of the Z-plane. As the poles move away from the location $(1 + j0)$ and on the unit circle, the frequency starts to increase. When the poles are in the right half of the unit circle, the frequency has increased even more, so as we travel from the location $(1 + j0)$ toward the location $(-1 + j0)$, we are continuously increasing the frequency in the response. While the poles lie on the unit circle, the amplitude of the oscillations stays constant. If the poles were inside the unit circle, then the amplitude of the oscillations would gradually decay to zero. If the complex conjugate poles were outside the unit circle, then the oscillations would gradually increase till the system output reaches an infinite value and this system is an unstable system.

The MATLAB script that we used to draw the diagram in Figure 4.8 is given in MATLAB 4.2.

MATLAB 4.2
```
n=0:0.2:20;      %Define the Time vector
a1=0.8+i*0.6;     %Location of the two poles on
a2=0.8-i*0.6;     %the unit circle.
x1=a1.^n+a2.^n;    %Compute the complex response,
x2=(-a1).^n+(-a2).^n;    %on the two opposite sides.
x3=n.*real(a1).^n;    %Compute the response for two,
x4=n.*real(-a1).^n;    %identical poles.
subplot(221);stem(n,x3); %plot the various responses.
title({ 'Response of two identical poles'; 'in the positive half
of'; 'the unit circle and less than 1'} );
subplot(222);stem(n,x1);
title({ 'Response of two complex conjugate'; 'poles on the positive
side';'of the unit circle'} );
subplot(223);stem(n,x4);
title({ 'Response of two identical poles'; 'in the negative half
```

```
of'; 'the unit circle and less than 1'});
subplot (224); stem (n, x2);
title ({ 'Response of two complex conjugate'; 'poles on the negative
side';'of the unit circle'});
```

4.7
Introduction to Two-Dimensional Z-Transform

In this chapter, we have studied the Z-transform in one dimension. There are occasions when we have a two-dimensional signal. The question that we want to address here is if we are able to extend the ideas of a one-dimensional Z-transform to the two-dimensional signal. In this section, we introduce the two-dimensional Z-transform. The definition of the two-dimensional Z-transform is similar to the definition of the one-dimensional Z-transform; however, we have to use a double summation, one for each dimension as shown in Eq. (4.37).

$$X(z_1, z_2) = \sum_{m=-\infty}^{\infty} \sum_{n=-\infty}^{\infty} x[m, n] z_1^{-m} z_2^{-n} \qquad (4.37)$$

In Eq. (4.37), the two variables z_1 and z_2 represent the two dimensions of the z-transform; they are both complex variables that can be expressed in their polar form as $z_1 = r_1 e^{j\omega_1}$ and $z_2 = r_2 e^{j\omega_2}$. With this substitution, Eq. (4.37) can be rewritten as

$$X(z_1, z_2) = \sum_{m=-\infty}^{\infty} \sum_{n=-\infty}^{\infty} x[m, n] \left(r_1 e^{j\omega_1}\right)^{-m} \left(r_2 e^{j\omega_2}\right)^{-n}$$

Just as we interpreted the one-dimensional z-transform as the DTFT of the modified sequence $x[n]r^{-n}$, we are now able to interpret the two-dimensional Z-transform as a two-dimensional DTFT of the sequence $x[m, n]r_1^{-m}r_2^{-n}$. When $r_1 = 1$ and $r_2 = 1$, the two-dimensional z-transform can be interpreted as the two-dimensional DTFT.

With the one-dimensional Z-transform, we were very aware that the Z-transform did not converge in the entire Z-plane and that there was a specific RoC. This is so even for the two-dimensional Z-transform. For convergence, we require that the double summation of the absolute value of the function $x[m, n]z_1^{-m}z_2^{-n}$ of Eq. (4.37) converge as shown in Eq. (4.38).

$$X(z_1, z_2) = \sum_{m=-\infty}^{\infty} \sum_{n=-\infty}^{\infty} \left|x[m, n]z_1^{-m}z_2^{-n}\right| < \infty \qquad (4.38)$$

A two-dimensional Z-transform is said to be separable if we can write $X(z_1, z_2)$ as a product of the two individual Z-transforms as $X(z_1, z_2) = X_1(z_1) X_2(z_2)$. The Z-transform will be separable only if the time domain sequence is separable, and the time domain sequence is separable when $x[m, n] = x_1[m] x_2[n]$. When the two-dimensional Z-transform is separable, the Z-transform of $x_1[m]$ is $X_1(z_1)$ and the Z-transform of $x_2[n]$ is $X_2(z_2)$.

Just as the one-dimensional Z-transform had some properties that we studied, the two-dimensional Z-transform also has similar properties. The proof of the properties of the two-dimensional Z-transform is easily extended from the one-dimensional case. Stability conditions of a two-dimensional system are interesting and complex. With the one-dimensional case, we were able to easily say that the system was stable if the poles of the system lie inside the unit circle. With the two-dimensional case, we can make a statement similar to the one-dimensional case; so, when the summation in Eq. (4.38) is finite, the system is stable. This can be extended to the bounded input–bounded output criterion to result in the requirement that the system will be stable when $|z_1| = |z_1| = 1$ and the summation of Eq. (4.38) converges.

When the Z-transform is separable, the criteria for stability is very similar to the one-dimensional case, which is to say that $X_1(z_1)$ has to have its poles in the unit circle of the z_1 complex plane and also $X_2(z_2)$ has to have its poles in the unit circle of the z_2 complex plane. In a general case when the Z-transform is not separable, the stability criterion is not so simple. A necessary and sufficient condition for a ratio of two-dimensional polynomials to represent a stable system that is also a causal system is that the denominator of the system cannot be equal to zero when both $|z_1|$ and $|z_2|$ are simultaneously greater than unity. This condition is consistent with the stability condition that we have written above for a separable system. That is when $X(z_1)$ has a pole outside its unit circle, that is, $|z_1|$ is greater than 1, at that point the denominator polynomial will be zero for all values of z_2 including values of $z_2 > 1$, and hence the system will be unstable.

One way to check for stability of a general two-dimensional system is to factor the denominator polynomial in either z_1 or in z_2. Say, we factor the polynomial in terms of z_2. In this case, the roots of the polynomial will be in terms of the variable z_1. Now, we require that the poles of this polynomial lie outside the unit circle of the z_2 plane only when the variable $|z_1| < 1$. This stability condition looks straightforward and simple in theory, but in practice it is difficult to apply.

4.8
Chapter Summary
4.8.1
Definition of Z-Transform

The first thing we did in this chapter was to define the Z-transform. The Z-transform is defined as an infinite double-sided summation. We say that the Z-transform exists when the summation over the infinite terms is finite. In making the infinite sum finite, we were able to define a region over which the sum would be finite, the sum did not have to be finite in the entire Z-plane. The region over which the sum would be finite was defined as the region of convergence. Here, we also pointed out that the Laplace transform and the Z-transform have a very similar relation to the Fourier transform and the DTFT. The DTFT does not exist for some time domain functions, but we saw that by defining the RoC the Z-transform can be made to exist for just about all the functions that we will come across in our studies. We further also saw that the expression for the Z-transform can be the same for more than one time domain sequence, but when we add the RoC to the expression, the functions are separated and there is no confusion any longer.

4.8.2
Inverse of the Z-Transform

The inverse transform of the Z-transform can be determined in one of the three different ways. Of the three ways, the inversion integral uses theory of complex numbers and evaluates the residues at the individual poles of the function. We also saw that if the number of poles changes with n, then this method is not very useful because with every change in the number of poles, we have to evaluate a new residue at every pole in the system. The next method uses a combination of look-up and the residue method. In the partial fraction expansion method, we first rewrite the function as partial fractions and evaluate the residue at each partial fraction. Then, looking at the RoC and the type of partial fraction, we can write the inverse transform directly from the table of simple inverse transforms. The last method to invert the Z-transform used long division. In this method, we divide the numerator by the denominator and the resulting coefficients of the powers of z correspond to the inverse z-transform. If the resulting sequence is in powers of z, then we have an anticausal sequence and if the sequence is in powers of z^{-1}, then we have a causal sequence.

4.8.3
Theorems and Properties of the Z-Transform

Just like all the other transforms that we have studied so far, the Z-transform also has some properties that are similar to the Laplace transform and the Fourier transform. These properties help us to understand what the relation between various sequences is and we are not required to determine the Z-transform all over again for minor changes. The properties that we studied are linearity, shift of a sequence, scaling, differentiation in the z-domain, and conjugation and convolution.

4.8.4
Application of Z-Transforms to Systems

In applying the Z-transform to systems, we saw that the system function determines the stability of the system. We further saw that the poles of the system function must lie inside the unit circle in the z-plane and that the unit circle in the z-plane must be part of the RoC of the system function.

4.8.5
Responses to Typical Pole–Zero Patterns of Systems

Now that we know that the poles of the system function must lie inside the unit circle and that system functions of real systems have real coefficients, we were able to identify the type of poles that we can expect in a system function. The real poles can be present as single poles and the complex poles always occur as complex conjugate pairs. MATLAB script 4.1 showed us how to compute and plot the response of typical pole patterns in the time domain.

4.8.6
Introduction to Two-Dimensional Z-Transform

The extension to two-dimensional Z-transform is not very simple especially when the two-dimensional Z-transform is not separable. The properties and the theorems of the two-dimensional Z-transform are an extension of the one-dimensional Z-transform, but the stability criteria are very difficult to apply.

5
Discrete Filter Design Techniques

5.1
Introduction

When we design filters to produce a specific frequency response we are trying to determine the best approximation that we can get for the desired ideal filter. What we get is only an approximation because we specify the filter, by specifying its magnitude or phase response that we want in the frequency domain, but implement the filter by determining the impulse response coefficients of the filter in the time domain. The choice of the impulse response coefficients is by some appropriate algorithm that will give us the frequency response that is as close to the desired ideal filter response. The translation between the two domains is not a perfect translation.

The specifications of the filter in the frequency domain generally refer to the specifications of the low-pass filter, but they could be for the high pass, band pass, or band stop filters. These are among the more common filters. The specifications may be given in either the analog domain or the discrete domain. If the sampling rate is known, then it is a straightforward matter to convert the specifications from the analog to the discrete domain. The discrete frequencies are given in either radians or in angle around the unit circle in the z-plane. The point $z = -1$ corresponds to $^1/_2$ the sampling frequency. In most design techniques that we will discuss in this chapter, the sampling rate plays no part in the approximation process. Therefore, it is least confusing to design discrete filters with the specifications given in terms of an angle around the unit circle rather than in terms of analog frequency.

In designing discrete filters, we will choose two major approaches. The first approach designs the filter in the analog domain and then converts the filter to its discrete equivalent. This method gives us filters that are known as infinite impulse response (IIR) filters. These filters are generally smaller in size. The other method of filter design is to design the filter directly in the discrete domain. This method gives us filters that are known as finite impulse response (FIR) filters. These type of filters can be designed to have a linear phase response in addition to approximating the desired magnitude characteristics. These filters are generally longer than the IIR filters, but they are very efficient filters, especially when utilized in multirate systems we will study in Chapter 9.

Digital Filters: Theory, Application and Design of Modern Filters, First Edition. Rajiv J. Kapadia.
© 2012 Wiley-VCH Verlag GmbH & Co. KGaA. Published 2012 by Wiley-VCH Verlag GmbH & Co. KGaA.

5.2
Design of Analog Filters: A Review

5.2.1
Filter Specifications

Since filter design is an approximation, most filter specifications are given in terms of tolerance such as the one shown in Figure 5.1. The curve represents the frequency response of the filter that we want approximated. The diagram and the curve are divided into three bands. The pass band is the band that allows the signal to go through the filter with almost no attenuation, the stop band ideally would completely block the signal in this frequency band, and the transition band is the frequency region where the filter response moves from the pass band to the stop band. The pass band specifications tell us the allowable error that we are willing to accept from unit magnitude transmission. This error is specified either as the absolute value or as a dB value. The pass band response is supposed to lie in the range shown in Eq. (5.1).

$$1-\delta_1 < \left|H(e^{j\Omega})\right| < 1+\delta_1 \qquad |\Omega| \le \Omega_p \tag{5.1}$$

In Eq. (5.1), Ω_p is the frequency till which the pass band extends from zero frequency. The stop band specifications are similarly given in terms of maximum allowable magnitude of the signal or the expected attenuation from the pass band in dB. So in the stop band, the magnitude of the response is supposed to be always less then δ_2 as shown in Eq. (5.2)

$$\left|H(e^{j\Omega})\right| \le \delta_2 \qquad \Omega_s \le |\Omega| \le \pi \tag{5.2}$$

The pass band cutoff frequency Ω_s and the stop band edge frequency are given in terms of angle around the unit circle in the Z-plane. For a low-pass filter, the stop band edge frequency is always greater than the pass band edge frequency. The transition band extends from the pass band to the stop band or $(\Omega_s - \Omega_p)$. It is generally assumed that the response in the transition band is monotonic and decreases smoothly from the pass band to the stop band.

One of the first things that we usually have to do is to be able to convert the specifications form absolute value to dB and back. For example, if we are told that the maximum error in the

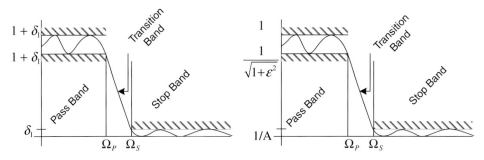

Figure 5.1   Filter bands and typical way of specifying filter parameters.

pass band should not exceed 1 dB, then the conversion to absolute value is explained in Eq. (5.3).

$$20\log_{10}\left(H(e^{j\Omega})\right) \geq -1$$

$$\left|H(e^{j\Omega})\right| \geq 10^{\frac{-1}{20}} \tag{5.3}$$

$$\left|H(e^{j\Omega})\right| \geq 0.891$$

Similarly, if we are told that the attenuation in the stop band has to be at least 15 dB, then the absolute value of the magnitude in the stop band will be as shown in Eq. (5.4).

$$20\log_{10}\left(\left|H(e^{j\Omega})\right|\right) \leq -15$$

$$\left|H(e^{j\Omega})\right| \leq 10^{-15/20} \tag{5.4}$$

$$\left|H(e^{j\Omega})\right| \leq 0.17782$$

So, the signal in the pass band in this example will never be less than 0.891 in magnitude and in the stop band it will never be greater than 0.17782 in magnitude. In the same way, if you are given the absolute value you can convert it to dB; for example, if you are given the minimum magnitude of 0.95 in the pass band, then this can be converted to dB as shown in Eq. (5.5).

$$20\log_{10}(0.95) = -0.4455 \text{ dB} \tag{5.5}$$

Similarly, if we are given the maximum magnitude of 0.015 in the stop band, then this translates to dB as shown in Eq. (5.6).

$$20\log_{10}(0.015) = -36.478 \text{ dB} \tag{5.6}$$

The second diagram in Figure 5.1 gives us the filter specifications in normalized form. In this form, the maximum magnitude is assumed to be an absolute value of 1, which is 0 dB, and the maximum allowable ripple is given by $1/\sqrt{1+\varepsilon^2}$ – this is the minimum value in the pass band. In the stop band, the maximum magnitude allowed is $1/A$, so the minimum attenuation provided in the stop band is given as $-20\log_{10}(1/A)$ dB.

In analog filter design, two more parameters are often determined before any work on the design of the filters begins. The two parameters aid enormously in simplifying the process of filter design. These parameters are the transition ratio and the discrimination parameter. The transition ratio is the ratio of the pass band frequency with the stop band frequency as shown in Eq. (5.7).

$$\text{Transition ratio} \quad \kappa = \frac{\Omega_p}{\Omega_s} \tag{5.7}$$

The discrimination parameter is also a ratio; it is a ratio of ε, which is a measure of the ripple allowed in the pass band to $\sqrt{A^2-1}$ that is a measure of the minimum attenuation provided in the stop band as shown in Eq. (5.8).

$$\text{Discrimination parameter} \quad \kappa_1 = \frac{\varepsilon}{\sqrt{A^2-1}} \tag{5.8}$$

The way we have defined the two parameters, they are both less than 1 for a low-pass filter with the discrimination parameter κ_1 being very much less than 1.

5.2.2
The Butterworth Approximation

This approximation gives us a smooth monotonic response in the frequency domain. The magnitude squared function of the Butterworth filter is given in Eq. (5.9). This function has a special property that the derivatives of the magnitude function $|H_a(j\Omega)|$ evaluated at $\Omega = 0$ are all zero; for this reason, this response is said to have a *maximally flat* response. In Eq. (5.9) the term Ω_c is the cutoff frequency of the filter.

$$|H_a(j\Omega)|^2 = \frac{1}{1+(\Omega/\Omega_c)^{2N}} \tag{5.9}$$

The cutoff frequency of a filter is defined as that frequency where the gain is $-3\,\text{dB}$ from the maximum in the filter response. In the Butterworth filter, the maximum occurs at $\Omega = 0$ so the gain at frequency Ω_c is $-3\,\text{dB}$. It is interesting to note that the gain at the $-3\,\text{dB}$ frequency is also $^1/_2$ the gain at zero frequency. A typical response for several orders of Butterworth filters is given in Figure 5.2. Note that as the order of the filter gets larger, the transition band gets narrower and the transition from the pass band to stop band is steeper.

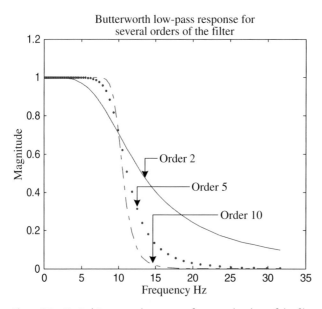

Figure 5.2 Typical Butterworth response for several orders of the filter.

Looking at Eq. (5.9), we can conclude that to design a Butterworth filter we need to determine two parameters from the filter specifications given. These two parameters are the filter order N and the cutoff frequency of the filter Ω_c. Once these two parameters are determined, the filter is completely determined. In order to determine these two parameters, we use the magnitude Eq. (5.9) at the two frequency points where we know the magnitude of the filter, specifically the pass band edge frequency and the stop band edge frequency as shown in Eq. (5.10) for the pass band edge and as shown in Eq. (5.11) for the stop band edge.

$$\left| H_a\left(j\Omega_p\right) \right|^2 = \frac{1}{1 + \left(\Omega_p/\Omega_c\right)^{2N}} = \frac{1}{1 + \varepsilon^2} \tag{5.10}$$

$$\left| H_a\left(j\Omega_s\right) \right|^2 = \frac{1}{1 + \left(\Omega_s/\Omega_c\right)^{2N}} = \frac{1}{A^2} \tag{5.11}$$

Solving Eqs. (5.10) and (5.11) simultaneously for the order N of the filter gives us Eq. (5.12). Notice that the order of the filter is completely determined by the transition ratio and the discrimination parameter.

$$N = \frac{1}{2}\frac{\log_{10}[(A^2-1)/\varepsilon^2]}{\log_{10}(\Omega_s/\Omega_s)} = \frac{\log_{10}(1/\kappa_1)}{\log_{10}(1/\kappa)} \tag{5.12}$$

The result in Eq. (5.12) will invariably be some value that is not an integer. Since the order of the filter must be an integer, the next higher integer larger than the value of N obtained from Eq. (5.12) is chosen. Once the filter order is determined, the cutoff frequency is determined from either Eq. (5.10) or Eq. (5.11). If Eq. (5.10) is used to determine the cutoff frequency, then the filter will exactly meet the response at the pass band edge frequency and will exceed the specifications at Ω_s, while if we use Eq. (5.11), the specification will be exactly met at the stop band edge and exceed the specifications at the pass band edge frequencies.

The MATLAB program can be effectively used to design the Butterworth filters. Some specific commands of the MATLAB program are given in Table 5.1.

Table 5.1 MATLAB commands to design Butterworth analog filters.

[z p k] = buttap(N)	Determines the pole, zero locations, and the gain for a Butterworth filter of order N
[[num den] = butter(N, Wn, 's')	Determines the numerator and the denominator polynomials for an analog Butterworth filter of order N and 3 dB cutoff frequency Wn. This is always a low-pass filter
[[num den] = butter(N, Wn, 'type', 's')	Same as above except you can specify the type of filter you want. Type is "low" or "high" or "band pass" or "band stop." In case of "band pass" or "band stop," Wn is a two-element vector
[[N Wn] = buttord(Wp, Ws, Rp, Rs, 's')	To determine the order of the Butterworth filter, use this command. Wp and Ws are the pass band and the stop band frequencies. Rp and Rs are the attenuations expected in the pass band and the stop band

5.2.3
The Chebyshev Type 1 Approximation

This approximation gives us ripples of equal magnitude in the pass band and a smooth monotonic response after that in the frequency domain. The ripples represent an error in the pass band from the ideal filter response. The magnitude squared function of the Chebyshev type 1 filter is given in Eq. (5.13).

$$|H_a(j\Omega)|^2 = \frac{1}{1 + \varepsilon^2 T_N^2(\Omega/\Omega_p)} \tag{5.13}$$

In Eq. (5.13), the function $T(\bullet)$ is the Chebyshev polynomial of order N and is given in Eq. (5.14). Notice that there are two definitions based on the value of the argument. If the argument $x \leq 1$, then the cos polynomial is used and when the argument $x > 1$, the cosh polynomial is used.

$$T_N(x) = \begin{cases} \cos(N\cos^{-1}(x)) & |x| \leq 1 \\ \cosh(N\cosh^{-1}(x)) & |x| > 1 \end{cases} \tag{5.14}$$

The Chebyshev polynomial is also defined recursively as shown in Eq. (5.15).

$$T_k(x) = 2xT_{k-1}(x) - T_{k-2}(x) \qquad k \geq 2$$
$$T_0(x) = 1 \quad \text{and} \quad T_1(x) = x \tag{5.15}$$

The typical plots of the Chebyshev type 1 filter are shown in Figure 5.3. Notice that the ripples across the entire pass band for a Chebyshev type 1 filter always have constant

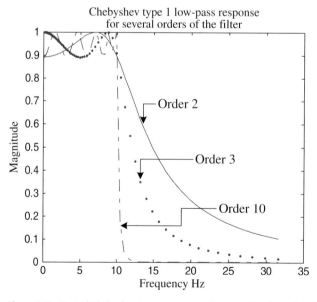

Figure 5.3 Typical Chebyshev type 1 response for several orders of the filter.

Table 5.2 MATLAB commands to design Chebyshev analog filters.

[[z p k] = cheb1ap(N, Rp)	Determines the pole, zero locations, and the gain for a Chebyshev type 1 filter of order N and ripple of Rp dB
[[num den] = cheby1(N, Rp, Wn, 's')	Determines the numerator and the denominator polynomials for an analog Chebyshev filter of order N and pass band edge frequency Wn and pass band ripple Rp
[[num den] = cheby1(N, Rp, Wn, 'type','s')	Same as above except you can specify the type of filter you want. Type is "low" or "high" or "band pass" or "band stop." In case of "band pass" or "band stop," Wn is a two-element vector
[[N Wn] = cheb1ord(Wp, Ws, Rp, Rs, 's')	To determine the order of the Chebyshev filter, use this command. Wp and Ws are the pass band and the stop band frequencies. Rp is the ripple in the pass band and Rs is the attenuations expected in the stop band

amplitude. The amplitude of the ripple is defined by the ripple parameter. The ripples for the three plots in Figure 5.3 are also equal because the ripple parameter for all the three plots was chosen to be equal for all the plots. For type 1 Chebyshev filter, the order N is determined shown in Eq. (5.16) as follows:

$$|H_a(j\Omega)|^2 = \frac{1}{1 + \varepsilon^2 T_N^2(\Omega_s/\Omega_p)} = \frac{1}{A^2}$$

$$|H_a(j\Omega)|^2 = \frac{1}{1 + \varepsilon^2 \left[\cosh\left(N\cosh^{-1}(\Omega_s/\Omega_p)\right)\right]} = \frac{1}{A^2} \tag{5.16}$$

Since we are in the process of designing a low-pass filter, the stop band frequency is larger than the pass band frequency and hence we have to use the cosh polynomial. The order N of the Chebyshev filter is easily obtained as the only unknown quantity in Eq. (5.16) is the order N. So by solving for N we get Eq. (5.17). Notice that the order of the filter is completely determined by the transition ratio and the discrimination parameter.

$$N = \frac{\cosh^{-1}\left(\left(\sqrt{A^2-1}\right)/\varepsilon\right)}{\cosh^{-1}(\Omega_s/\Omega_p)} = \frac{\cosh^{-1}(1/\kappa_1)}{\cosh^{-1}(1/\kappa)} \tag{5.17}$$

Here also, we are able to use the discrimination and the transition parameters to simplify our work in determining the order of the filter. Here also, like in the case of Butterworth filter, the filter order has to be an integer, so we will choose an integer higher than the value we get in Eq. (5.17).

The MATLAB program can be effectively used to design the Chebyshev filters. The MATLAB program has some specific commands that are given in Table 5.2.

5.2.4
The Chebyshev Type 2 Approximation

This approximation gives us ripples of equal magnitude in the stop band and a smooth monotonic response in the pass band in the frequency domain. The magnitude

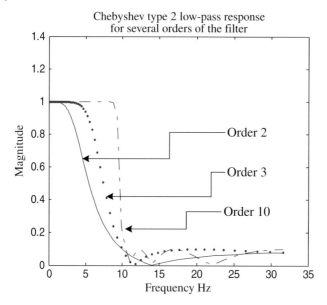

Figure 5.4 Typical Chebyshev type 2 response for several orders of the filter.

squared function of the Chebyshev type 2 filter is given in Eq. (5.18). Both the Butterworth and the Chebyshev type 1 filters are all pole filters, but the Chebyshev type 2 filter will have the Chebyshev polynomial in the numerator also, so this is not any more an all-pole filter. The order of the Chebyshev type 2 filter is the same as the order of the type 1 filter and hence it is calculated in the same way as the type 1 filter and given in Eq. (5.17).

$$|H_a(j\Omega)|^2 = \frac{1}{1 + \varepsilon^2 \left((T_N(\Omega_s/\Omega_p)) / (T_N(\Omega_s/\Omega)) \right)^2} \tag{5.18}$$

The typical plots of the Chebyshev type 2 filter are shown in Figure 5.4. Notice that the ripple is in the stop band for each plot. The ripple for the three plots is also equal because that was the ripple parameter chosen for the plots. For this type of filter, the order N is determined as shown in Eq. (5.16). Notice that the order of the filter is completely determined by the transition ratio and the discrimination parameter.

The MATLAB program can be effectively used to design the Chebyshev filters. The MATLAB program has some specific commands that are given in Table 5.3.

5.2.5
Scaling the Filters

Design of analog filters is generally carried out with the assumption that the cutoff frequency of the filter is unity irrespective of the desired cutoff frequency of the filter. This is because it is easy to scale the filter to the desired frequency region once we are satisfied with the magnitude response. Another reason to work with cutoff frequency of 1 is that it is much

Table 5.3 MATLAB commands to design Chebyshev analog filters.

[[z p k] = cheb2ap(N, Rs)	Determines the pole, zero locations, and the gain for a Chebyshev type 2 filter of order *N* and ripple of Rs dB
[[num den] = cheby2(N, Rs, Wn, 's')	Determines the numerator and the denominator polynomials for an analog Chebyshev filter of order *N* and pass band edge frequency Wn and pass band ripple Rs
[[num den] = cheby2(N, Rs, Wn, 'type', 's')	Same as above except you can specify the type of filter you want. Type is "low," "high," "band pass," or "band stop." In case of "band pass" or "band stop," Wn is a two-element vector
[[N Wn] = cheb2ord(Wp, Ws, Rp, Rs, 's')	To determine the order of the Chebyshev filter, use this command. Wp and Ws are the pass band and the stop band frequencies. Rp is the attenuations expected in the pass band and Rs is the ripple in the stop band

easier to work with small numbers rather than with large numbers. Scaling of the filter to the desired location is not a difficult operation. We first determine the scaling constant, which is just a ratio of the cutoff frequency of the filter that we have to the cutoff frequency of the filter that we want. This is shown in Eq. (5.19). Once the scaling constant is determined, we replace the *s* parameter in the filter that we have with a scaled *s* parameter to get the filter that we want. This is also shown in Eq. (5.19). Equation (5.19) explains how the scale factor is calculated and how the parameter *s* is replaced. Examples 5.1 and 5.2 will clear up any issues that may be present.

$$K_{freq} = \frac{\Omega_{old}}{\Omega_{new}} \quad \text{and} \quad s_{new} = K_{freq}s_{old} \tag{5.19}$$

Example 5.1

Design a low-pass filter with a pass band frequency of 30 Hz, pass band attenuation of no more than 0.5 dB, stop band attenuation of 35 dB, and stop band cutoff frequency of 90 Hz. Design the filter as a prototype filter first and then scale the filter to its proper frequency range. Use the Butterworth approximation for this filter.

Solution 5.1: For the prototype filter, we are going to choose the pass band edge frequency to be 1 rad so the stop band frequency will be $\Omega_s = (2\pi \times 90/(2\pi \times 30) = 3$ rad.

We will first use the command *buttord* to determine the order of the Butterworth filter. This command in MATLAB will look like

$$[N, Wn] = buttord(1, 3, 0.5, 35, 's')$$

The order of the filter and the cutoff frequency can also be obtained by using Eq. (5.12) to determine the order of the filter and Eq. (5.10) to determine the cutoff frequency of the filter.

By using this command, we get the filter order to be $N = 5$ and the 3 dB cutoff frequency to be 1.3401. Next, we will use the command *butter* to determine the transfer function of the order 5 filter. For this the MATLAB command will look like

$$[Num\ Den] = butter(N, Wn, \text{'s'})$$

The transfer function can be obtained by displaying the numerator and the denominator coefficients. The transfer function for our filter is given in Eq. (5.20).

$$H_{LP}(s) = \frac{4.3219}{s^5 + 4.34s^4 + 9.4s^3 + 12.6s^2 + 10.44s + 4.3219} \tag{5.20}$$

This filter has a pass band edge frequency of 1 rad/s and stop band edge frequency of 3 rad/s. We can see this from the magnitude plot of this filter that the pass band and the stop band edges are exactly as we designed them. Next, we have to scale this filter. To scale the filter, the scaling factor is $K_{freq} = 1/30$; with this scaling factor, the filter transfer function will now become

$$H_{LP}(s) = \frac{105022170}{s^5 + 130s^4 + 8460s^3 + 340200s^2 + 8456400s + 105022170}$$

We can now draw the frequency response plot of this filter to see what the response looks like and compare it with the prototype response that we drew. The two frequency response plots are drawn in Figure 5.5. Notice that the two plots are identical, only shifted in frequency. The response of the scaled filter has a pass band edge of 30 Hz and a stop band edge of 90 Hz just as we required.

5.2.6
Transforming Filters

Transforming one type of prototype filter to a different type of prototype filter is just as easy as scaling the filter; for this reason, we almost always design prototype low-pass filters to begin with. Once we have the prototype filter, this filter is transformed to whatever type of filter is desired. The transformation process is just a matter of substituting the variable s in the low-pass filter by the variable shown in Table 5.4 for the appropriate filter. The transforming equations are given in Table 5.4.

Example 5.2

First, transform the prototype filter designed in Example 5.1 to a prototype high-pass filter. Then, scale the prototype high-pass filter, so that its stop band frequency is 50/3 Hz and its pass band frequency is 50 Hz.

Solution 5.2: To transform the filter from low pass to high pass, we replace s in the low-pass filter by $1/s$ to get the high-pass filter. When we substitute s by $1/s$ in Eq. (5.20), we get Eq. (5.21).

$$H_{HP}(s) = \frac{4.3219s^5}{1 + 4.34s + 9.4s^2 + 12.6s^3 + 10.44s^4 + 4.3219s^5} \tag{5.21}$$

This is the prototype high-pass filter with pass band frequency of 1 Hz. The frequency response plot of this filter is given in Figure 5.6. Next, we want to scale this filter so that its

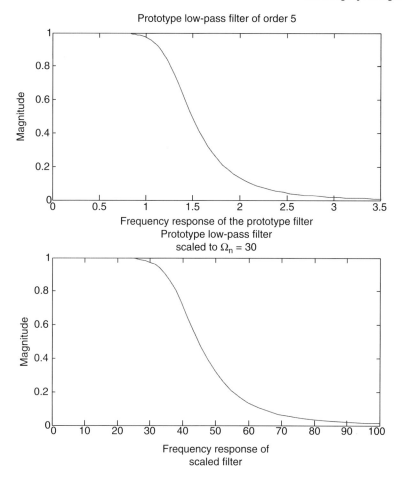

Figure 5.5 Frequency response of filters from Example 5.1.

pass band frequency is 50 Hz and its stop band frequency is 50/3 Hz. Scaling a filter is done in the same way no matter what type of filter you have. We have seen how a filter is scaled in Eq. (5.19), so we will scale this filter in exactly the same way. The scaling factor this time is 1/50. After the filter is scaled the transfer function becomes

Table 5.4 Equations to be used to transform low-pass filter to another type.

Transform from low pass to	Substitute s by
High pass	$1/s$
Band pass	$s + 1/s$
Band stop	$1/(s + 1/s)$

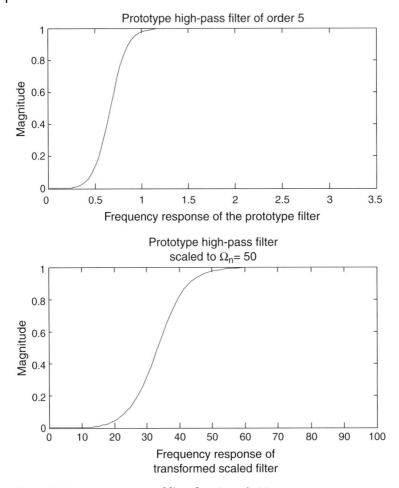

Figure 5.6 Frequency response of filters from Example 5.2.

$$H_{\mathrm{HP}}(s) = \frac{4.3219s^5}{312.5 \times 10^6 + 27.1 \times 10^6 s + 1.175 \times 10^6 s^2 + 31.5 \times 10^3 s^3 + 512 s^4 + 4.3219 s^5}$$

$$(5.22)$$

The plot of the transformed prototype high-pass filter derived from the low-pass filter and from Eq. (5.21) is given in Figure 5.6. The scaled filter frequency response for the high-pass filter with pass band frequency of 50 Hz is also drawn in Figure 5.6.

All the work that we have shown in the two examples can be done directly in MATLAB with no intervention by you. Examine the MATLAB script MATLAB 5.1.

MATLAB 5.1. Design a prototype low-pass filter. Scale the prototype filter to a higher frequency. Transform the low-pass prototype filter to a high-pass prototype filter. Scale the high-pass prototype filter to a different frequency.

```
%% First design the prototype filter.
[N Wn] = buttord(1,3,0.5,35,'s'); %Determine the order of
[Nump Denp] = butter(N,Wn,'s');    %the prototype filter
w = logspace(-2,.6,200);
[Hp Wp] = freqs(Nump,Denp,w);        %Compute and Draw its Freq
subplot(211);plot(Wp,abs(Hp));axis([0 3.5 0 1) %response.
title(['Prototype Low Pass Filter of order ',int2str(N)]);
xlabel('Frequency Hz');ylabel('Magnitude');
%% Now scale the filter to the frequency range of interest.
Kf = 1/30; %Scale factor for Lp to Lp
Nums = (1/Kf)^N* Nump; %Scale the Numerator
sv = (1/Kf).^(0:N);    %Vector to scale the
Dens = Denp.* sv; %denominator. Scale it.
w = logspace(-2,2,200);
[Hnew Wnew] = freqs(Nums,Dens,w); %Frequency response of the
subplot(212);plot(Wnew,abs(Hnew)) %Scaled filter.
title(['Prototype filter scaled to \omega_n = ',int2str(30)]);
xlabel('Frequency Hz');ylabel('Magnitude');
%% Now convert to High pass with stop band freq 50/3 and
%% pass band freq 50. First the prototype.
for i = 0:N %Substitute 1/s for the s
   Numh(i + 1) = Nump(N-i + 1); %in the numerator
   Denh(i + 1) = Denp(N-i + 1); %and the denominator.
end
figure;[Hh Wh] = freqs(Numh,Denh,w); %Frequency Response of the
subplot(211);plot(Wh,abs(Hh));
axis([0 3.5 0 1]) %transformed filter.
title(['Prototype High Pass Filter of order ',int2str(N)]);
xlabel('Frequency Hz');ylabel('Magnitude');
%% Now scale the High pass filter
Kfh = 1/50; %Scale factor to scale the
Numsh = Numh; %High pass filter.
svh = (1/Kfh).^(0:N); %Scaling vector to scale
Densh = Denh.* svh; %the denominator. Scale it.
w = logspace(-2,2,200);
[Hnewh Wnewh] = freqs(Numsh,Densh,w); %Compute and Plot the
subplot(212);plot(Wnewh,abs(Hnewh)); %frequency Response.
title({ 'Prototype High Pass Filter'; ['scaled to \omega_n= ',
int2str(50)]} );
xlabel('Frequency Hz');ylabel('Magnitude');
```

5.3
Design of IIR Filters from Analog Filters

Traditionally, the approach of designing discrete filters has been to first design the required analog filter and then to transform it to its discrete equivalent filter. This approach has been followed as the art of analog filters is very well known and very good results in designing analog filters are easily obtained. The analog design methods give good closed form formulas, so discrete equivalent filters based on these closed formulas are easy to implement.

The analog filters are usually specified as transfer functions in the frequency domain as shown in Eq. (5.23).

$$H_a(s) = \frac{Y_a(s)}{X_a(s)} \tag{5.23}$$

The transfer function relation can also be written as an input–output relation of such a system and this is given by a convolution integral in the time domain as shown in Eq. (5.24)

$$y_a(t) = \int_{-\infty}^{\infty} x_a(\tau)h_a(t-\tau)d\tau \tag{5.24}$$

or as a differential equation shown in Eq. (5.25).

$$\sum_{r=0}^{R} a_r \frac{d^r y(t)}{dt^r} = \sum_{k=0}^{K} b_k \frac{d^k x(t)}{dt^k} \tag{5.25}$$

The corresponding filter for the discrete time can also be specified as a transfer function and it has the form as shown in Eq. (5.26).

$$H(z) = \frac{Y(z)}{X(z)} \tag{5.26}$$

The transfer function relation of the discrete domain can also be written as an input–output relation of such a system and this is given by a convolution summation in the time domain as shown in Eq. (5.27)

$$y[n] = \sum_{r=-\infty}^{\infty} x[r]h[n-r] \tag{5.27}$$

or as a difference equation shown in Eq. (5.28).

$$\sum_{r=0}^{R} a_r y[n-r] = \sum_{k=0}^{K} b_k x[n-k] \tag{5.28}$$

Our goal here is to determine the impulse response of the discrete filter $h[n]$, so that the discrete impulse response resembles the impulse response of the analog filter $h_a(t)$ in some manner while preserving some desirable property of the analog filter.

So when we are trying to transform an analog filter to its discrete equivalent, we are interested in obtaining $h[n]$. In performing the transformation, we require that the general properties of the analog filter should be preserved. The most basic of these requirements is that a stable filter in the analog domain transform to a stable filter in the discrete domain. This implies that the entire left half of the s-plane transforms to the interior of the unit circle in the z-plane. The basic idea of the filter is to have a certain response in the frequency domain. So, this second characteristic of the analog filter must also be preserved when we transform the filter to the discrete domain. The frequency response of the discrete filter must match the frequency response of the analog filter.

5.3.1
Impulse Invariance Method

One approach to designing IIR filters from analog filters begins with the design of the analog filter whose impulse response is adequate. Once we have the impulse response $h_a(t)$ of the analog filter that we need, we will try to match the impulse response $h[n]$ of the discrete filter to the impulse response $h_a(t)$ of the analog filter. Our goal is to match the impulse response of the discrete filter to the impulse response of the analog filter that we have. If we were to sample this analog impulse response and get a discrete impulse response, it should resemble the impulse response of the analog filter. In principle then, the sampled impulse response $h[n]$ of the discrete filter should match the impulse response of the analog filter; this is shown in Eq. (5.29).

$$h[n] = h_a(t)_{t=nT}| = h_a(nT) \tag{5.29}$$

The Z-transform of $h[n]$ and the Laplace transform of $h_a(t)$ are related to each other by the sampling of the analog signal and the result is similar to the relation given in Eq. (2.36). In Eq. (2.36), we saw that the spectrum of $H(z)$ was a periodic repeat of the spectrum of $H_a(s)$. At that time, we said that if $H_a(s)$ is band limited, then there will be no aliasing in the spectrum of $H(z)$. Here, however, the spectrum of $H_a(s)$, which is a filter, is not band limited; it becomes very small in the stop band region, but it is never zero. So the impulse invariant filter design method will definitely suffer some aliasing.

In Chapter 3, we also saw that one way to reduce the effect of aliasing was to increase the sampling rate. Even that technique does not help us here. The reason for this is that when we have the critical frequency in the discrete domain, we have to first convert it to the analog domain. This relation between the frequency in the analog domain and the discrete domain is given in Eq. (5.30).

$$\omega_c = \Omega_c T \tag{5.30}$$

So as long as ω_c stays constant, making T smaller results only in making Ω_c larger and the aliasing stays the same. This is shown in Figure 5.7. With the knowledge that there will be some aliasing, we demonstrate the design process for impulse invariance for converting an analog filter to a discrete filter and always begin with the analog filter in its factored form as shown in Eq. (5.31).

$$H_a(j\Omega) = \sum_{r=1}^{N} \frac{A_r}{(s-s_r)} \tag{5.31}$$

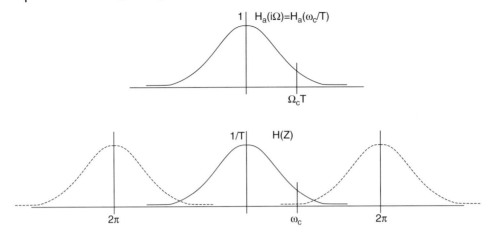

Figure 5.7 Representation of aliasing in an impulse invariance design procedure irrespective of the sampling time interval.

Since this filter is in its partial factored form, we can determine the inverse Laplace transform of the filter to get the impulse response of the analog filter as shown in Eq. (5.32).

$$h(t) = \mathcal{L}^{-1}(H_a(s)) = \sum_{r=1}^{N} A_r e^{s_r t} \tag{5.32}$$

Next, we will sample $h(t)$ in Eq. (5.32) to get the impulse response in the discrete domain as shown in Eq. (5.33).

$$h[n] = h(t)|_{t=nT} = \sum_{r=1}^{N} A_r e^{s_r n T} \tag{5.33}$$

This is the filter that we want. But we get some interesting insight if we go one step further and obtain the Z-transform of the discrete sequence $h[n]$ in Eq. (5.33).

$$H(z) = \mathcal{Z}[h[n]] = \sum_{n=-\infty}^{\infty} \left[\sum_{r=1}^{N} A_r e^{s_r n T} \right] z^{-n}$$

$$H(z) = \sum_{r=1}^{N} A_r \sum_{n=-\infty}^{\infty} \left(e^{s_r T} z^{-1} \right)^n = \sum_{r=1}^{N} \frac{A_r}{1 - (e^{s_r T} z^{-1})} \tag{5.34}$$

The result in Eq. (5.34) is interesting. Compare Eq. (5.31) with Eq. (5.34). They are both in partial fraction form, they both have N poles and the residue at these poles is the same in both the analog and the discrete domain. So, to get the Z-transform of the filter designed using impulse invariance, we first determine the pole locations of the filter and then transform the pole location using the transform $s_k = e^{s_k T}$ to get the Z-transform of the filter. It is important to note that while we are mapping the pole locations, we are not mapping the function $H_a(s)$. In Eq. (5.34), we have directly mapped the pole locations, but there is no mention of what happens to the zeros of $H_a(s)$. Consider the following example of mapping a second-order Butterworth filter.

Example 5.3

Transform a second-order Butterworth filter with cutoff frequency 10 Hz to its discrete equivalent using the impulse-invariant method.

Solution 5.3: A second-order Butterworth filter in its factored form is given in Eq. (5.35).

$$H_a(s) = \frac{7.1i}{s+7.1+7.1i} - \frac{7.1i}{s+7.1-7.1i} \tag{5.35}$$

From this we can directly write the filter transfer function in the Z-domain as shown in Eq. (5.36).

$$H(z) = \frac{7.1i}{1-e^{-7.1T-7.1iT}z^{-1}} - \frac{7.1i}{1-e^{-7.1T+7.1iT}z^{-1}} \tag{5.36}$$

By rearranging Eq. (5.36), we get

$$H(z) = \frac{14.2\sin(7.1T)e^{-7.1T}z^{-1}}{1-2\cos(7.1T)e^{-7.1T}z^{-1} + e^{2(7.1T)}z^{-2}}$$

The basis or the motivation for the impulse invariance method is not so much to maintain the shape of the response but the knowledge that if the analog filter is band limited, then the digital filter frequency response will closely approximate the analog frequency response. Sometimes, however, the primary objective may be to control some aspect of the time response such as the peak overshoot or the rise time. In such cases, a natural approach to the design is by impulse invariance. If we discount aliasing, then the shape of the frequency response is preserved and the relation between the analog and the discrete frequencies is linear. This filter design method cannot be used for high-pass or band stop filters. Extra care should be taken if this method is to be used for band pass filters.

5.3.2
Bilinear Transform Method

The impulse invariance method is good when you want to match some characteristics like the rise time or the peak overshoot. There is another method that is superior to the impulse invariance method when other properties are more important. This is the bilinear transform method. The bilinear transform method is more commonly used to design IIR filters when we want to convert an analog filter to its discrete equivalent. The bilinear transform is a simple substitution of all the s terms in the analog filter by a z-domain function to convert the analog filter to its discrete equivalent by the substitution shown in Eq. (5.37) as follows:

$$s = \frac{2}{T}\left(\frac{1-z^{-1}}{1+z^{-1}}\right) \tag{5.37}$$

The transform given maps the entire left half of the s-plane into the interior of the unit circle in the z-plane, it is a one-to-one transform and it maps a stable filter from the s-plane to a stable filter in the z-plane. The relation between the s-domain and the z-domain transfer functions is given by Eq. (5.38).

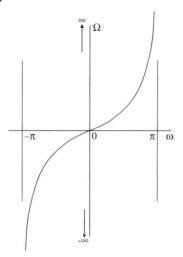

Figure 5.8 Mapping of the analog frequency to the discrete frequency using bilinear transform.

$$H(z) = H_a(s)|_{s=(2/T)[(1-z^{-1})/(1+z^{-1})]} \tag{5.38}$$

It is interesting to note that the bilinear transform is derived by approximating the integral in the analog domain using the trapezoidal method.[1] In arriving at the bilinear method, the parameter T is the step size of the approximation method. In the following section, we examine how the transform will map the poles and the zeros of the s-plane to the z-plane by the bilinear transformation method.

$$s = \frac{2}{T}\frac{1-z^{-1}}{1+z^{-1}}\bigg|_{z=re^{j\omega}} = \frac{2}{T}\frac{1-r^{-1}e^{-j\omega}}{1+r^{-1}e^{-j\omega}} \tag{5.39}$$

By evaluating Eq. (5.39) on the imaginary axis, we get the result that we are looking for. To do this, we set $s = \sigma + j\Omega = 0 + j\Omega$ since we are evaluating on the imaginary axis. Similarly, we will replace $z = re^{j\omega} = 1e^{j\omega}$. This gives us the relation to map the frequency from the analog domain to the frequency in the discrete domain. This is shown in Eq. (5.40) and in Figure 5.8.

$$0 + j\Omega = \frac{2}{T}\frac{1-e^{-j\omega}}{1+e^{-j\omega}} = j\frac{2}{T}\tan^{-1}\left(\frac{\omega}{2}\right)$$

$$2\tan\left(\frac{T\Omega}{2}\right) = \omega \tag{5.40}$$

As can be clearly seen from Figure 5.8, the mapping is not linear. As the frequencies in the analog domain approach infinity, the frequencies in the discrete domain approach π, but this approach is very crammed near the region $\pm\pi$ in the discrete domain. This is

---

1) Refer to Problem 4.10 for more details.

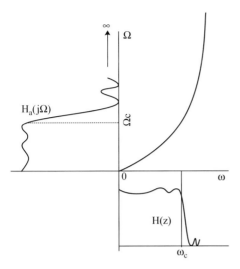

Figure 5.9 Distortion of the analog frequency to the discrete frequency by bi-linear transform.

known as the *warping* of the frequency, and to see the effect of warping, examine Figure 5.9. In Figure 5.9, we see that the low-frequency region gets stretched out, while the high-frequency region is crunched together when we use the curve to map the analog frequencies to the discrete frequencies. Since this warping effect is present in the mapping, we must take precaution before designing the analog filter by *prewarping* the frequency from the discrete domain to the analog domain by using the prewarping equation. What the prewarping equation does is that it recognizes that during bilinear transform warping will take place. To account for this warping, we will prewarp the critical frequencies from the discrete domain to the analog domain as shown in Eq. (5.41). Once we know what the warped frequency is, we can design the analog filter with these prewarped critical frequencies and then use the bilinear transform to transform the analog filter to its discrete equivalent that will then unwarp the critical frequencies and place them exactly where we wanted them to be in the first place.

$$\Omega = \frac{2}{T}\tan\left(\frac{\omega}{2}\right) \tag{5.41}$$

We should remember that the bilinear transform preserves the magnitude response of filters that can be described by piece-wise linear response, that is, the shape of the magnitude response in both the analog and the discrete domain is very similar. The phase of the response is, however, not preserved during the response. Hence, this method of mapping should be used only when it is the magnitude portion of the response that we consider as critical and we are not concerned by the alteration of the phase of the response.

Example 5.4

Transform a second-order Butterworth filter with cutoff frequency 10 Hz to its discrete equivalent using the bilinear transform method.

Solution 5.4: A second-order Butterworth filter is given in given in Eq. (5.42).

$$H_a(s) = \frac{100}{s^2 + 14.2s + 100} \tag{5.42}$$

In this equation, when we substitute for s using Eq. (5.39), we get Eq. (5.43) as follows:

$$H(z) = H_a(s)|_{s=(2/T)[(1-z^{-1})/(1+z^{-1})]}$$

$$= \frac{100}{((2/T)[(1-z^{-1}/1+z^{-1})])^2 + 14.2((2/T)[(1-z^{-1}/1+z^{-1})]) + 100} \tag{5.43}$$

On simplifying, we get the required discrete filter as shown in Eq. (5.44). The method itself is quite simple and straightforward; however, it is very tedious and takes up a lot of time to keep the algebra straight.

$$H(z) = \frac{T^2 100(1 + 2z^{-1} + z^{-2})}{(4 + 28.4T + 100T^2) + (200T^2 - 8)z^{-1} + (4 - 28.4T + 100T^2)z^{-2}} \tag{5.44}$$

In this example, we were given the cutoff frequency of the analog filter and hence now the sampling interval plays a very critical role. Let us change the requirement slightly and work on a more realistic example. When we have to design a discrete filter, we will be given the specifications in the discrete domain. Specifically, consider the design shown in Example 5.5. We have two specific purposes in this example. First, it will demonstrate the design method and second, to show that when you start with the specifications in the discrete domain and end up with the filter in the discrete domain, the approximation step size parameter T plays no part in the design and hence can be chosen to be any value that is convenient, which is usually 2, so the multiplier $2/T$ vanishes.

Example 5.5

Design a Butterworth response for a filter with pass band edge frequency of 0.15π rad and the stop band edge frequency of 0.35π rad. It is also required that the droop in the pass band is not more than 0.75 dB and the minimum attenuation in the stop band is at least 20 dB.

Solution 5.5: Here, we are given the specifications in the discrete domains, so we will first convert the specifications to the analog domain. This is where we prewarp the critical frequency from the discrete domain to the design frequencies in the analog domain as shown in Eq. (5.45).

$$\Omega_p = \frac{2}{T}\tan\left(\frac{\omega_p}{2}\right) = \frac{2}{T}\tan\left(\frac{0.15\pi}{2}\right) = \frac{2}{T}(0.24)$$

$$\Omega_s = \frac{2}{T}\tan\left(\frac{\omega_s}{2}\right) = \frac{2}{T}\tan\left(\frac{0.35\pi}{2}\right) = \frac{2}{T}(0.73) \tag{5.45}$$

Next, we determine the parameter ε^2 from the pass band droop of 0.75 dB and A^2 from the minimum attenuation in the stop band of 20dB, as done in Eq. (5.46).

$$20 \log_{10}\left(\sqrt{1+\varepsilon^2}\right)dB = 0.75 \qquad \text{this gives } \varepsilon = 0.4342$$

$$20 \log_{10}(1/A)dB = 20 \qquad \text{this gives } A^2 = 100 \tag{5.46}$$

From Eqs. (5.45) and (5.46), we can determine the transition ratio and the attenuation parameter required to compute the order of the filter. Refer to Eqs. (5.7) and (5.8). The two parameters are determined in Eq. (5.47).

$$\kappa = \Omega_p/\Omega_s = ((2/T)0.24)/((2/T)0.7265) = 0.3304$$

$$\kappa_1 = \frac{\varepsilon}{\sqrt{A^2-1}} = \frac{0.4342}{9.9499} = 0.0436 \tag{5.47}$$

Next, we determine the required order of the filter as shown in Eq. (5.48), so we need a third order Butterworth filter.

$$N = \frac{\log_{10}(1/\kappa_1)}{\log_{10}(1/\kappa)} = \frac{\log_{10}(22.9358)}{\log_{10}(3.0266)} = 2.8283 = 3 \tag{5.48}$$

Also, note that the approximation time T plays no part in selecting the filter size. Now that we have the filter size, we will choose a third-order prototype filter to begin with. This filter is given in Eq. (5.49). We have to first scale this filter so that it has its cutoff frequency where we need it. To determine where we need the cutoff frequency to be, we will use Eq. (5.9).

$$H_{\text{pro}}(s) = \frac{1}{(s+1)(s^2+s+1)} \tag{5.49}$$

By using Eq. (5.9) with the specified gain at the pass band edge (or at the stop band edge), we get the cutoff frequency to be

$$|H_a(\Omega_p)|^2 = \frac{1}{1+\left(\Omega_p/\Omega_c\right)^{2*3}} = \frac{1}{1+\varepsilon^2} \Rightarrow \Omega_c = \Omega_p * 1.3206$$

$$|H_a(\Omega_s)|^2 = \frac{1}{1+\left(\Omega_s/\Omega_c\right)^{2*3}} = \frac{1}{A^2} \Rightarrow \Omega_c = \Omega_s * 0.4649 \tag{5.50}$$

Depending on which one of the two equations you choose you will get the cutoff frequency to be either 0.3171 or 0.3378. Now that we know the cutoff frequency of our filter and the cutoff frequency of the prototype (which is 1), we need to first scale the prototype as shown in Eq. (5.51).

$$H_a(s) = H_{\text{pro}}\left(\frac{s}{(2/T)0.3171}\right)$$

$$= \frac{(8/T^3)(0.03189)}{(s+(2/T)0.3171)\left(s^2+((2/T)0.3171)s+(2/T)^20.1006\right)} \tag{5.51}$$

To the filter of Eq. (5.51) we will apply the bilinear transform by substituting for the variable s as given in Eq. (5.39). Doing this gets us the Z-domain function as shown in Eq. (5.52).

$$H(z) = \frac{(8/T^3)(0.03189)}{\left((2/T)\left(\frac{1-z^{-1}}{1+z^{-1}}\right) + (2/T)0.3171\right)}$$ (5.52)

$$\left\{(2/T)^2\left(\frac{1-z^{-1}}{1+z^{-1}}\right)^2 + 0.3171(2/T)^2\left(\frac{1-z^{-1}}{1+z^{-1}}\right) + (2/T)^2 0.1006\right\}$$

The first thing that we notice in Eq. (5.52) that the factor $2/T$ just cancels out from the numerator and the denominator, that is, T plays no part in converting the analog filter to its discrete equivalent. By completing the required algebra, we get the discrete filter as given in Eq. (5.53).

$$H(z) = \frac{0.09823(1+z^{-1})^3}{(1.3171 - 0.6829z^{-1})(1.4177 - 1.7988z^{-1} + 0.7835z^{-2})}$$ (5.53)

The plot of the magnitude and the gain response of the filter are given in Figure 5.10. The MATLAB program used to compute the various parameters is given in MATLAB 5.2

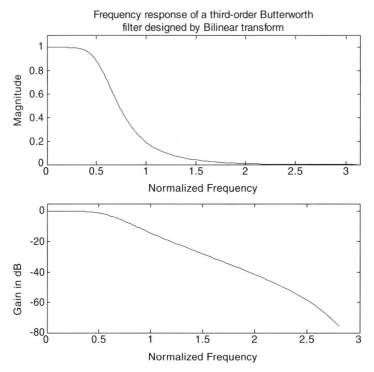

Figure 5.10 Magnitude and gain response of a third-order Butterworth filter designed using bilinear transform with prewarping.

MATLAB 5.2. Design a prototype low-pass filter. Scale the prototype filter to a higher frequency. Transform the low-pass prototype filter to a high-pass prototype filter. Scale the high-pass prototype filter to a different frequency.

```
%% Calculate the critical frequencies by pre-warping
Op = tan(0.15*pi/2); %pre warp the pass band edge
Os = tan(0.4*pi/2); %pre warp the stop band edge
%% Compute the two filter parameters required for filter order.
ep = sqrt(1/10^(-0.75/10)-1); %Compute the required constants
A2 = (1/10^(-20/20))^2; %from the filter specifications
k = Op/Os; %The transmission parameter
k1 = ep/sqrt(A2-1); %The discrimination parameter
N = log10(1/k1)/log10(1/k); %The required filter order
%% Determine the cutoff frequency of the required filter
Oc1 = Op/(ep)^(1/3); %Filter cutoff frequency
Oc2 = Os/(A2-1)^(1/6); %Alternate cutoff frequency
%% Define the Numerator and the denominator of the filter
Numd = 0.06123*[ 1 3 3 1]; %Numerator and denominator
Dend = [1 -1.7873 1.2105 -0.2865];    %polynomials of the filter
%% Compute the response and draw the plots.
[Hd Wd]  = freqz(Numd,Dend,logspace(-2,pi,250)); %Compute the
response
Max = max(abs(Hd));
subplot(2,1,1);plot(Wd,1/Max*abs(Hd)); %Plot the Magnitude
axis([ 0,pi,0,1.1] );
xlabel('Frequency in Radians');ylabel('Magnitude');
title({ 'Frequency response of a third order Butterworth';'filter
designed by Bilinear transform'} )
subplot(2,1,2);plot(Wd(1:245),20*log10(1/Max*abs(Hd
(1:245)))));
axis([0,3,-80,5] ); %Plot the gain response.
xlabel('Frequency in Radians');ylabel('Gain in dB');
```

5.3.3
Frequency Transformations of IIR Discrete Filters

In transforming analog filters into discrete filters, we have always designed the analog filter as a low-pass filter and then transformed it to a discrete low-pass filter. The commonly used frequency-selective filters come in several types such as high pass, band pass, and band stop. The traditional method of design for analog filters has always been to design a normalized prototype low-pass filter and then to scale the filter to the frequency region of interest and transform it to be the desired type. This transformation is done by simple algebraic substitution as shown in Table 5.4.

In the case of digital frequency-selective filters, we can also design an analog prototype normalized low-pass filter and transform this filter to its discrete equivalent by either the impulse-invariant or the bilinear transform method. We have seen that when we transform a low-pass filter, we do not suffer from a very severe case of aliasing for either of these methods. Once we have a discrete low-pass normalized filter, we can scale it to the frequency range of interest and then transform it to be the desired type of filter. This transformation of a digital filter to a different type of digital filter can be applied to any digital filter no matter how we obtained the filter in the first place.

To understand how the mapping is done, let us denote $H_l(z)$ and z as the low-pass function and the variable of the low-pass filter that we have. Let us also associate $H_d(Z)$ and Z as the desired filter function and the variable of the desired filter that we want. Now define the mapping from z to Z as shown in Eq. (5.54).

$$z^{-1} = T(Z^{-1}) \tag{5.54}$$

So with this mapping, the desired filter $H_d(Z)$ can be written in terms of the filter $H_l(z)$ as shown in Eq. (5.55) and $T^{-1}()$ is the mapping of z^{-1} to Z^{-1}.

$$H_d(Z) = H_l(T^{-1}(z^{-1})) \tag{5.55}$$

The requirement of the mapping function that maps one discrete filter to another discrete filter are simple; first, it should map the entire region from inside the unit circle to a region inside the unit circle. This will ensure that a stable filter always maps to be a stable filter. That the mapping function $T(Z^{-1})$ be a rational function so that a causal filter will always map to be a causal filter. It has been shown that the most general form of the required function $T(Z^{-1})$ is an all-pass filter of the type shown in Eq. (5.56).

$$T(Z^{-1}) = \pm \prod_{r=1}^{R} \frac{Z^{-1} - a_r}{1 - a_r Z^{-1}} \qquad |a_r| < 1 \tag{5.56}$$

The requirement that $|a_r| < 1$ implies that the transformation leads always to a stable filter. By choosing appropriate values of R, the numbers of all pass sections and a_r, the parameter that determines the pole and the zero locations, a variety of mappings can be obtained, and the simplest of these mappings is to transform one low-pass filter to another low-pass filter at a different frequency. Examining this mapping will give us an insight into all the other mappings. The simplest mapping then is when $R = 1$, so we get Eq. (5.57).

$$z^{-1} = T(Z^{-1}) = \frac{Z^{-1} - a}{1 - a Z^{-1}} \tag{5.57}$$

In Eq. (5.57), if we let $z = e^{i\theta}$ and $Z = e^{i\omega}$, we get the relation for ω that we want to obtain as shown in Eq. (5.58). The relation between θ and ω for several values of α is plotted in Figure 5.11.

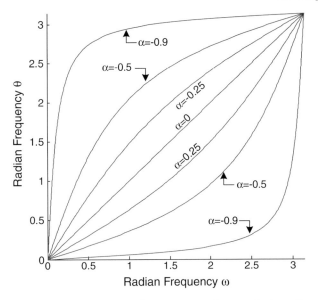

Figure 5.11 Frequency transformation curves for several values of α.

$$e^{-j\theta} = \frac{e^{-j\omega} - \alpha}{1 - \alpha e^{-j\omega}}$$

$$\omega = \tan^{-1}\left(\frac{(1 - \alpha^2)\sin(\theta)}{2\alpha + (1 + \alpha^2)\cos(\theta)}\right)$$

$$(5.58)$$

Figure 5.11 shows a warping of the frequency scale. Even though there is warping, the low-pass filter with the pass band till θ and a piece-wise linear magnitude response will be transformed into a low-pass filter with the pass band till ω and a piece-wise linear magnitude. The transformation suggested by Eq. (5.57) will be known and will be complete as soon as we determine the parameter α. This parameter α is determined by solving for it in Eq. (5.58). When we solve for α in Eq. (5.58) we get Eq. (5.59).

$$\alpha = \frac{\sin\big((\theta_p - \omega_p)/2\big)}{\sin\big((\theta_p + \omega_p)/2\big)}$$

$$(5.59)$$

Once the parameter α is determined, the low-pass filter that you have can be transformed to the frequency range of interest by simple substitution as shown in Eq. (5.60).

$$H_d(Z) = H_l(z)\big|_{z^{-1} = (Z^{-1} - a)/(1 - \alpha Z^{-1})}$$

$$(5.60)$$

The other transforms from low pass to high pass, band pass, and band stop all use a similar procedure where we solve for one or two constants and once we have determined the required number of constants, the filter is transformed by a simple substitution. The transforms for other types of filters are given in Table 5.5. As an example of this method,

Table 5.5 Equations to scale and transform discrete filters from one type to another.

Filter type	Transformation equation	Design parameters
Low pass	$z^{-1} = \dfrac{Z^{-1}-\alpha}{1-\alpha Z^{-1}}$	$\alpha = \dfrac{\sin\left(\dfrac{\theta-\omega}{2}\right)}{\sin\left(\dfrac{\theta+\omega}{2}\right)}$ θ = cutoff frequency of filter we have ω = cutoff frequency of filter we want
High pass	$z^{-1} = -\dfrac{Z^{-1}+\alpha}{1+\alpha Z^{-1}}$	$\alpha = -\dfrac{\cos\left(\dfrac{\theta+\omega}{2}\right)}{\cos\left(\dfrac{\theta-\omega}{2}\right)}$ θ = cutoff frequency of filter we have ω = cutoff frequency of filter we want
Band pass	$z^{-1} = -\dfrac{Z^{-2}-(2\alpha\delta/\delta+1)Z^{-1}+(\delta-1/\delta+1)}{(\delta-1/\delta+1)Z^{-2}-(2\alpha\delta/\delta+1)Z^{-1}+1}$	$\alpha = \dfrac{\cos((\omega_2+\omega_1)/2)}{\cos((\omega_2-\omega_1)/2)}$ $\delta = \cot((\omega_2-\omega_1)/2))\tan\left(\dfrac{\theta}{2}\right)$ $\omega_1;\omega_2$ = desired lower and upper cutoff frequencies of the filter we want
Band stop	$z^{-1} = -\dfrac{Z^{-2}-(2\alpha/\delta+1)Z^{-1}+(1-\delta/1+\delta)}{(1-\delta/1+\delta)Z^{-2}-(2\alpha/\delta+1)Z^{-1}+1}$	$\alpha = \dfrac{\cos((\omega_2+\omega_1)/2)}{\cos((\omega_2-\omega_1)/2)}$ $\delta = \tan((\omega_2-\omega_1)/2))\tan\left(\dfrac{\theta}{2}\right)$ $\omega_1;\omega_2$ = desired lower and upper cutoff frequencies of the filter we want

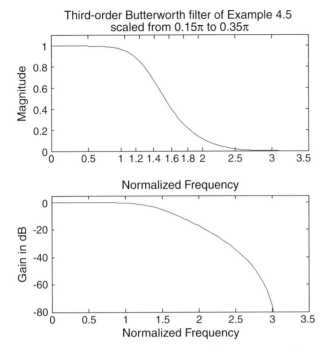

Figure 5.12 Filter scaled from one pass band frequency to another pass band frequency in the discrete domain.

translate the filter of Example that was obtained by using bilinear transform. This filter had a pass band edge at 0.15π rad. We would like to scale the filter so that it has a pass band edge at 0.35π rad.

We first determine the parameter α; using Eq. (5.59) we get $\alpha = -0.437$. With this value of α, the transformation will be given by $z^{-1} = (Z^{-1} + 0.437)/(1 + 0.437Z^{-1})$. On substituting the required function, we get the desired filter as shown in Eq. (5.61).

$$H_d(Z) = \frac{0.1412(1 + Z^{-1})^3}{(1.0187 - 0.1073Z^{-1})(0.7813 - 0.2185Z^{-1} + 0.2682Z^{-2})} \qquad (5.61)$$

A frequency response plot of the filter of Eq. (5.61) confirms that this filter indeed has the same shape as the filter of Example 5.5 and Figure 5.10, but the pass band has been shifted to 0.35π rad. The plot of this filter is created by a program similar to the script MATLAB 5.1. The plot of the filter is given in Figure 5.12.

5.4
Design of FIR Filters

The IIR filters that we have studied in the previous section provide excellent magnitude response, but this comes at the cost of the phase response. When the phase of the input signal has to be preserved, then the FIR filter will most often be used. The FIR filters are designed mostly with the condition that the phase response of these filters be linear. These filters are also very efficient as we are able to take advantage of very efficient computing algorithms for the DFT to use these filters. These filters are very well suited for multirate signal processing we will study in Chapter 9. For all these reasons, design techniques for FIR filters have aroused considerable interest.

The FIR filter is defined by a finite summation as shown in Eq. (5.62).

$$H(z) = \sum_{n=0}^{N-1} h[n]z^{-n} \tag{5.62}$$

So $H(z)$ is the discrete filter and, as shown in Eq. (5.62), it is represented as a polynomial in z^{-1} and it is of order N. A polynomial with order N has $N-1$ roots, so the discrete FIR filter will have $N-1$ zeros in the Z-plane and it will have $N-1$ poles at $z=0$. The definition of the FIR filter can also be written as shown in Eq. (5.63).

$$H(e^{j\omega}) = \sum_{n=0}^{N-1} h[n]e^{-j\omega n} \tag{5.63}$$

Equation (5.63) looks very much like the Fourier series representation. This representation has only N distinct frequency representations, so the sequence is completely specified by N terms that are the N samples of its Fourier transform. So the design method of FIR filters reduces to finding either N impulse response coefficients $h[n]$ or equivalently N samples of its frequency response $H(e^{j\omega})$ around the unit circle. In the next section, we will discuss both these methods individually.

5.4.1
Linear Phase Transfer Functions

Earlier we stated that the FIR filters are designed with the express purpose that they have linear phase. This way we avoid phase distortion when the signal is filtered with such filters. Here, we examine the characteristics of a linear phase transfer function. If the output from the filter is an exact replica of the input but delayed by n_0 samples, we can say that there is no phase distortion. So the relation for no phase distortion can be written as $y[n] = x[n - n_0]$. Next, we determine the Fourier transform of both sides to get Eq. (5.64).

$$Y(e^{j\omega}) = e^{-j\omega n_0}X(e^{j\omega})$$
$$H(e^{j\omega}) = \frac{Y(e^{j\omega})}{X(e^{j\omega})} = e^{-j\omega n_0} \tag{5.64}$$

Equation (5.64) shows us that the magnitude response of the filter $H(e^{j\omega})$ is 1 and the phase is linear. In this filter, all the frequencies in the pass band will pass through undistorted as the magnitude is unity and the phase is linear. The magnitude and phase response of this filter are drawn in Figure 5.13. In the figure, we only see the phase response of the filter in the pass band; no mention has been made of the phase response in the stop band as no signal will go through this region.

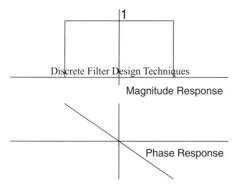

Figure 5.13 Magnitude and phase response of a filter that will give you a distortionless output in the pass band.

5.4.2
Requirements of a Linear Phase Transfer Function

In Section 5.3.1, we have just seen that if we want to avoid phase distortion, then we must have a linear phase transfer function. Here, we examine the criteria that will always give us a linear phase transfer function $H(z)$ with a real impulse response $h[n]$. Since we want $H(z)$ to have linear phase, we will factor $H(z)$ into its amplitude response that has zero phase and its phase response as shown in Eq. (5.65).

$$H(z) = \sum_{n=0}^{N-1} h[n]z^{-n} = e^{j(c\omega + \beta)} H_{ZP}(\omega) \tag{5.65}$$

In Eq. (5.65), the phase is represented as $e^{j(c\omega + \beta)}$ where β represents a constant phase shift and $c\omega$ represents the linear phase shift, the amplitude response $H_{ZP}(\omega)$ represents the amplitude of the filter response that represents a zero phase function. For the impulse response $h[n]$ to be real, the magnitude response of the filter $|H(e^{j\omega})|$ must be an even function or $|H(e^{j\omega})| = |H(e^{-j\omega})|$. Since $|H(e^{j\omega})| = |H_{ZP\setminus}(\omega)|$, the amplitude function $H_{ZP}(\omega)$ must be either an even function or an odd function, so we can write $H_{ZP}(\omega) = \pm H_{ZP}(-\omega)$. With this the frequency response $H(e^{j\omega})$ will satisfy the relation $H(e^{j\omega}) = H^*(e^{-j\omega})$. By substituting this equality in Eq. (5.65), we get Eq. (5.66).

$$e^{j(c\omega + \beta)} H_{ZP}(\omega) = e^{-j(c(-\omega) + \beta)} H_{ZP}(-\omega) \tag{5.66}$$

In Eq. (5.66) when $H_{ZP}(\omega)$ is an even function, we can obtain a condition on β as we must have $e^{j\beta} = e^{-j\beta}$ for the equality to exist. This happens only when $\beta = 0$ or $\beta = \pi$. Setting $\beta = 0$, we get Eq. (5.67).

$$H_{ZP}(\omega) = \pm e^{-jc\omega} H(e^{j\omega}) = \pm \sum_{n=0}^{N} h[n] e^{-jc\omega(c+n)} \tag{5.67}$$

We get another relation from the fact that $H_{ZP}(\omega)$ is an even function, so $H_{ZP}(\omega) = H_{ZP}(-\omega)$, which we can write as shown in Eq. (5.68).

$$H_{ZP}(-\omega) = \pm e^{jc\omega} H(e^{j\omega}) = \pm \sum_{r=0}^{N} h[r]e^{j\omega(r+c)} \tag{5.68}$$

We can rewrite Eq. (5.68) after we substitute $r = N - n$ to get Eq. (5.69).

$$H_{ZP}(-\omega) = \pm \sum_{n=0}^{N} h[N-n]e^{j\omega(c+N-n)} \tag{5.69}$$

$H_{ZP}(\omega)$ is an even function and we can equate the RHS of Eq. (5.67) to the RHS of Eq. (5.69). By doing this we get the condition that we must have the equalities shown in Eq. (5.70).

$$h[n] = h[N-n] \qquad c = \frac{-N}{2} \quad \text{and} \quad \beta = 0 \tag{5.70}$$

The condition in Eq. (5.70) says that for $h[n]$ to have linear phase, it must be a symmetric impulse response. In arriving at the conditions of Eq. (5.70), we chose $H_{ZP}(\omega)$ to be an even function, when $H_{ZP}(\omega)$ is to be an odd function, we can follow an almost identical argument to arrive at the condition given in Eq. (5.71).

$$h[n] = -h[N-n] \qquad c = -N/2 \quad \text{and} \quad \beta = \pi/2 \tag{5.71}$$

So for an FIR filter to have linear phase, we conclude that the filter must be either even symmetric as suggested by Eq. (5.70) or it must be odd symmetric as suggested by Eq. (5.71).

5.4.3
Design of FIR Filters Using Windows

One obvious way to determine the impulse response coefficients is to start with a piece-wise linear frequency response of the desired ideal filter in the frequency domain. Next, we determine the IDTFT of this piece-wise linear frequency response of the desired ideal filter. This in general will give us an infinite sequence $h_d[n]$. For the filter to have a finite impulse response, we will truncate the filter to be of length N. This truncation of the infinite sequence in the time domain is the same as multiplying the sequence $h_d[n]$ by a window also of infinite length, but with only a finite number (specifically N) of terms that are not zero. Such a window function is shown in Figure 5.14. Thus, the product of the infinite sequence $h_d[n]$ with a window sequence $w[n]$ gives us a finite length sequence as shown in Figure 5.14.

Now that we have multiplied the desired filter sequence with the window sequence, the frequency response of the product of two sequences will not be exactly the same as the desired ideal filter. This is because product in the time domain results in convolution in the frequency domain. So the frequency response of the product sequence will be the convolution of the frequency response of the filter with the frequency response of the ideal desired filter. Figure 5.15a shows the frequency response of the ideal filter that we want to design. Figure 5.15b represents the frequency response of the rectangular window that we have used to limit the impulse response to a finite length. The result of the windowing process is the convolution of the two frequency responses. The result of this convolution is shown in Figure 5.15c. Notice that the resulting frequency response resembles the ideal frequency response, but the edges are not sharp, as they appear to be smeared. The smearing occurs because the main lobe of the window frequency response has some width.

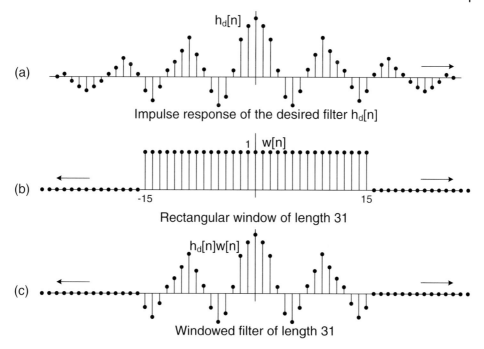

Figure 5.14 Demonstrating the windowing operation in the time domain.

If the frequency response of the window is an impulse in the frequency domain, then the convolution will not smear the frequency response of the desired ideal filter. The frequency response of the window is never an impulse, so there will be some smearing and this is how the well-known Gibbs phenomenon arises when we try to implement FIR filters. This effect is shown in Figure 5.14.

Since we will be using a window to limit the length of the infinite filter, a question that often comes up here is the choice of the length of the window. How long should the window be? When we examine the relation between the length of the window and the width of the main lobe, we find that longer the window is narrower the width of the main lobe will be. When the width of the main lobe is narrow, the smearing that occurs will be limited to a very narrow frequency band when the window and the filter frequency responses are convolved together. We obviously do not want to make the length of the window infinite; we want to make it as short as possible. If we make the window too short, then the main lobe of the frequency response of the window will be too wide. When the main lobe of the window is wide, the transition band of the filter will also be wide. Deciding on the length of the window is not an exact science. Window lengths are decided on either by empirical relations or by trial and error. Some of the more popular empirical relations to determine the window length are Kaiser's formula and Bellanger's formula. Both the formulas take the width of the required transition band into account in determining the length of the window. The empirical relations are given in Eq. (5.72).

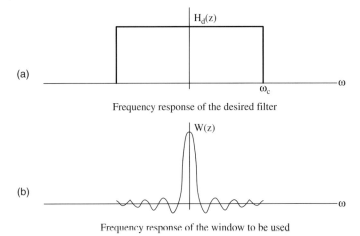

(a)

Frequency response of the desired filter

(b)

Frequency response of the window to be used

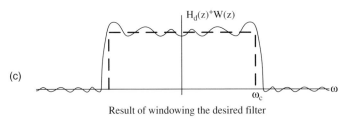

(c)

Result of windowing the desired filter

Figure 5.15 Smearing of the desired filter response due to the convolution of the window and the filter.

Kaiser's formula $\quad N = \dfrac{-20\log_{10}\left(\sqrt{\delta_p\delta_s}\right)-13}{14.6\left(\omega_s-\omega_p\right)/2\pi}$

Bellanger's formula $\quad N = \dfrac{2\log_{10}\left(10\delta_p\delta_s\right)}{3\left(\omega_s-\omega_p\right)/2\pi}$

(5.72)

These two empirical relations are interesting in that they both show that the length of the window is inversely proportional to the transition width $(\omega_s - \omega_p)$. In these empirical relations, some attempt has been made to include the pass band and the stop band ripple. The Bellanger's formula slightly overestimates, while the Kaiser's formula slightly under-estimates. Later when we talk about specific windows, we will look at some other formulas that pertain to the window itself.

The windowing operation makes an infinite filter into a finite length filter. This filter is still not usable as it is a not a causal filter. This is seen in Figure 5.14c. In order to be able to use the filter, we will shift (delay) the filter by half the length of the filter or $(N - 1)/2$ when N is odd or $(N/2)$ sample points when N is even and then use the filter. The effect of the delay in the time domain is to cause a rotation in the frequency domain. So when we consider an ideal low-pass filter with cutoff frequency of ω_c, the design process for this filter will begin with the fact that we expect the filter to be delayed by half the sample units, so we will write the filter transfer function to be as shown in Eq. (5.73).

$$H_{\mathrm{d}}(\omega) = \begin{cases} e^{-j\omega(N/2)} & 0 \leq |\omega| \leq \omega_{\mathrm{c}} \\ 0 & \omega_{\mathrm{c}} \leq |\omega \leq \pi| \end{cases} \tag{5.73}$$

Now, we can determine the IDTFT of the desired filter to get the filter in the time domain as shown in Eq. (5.74). By performing integration, we get the sequence $h_{\mathrm{d}}[n]$, which is the sequence we want, and the sequence is shifted by $N/2$ sample points, but it is of infinite length as shown in Eq. (5.74).

$$h_{\mathrm{d}}[n] = \frac{1}{2\pi} \int_{-\omega_c}^{\omega_c} e^{j\omega(n-(N/2))} d\omega = \frac{e^{j\omega(n-(N/2))}}{2\pi j(n-(N/2))} \bigg|_{-\omega_c}^{\omega_c} = \frac{\sin(\omega_{\mathrm{c}}(n-(N/2)))}{\pi(n-(N/2))} \tag{5.74}$$

To make this sequence a finite length sequence from an infinite length sequence, we window the sequence. Windowing the sequence involves multiplying the entire sequence $h_{\mathrm{d}}[n]$ with the entire window sequence that is also shifted by $N/2$ sample points and it is nonzero in the range 0–$N-1$ and zero everywhere else as shown in Figure 5.14. The windowing process is shown in Eq. (5.75).

$$h_{win}[n] = w[n] \cdot h[n] \tag{5.75}$$

In Eq. (5.75), $w[n]$ is the window function with only a finite number of nonzero values. The sequence $h[n]$ is the finite length filter that we will use. The design process to design the required FIR filter is as follows. We begin with an ideal filter in the frequency, which is piecewise linear, and determine the approximate length of the filter using the empirical relations given in Eq. (5.72). Now that we know the length, we shift the filter by half the length of the filter and then determine the IDTFT of the filter. This will give us an infinite length-shifted filter. We will now truncate this filter by windowing the filter as shown in Eq. (5.75) using an appropriate window. Example 5.6 will make this process clear.

Example 5.6

Design an FIR filter to resemble the ideal filter shown in Figure 5.16. Determine the length of the filter required by both the empirical formulas given in Eq. (5.72). Use $\delta_{\mathrm{p}} = \delta_{\mathrm{s}} = \delta = 0.001$ and a transition width of 0.05π and $\omega_{\mathrm{c}1} = 0.3\pi$, $\omega_{\mathrm{c}2} = 0.45\pi$.

Solution 5.6: The ideal filter given to us is not any standard filter but that does not matter. The procedure stays the same, so we will first determine the IDTFT of the desired filter that we are given. This is done in Eq. (5.76).

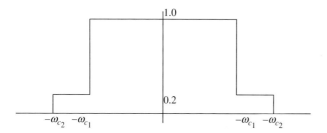

Figure 5.16 Multiple pass band filter to be designed.

$$h_{\rm d}[n] = \frac{1}{2\pi}\int_{-\omega_{c_1}}^{\omega_{c_1}} e^{j\omega(n-(N/2))}\,{\rm d}\omega + \frac{1}{2\pi}\int_{-\omega_{c_2}}^{-\omega_{c_1}} 0.2e^{j\omega(n-(N/2))}\,{\rm d}\omega + \frac{1}{2\pi}\int_{\omega_{c_1}}^{\omega_{c_2}} 0.2e^{j\omega(n-(N/2))}\,{\rm d}\omega$$

$$h_{\rm d}[n] = \begin{cases} \dfrac{\omega_{c_1}-\omega_{c_2}}{\pi} & n = \dfrac{N}{2} \\[4mm] & 0 \leq n \leq N-1 \\[2mm] \dfrac{0.8\sin(\omega_{c_1}(n-(N/2))) + 0.2\sin(\omega_{c_2}(n-(N/2)))}{\pi(n-(N/2))} & n \neq \dfrac{N}{2} \end{cases}$$

(5.76)

Next, we determine the length of the filter and the window by the two formulas in Eq. (5.72). This is done in Eq. (5.77).

Kaiser $\quad \dfrac{-20\log_{10}\left(\sqrt{(0.01)(0.01)}\right) - 13}{14.6(0.05\pi)/2\pi} = 73.972 = 74$

(5.77)

Ballanger $\quad \dfrac{-2\log_{10}(10(0.01)(0.01))}{3(0.05\pi)/2\pi} = 80 = 80$

We now have to choose which one of the many available windows we will use. Since we have not yet discussed which windows are possible, we will choose the rectangular window with the window opening height 1 and window length 75 or 81 (we want to choose a window of odd length here). Using the rectangular window to filter the desired sequence gives us the windowed filter sequence $h[n]$ that is the same as $h_{\rm d}[n]$ that we have obtained in Eq. (5.76) except that the windowed sequence will be limited in size to a length of 75 or 81 depending on which window length method you decide to choose. A plot of the response from Eq. (5.76) is given in Figure 5.17. The plot is drawn for three different window lengths. When we compare the three plots, we see the effect of altering the window length. From the plot we can conclude that increasing the window length reduces the transition width.

5.4.3.1 Windows Used for FIR Filters

The windows that we use to limit the length of the IDTFT of the desired filter can be varied and can come in different forms but the goal is the same, we would like the frequency response of the window be as close to an impulse as possible. Over the years, many different windows have been developed and used in practice. The most popular among these are given in Table 5.6.

The plot of the popular windows used in FIR filter design in the time domain is given in Figure 5.18. Notice that all the windows are very similar in that they are at a peak in the center and gradually taper to zero at the ends (all except the rectangular window). The plot of the windows in the frequency domain is given in Figure 5.19. From the plot in Figure 5.19, we see that different windows, even though they are the same size, provide different amount of attenuation. Another remarkable fact of these windows is that the window that provides a larger attenuation in the stop band also has a wider main lobe width. The rectangular window has the narrowest main lobe, but it provides the least amount of attenuation (only about 12–13 dB), while the Blackman window provides the most amount of attenuation (almost 55 dB)

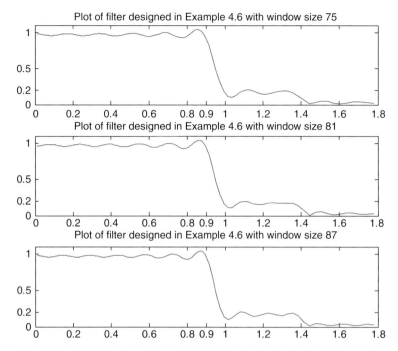

Figure 5.17 Plot of the filter of Example 5.6. Notice that the window length gives us the required transition width. Increasing the window reduces the transition width.

but the width of the main lobe is very wide compared to the width of the main lobe of the rectangular window. The attenuation provided by the various windows is given in Table 5.7.

In Figure 5.20, we see the plot of the Hamming window for several different window lengths. For each window length, we also see the frequency response of the window. From the plot of the frequency responses, we can clearly see that extending the window length

Table 5.6 Some of the popular windows used for FIR filter design.

Window name	Window expression
Rectangular	$W[n] = 1 \quad 0 \le n \le N-1$
Triangular	$W[n] = \begin{cases} 2n/(N-1) & 0 \le n \le \dfrac{N}{2} \\ 2-(2n/(N-1)) & \dfrac{N}{2} \le n \le N-1 \end{cases}$
Hamming	$W[n] = 0.54 - 0.46 \cos\left(\dfrac{2\pi n}{N-1}\right) \quad 0 \le n \le N-1$
Hanning	$W[n] = 0.5 + 0.5 \cos\left(\dfrac{2\pi n}{N-1}\right) \quad 0 \le n \le N-1$
Blackman	$W[n] = 0.42 + 0.5 \cos\left(\dfrac{2\pi n}{N-1}\right) + 0.08 \cos\left(\dfrac{4\pi n}{N-1}\right) \quad 0 \le n \le N-1$

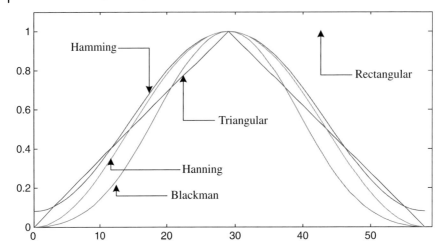

Figure 5.18 Plot of the popular windows for FIR filter design. Window length 59.

reduces the width of the main lobe in the frequency response, but it does not alter the attenuation provided in the stop band. From this we can conclude that the window length controls the main lobe width of the window, the window length this way also controls how wide the transition band will be. Using the empirical formulas given in Table 5.7, we can calculate the width of the main lobe and compare the result with the main lobe width in the plot in Figure 5.20. From this we can say that when designing the filters, we usually make an informed estimate of the window length by using one of the formulas from Eq. (5.72) or as an alternative expression we can use the last column of Table 5.7 to estimate the window length for a required main lobe width. With the window length determined, we first draw the frequency response of the entire windowed filter. If the window length used provides us with a required transition band, then we are done, or else we increase or decrease the window length till we get the required transition band. The following example will show the complete process of filter design.

Example 5.7

Design a low-pass filter with a cutoff frequency of 0.4π, a transition band no wider than 0.1π, and stop band attenuation to be at least 30 dB down.

   Solution 5.7: We begin with the understanding that we will have to shift the filter so we will introduce the shift in the pass band of the ideal filter. This gives us the desired filter as shown in Eq. (5.78).

$$h_d[n] = \int_{-\pi}^{\pi} e^{-j\omega(N/2)} e^{j\omega n} d\omega = \int_{-\omega_c}^{\omega_c} e^{j\omega(n-(N/2))} d\omega = \frac{\sin(\omega_c(n-(N/2)))}{\pi(n-(N/2))} \qquad (5.78)$$

   Next, we will limit this filter with a window function. Since we need at least 30 dB of attenuation, we are limited to using the Hamming, the Hanning, or the Blackman window from the list of popular windows. Then, we will try to estimate the size of the filter. To do this

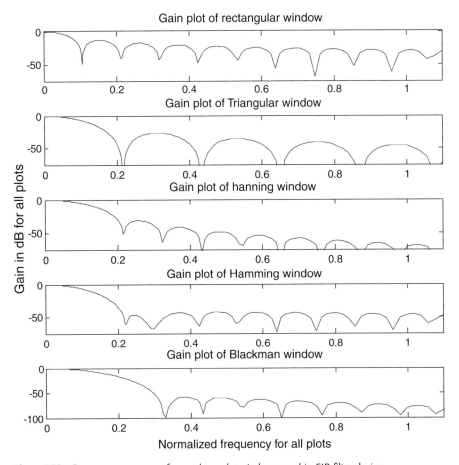

Figure 5.19 Frequency response of several popular windows used in FIR filter design.

we will use the information about the windows given in Table 5.5. The length of the filter with these three windows and the required transition width is calculated in Eq. (5.79).

$$\text{Hamming} \quad N = \frac{6.22 * \pi}{0.1 * \pi} = 63$$

$$\text{Hanning} \quad N = \frac{6.65 * \pi}{0.1 * \pi} = 67 \tag{5.79}$$

$$\text{Blackman} \quad N = \frac{11.1 * \pi}{0.1 * \pi} = 111$$

The frequency response plot of this filter with the three windows and their respective lengths is drawn using MATLAB and is given in Figure 5.21. From the figure we can see that all the windows meet the required specifications as the transition band of 0.1π is 0.3141 rad of normalized frequency. The Hamming window has a transition band from about 1.1 to about 1.4, the Hanning window has a transition band from about 1.15 to about 1.4, and the Blackman window has a transition band from 1.2 to about 1.4.

Table 5.7 Attenuation provided by the window and the window transition width in number of samples in the window.

Window name	Attenuation in the stop band	Width of the transition band
Rectangular window	−13.3 dB	$\dfrac{1.85\pi}{N}$
Triangular window	−26.5 dB	$\approx \dfrac{4.15\pi}{M}$
Hamming window	−31.5 dB	$\dfrac{6.22\pi}{N}$
Hanning window	−42.7 dB	$\dfrac{6.65\pi}{N}$
Blackman window	−58.1 dB	$\dfrac{11.1\pi}{N}$

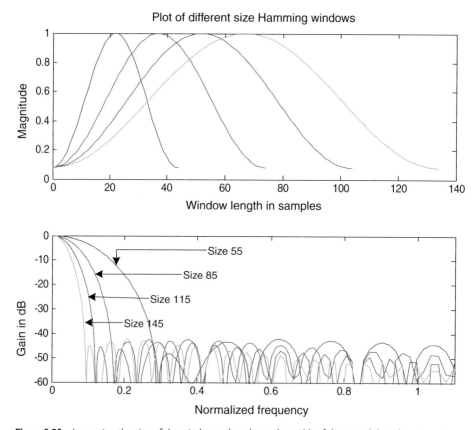

Figure 5.20 Increasing the size of the window only reduces the width of the main lobe. The attenuation provided by the window is fixed by the window shape.

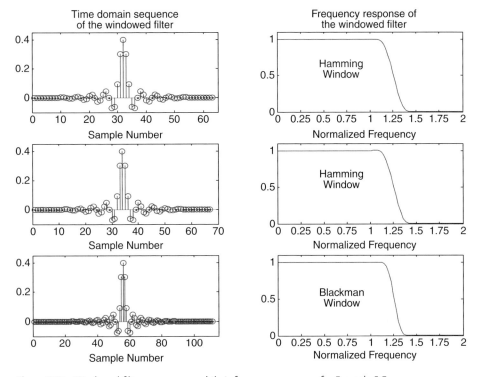

Figure 5.21 Windowed filter sequences and their frequency response for Example 5.7.

Since all the filters exceed the specifications, you would go back and evaluate the filters with a shorter length, as the length of the filter and the window determines how much computational effort will be required. A longer filter requires greater computational effort. So in designing the FIR filters we would first estimate the length as we did for this example and then verify that the length is just right and if it is not then we adjust the length as required till we get it just right.

5.4.4
Design of Filters Using Frequency Sampling

Earlier we showed that a finite duration sequence can be represented by its DFT. We have also seen that an FIR filter is a finite duration sequence, so it should be possible to represent the FIR filters $h[n]$ by the samples of its frequency response. We can view this as shown in Eq. (5.80).

$$H[k] = H(z)|_{z=e^{(2\pi k/N)}} = \sum_{n=0}^{N-1} h[n]e^{(-j2\pi kn)/N} \qquad (5.80)$$

Equation (5.80) says that we can sample the Z-transform of the filter on the unit circle to get $H[k]$ from which we can then get the filter $h[n]$. We of course have to take special care to make sure that the magnitude and the phase of the samples satisfy the conditions required to give us a causal real filter $h[n]$. We can specify the samples as either the combination of magnitude and the phase or the real and the imaginary parts of the samples. It is much easier to specify the magnitude and the phase rather than the real and the imaginary parts. It should be remembered that the phase has to be consistent with the length of the filter (odd or even) as seen in Eq. (5.70) or in Eq. (5.71). Since the frequency sampling method interpolates between sample points, the choice of inappropriate phase specifications will show up as undesirable behaviors between samples. Since we expect the filter to have linear phase, let us write the filter in terms of its magnitude and phase as shown in Eq. (5.81).

$$H[k] = |H[k]|e^{j\theta(k)} \tag{5.81}$$

In this filter, we want the phase $\theta(k)$ to be linear, so we will force it to be as shown in Eq. (5.82) where we have to determine the constant α.

$$\theta(k) = -\alpha k \tag{5.82}$$

To determine the coefficient α, let us determine the filter coefficients in the time domain by computing the IDFT of the filter given in Eq. (5.81) with $\theta(k) = -\alpha k$. We do this in Eq. (5.83).

$$h[n] = \frac{1}{N}\sum_{k=0}^{N-1}|H[k]|e^{-j\alpha k}e^{(j2\pi nk/N)} = \frac{1}{N}\sum_{k=0}^{N-1}|H[k]|e^{-jk(\alpha - 2\pi n/N))} \tag{5.83}$$

Equation (5.83) is in terms of magnitude and phase, and if we rewrite the equation in terms of the real part and the imaginary parts, we get the result that we want as shown in Eq. (5.84). For the imaginary part to be zero, we must have $\alpha = (2\pi n/N)$. From Eq. (5.84), we see that the magnitude is simply the absolute value of the frequency response magnitude at the appropriate sample point.

$$h_{\text{Re}}[n] = \frac{1}{N}\sum_{k=0}^{N-1}|H[k]|\cos\left(k\left(\alpha - \frac{2\pi n}{N}\right)\right)$$

$$h_{\text{Im}}[n] = 0 = \frac{1}{N}\sum_{k=0}^{N-1}|H[k]|\sin\left(k\left(\alpha - \frac{2\pi n}{N}\right)\right) \tag{5.84}$$

We also have to have this filter as a causal filter, so to make this filter a causal filter we must shift the filter by half the sample points. This is easily done as the shifting introduces rotation and rotation is represented by a phase, so we make the parameter α greater by the shift amount as shown in Eq. (5.85).

$$\alpha = \frac{2\pi n}{N}\frac{N-1}{2} = \frac{\pi n(N-1)}{N} \tag{5.85}$$

So the method of designing FIR filters with frequency sampling can be stated as follows. Begin with a piece-wise linear ideal filter in the frequency domain. Sample this filter from 0 to 2π using periodic sampling (samples are spaced linearly around the unit circle). If the filter order desired is N, remember to take N samples. The phase of the filter has to be linear with a delay of $N/2$ samples. So add this phase to the magnitude samples that you obtained. This gives you a length N sequence in the frequency domain. Determine the IDFT of this sequence to get a sequence in the time domain that is the impulse response of the desired filter. Example 5.8 demonstrates the method.

Example 5.8

A filter of length 33 having a cutoff frequency of 0.4π is to be designed using the frequency sampling method.

    Solution 5.8: The plot of the filter along with the sample points is shown in Figure 5.22. When we sample the filter at 33 points, we get the samples as shown in Figure 5.22. The first 14 samples are 1 and the remaining 19 samples are 0. To these samples we have to add the phase of the filter. Since the filter length is odd, β of Eq. (5.70) will be zero and the linear phase delay will be as shown in Eq. (5.85). The MATLAB script for the design of the filter is given in MATLAB 5.3 and the plot of the response of the filter is given in Figure 5.22.

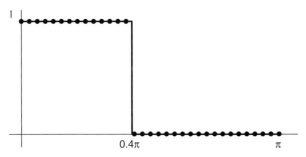

Low-pass filter to be designed by frequency sampling

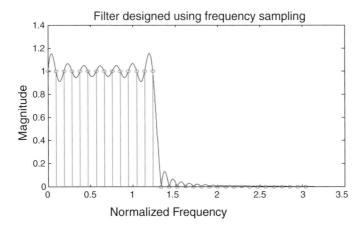

Figure 5.22 Low-pass filter designed using frequency sampling method.

MATLAB 5.3. Script for FIR filter using frequency sampling.

```
%% Build a filter using Frequency sampling method.
N = 66;                          %The length of the filter
n = 0:(N-1);                     %The sample number vector
p = ceil(0.4*N/2);              %Number of points in pass band
Hm =[ones(1,p) zeros(1,N-p)] ; %Define the Magnitude response
ph = exp(-j*pi*n*(N-1)/N);      %Define the Phase response
H = Hm.*ph;                      %Combine the two responses
h = ifft(H);                     %The filter in the time domain
[B W]  = freqz(h,1,16*N);        %Determine its frequency response
plot(W,abs(B));hold             %Draw the plots.
stem(W(1:32:N*16),abs(B(1:32:16*N)),'g');
title('Frequency Sampling Filter Design');
```

Some points of interest about the frequency sampling method: Note from the solid curve and the stems in Figure 5.22 that the stems are the frequency samples we started with. That the frequency response (the solid curve) passes through each one of them so at the frequency where we took the sample there is no error. In between the sample points, the frequency sampling method approximates and interpolates the response. That is why you get the oscillations between the sample points.

Earlier, we saw that using windows on the FIR filters reduced the Gibbs phenomenon, hence smoothing out the oscillations. We can take advantage of that here also to reduce the oscillations. So once we have the desired filter using the frequency sampling method, we can window the filter with a window of appropriate length. This will have the same effect and reduce the oscillations between the sample points.

In the frequency sampling method, we knew that the cutoff frequency had to be 0.4π, but when we took the samples of the response, none of the samples fell on the transition, the transition occurs between the 14 and the 15 samples that are located at frequencies of 0.39π and 0.42π. This is the biggest drawback of this filter design method. You cannot place a sample exactly on the transition frequency.

Another major drawback of this method is that the error or the overshoot is maximum near the transition band. We can effectively reduce this error if we took one sample between the pass band and the stop band. In the design that we just completed, we have no points in the transition band. Of course, choosing to place a sample in the transition band effectively doubles the transition band as at present the transition band is just one sample and after we place one sample in the transition band the transition band will become two samples.

5.4.5
A Comparison of the FIR and the IIR Filters

Often, I am asked the questions: Which type of filter is the best to use? Which filter design method should we use? Which window is the best to use for my filters? The answer to all these questions is simple: no single type of filter or design method is the best. Each type of filter and design method has its own merits, limitations, and peculiarities. Pay attention to them and use them to your advantage.

The choice between FIR and the IIR filters: The advantage of the IIR filters is that a variety of filters can be designed with a closed form formula. With a closed form formula, obtaining the filter is a simple matter, scaling and translating the filter is also just an exercise in algebra, and the pole–zero locations are explicitly known. This is a distinct advantage when computing power is expensive or not available.

With FIR filters, a closed form formula is not possible. These filters also suffer from the well-known Gibbs phenomenon, although windowing eliminates most of the Gibbs effect. To get a good FIR filter, the engineer may go through several iterations to adjust the ripple and the transition width when using fixed windows. Two-parameter windows (discussed in Section 5.4) help tremendously, but some iteration is still required.

With the IIR filters, we are limited to the standard classical filter types such as the low pass, high pass, and so on. The design methods of the IIR filters concentrate on getting the magnitude response right, but this happens at the cost of the phase response. The phase response of most of the IIR filters is somewhat linear in the pass band region, but it is highly nonlinear in and near the transition band region.

The FIR filters on the other hand are designed with the express purpose of having exactly linear phase and in addition can be designed to have a magnitude response that is piece-wise linear and not just limited to the classical filter types such as the low pass, high pass, and so on. There is also a possibility of (or an appearance of) more control with the FIR filters, as there are optimization theorems that allow you to move the error energy from one area to a different area.

Finally, there is also the economic consideration. In general, a longer filter is more expensive than a shorter filter in both the hardware required and the software used (memory space and computational power required) to implement the filter. Since the FIR filters are at least an order of magnitude greater than the IIR filters, they will incur an additional cost because of their size. This additional cost is more than returned when the phase of the filter is required to be linear. To meet this requirement with IIR filters, you will have to use one or more all pass filters just to adjust the phase of the filter.

So many trade-offs must be considered in designing digital filters. The final choice is mostly made in terms of engineering judgment on issues such as filter specifications, method of implementation, computational facilities available, and the experience and the expertise of the design engineer.

5.5
Design of Windows

The different windows that we used in Section 5.3 all have been fixed windows. In these windows, the only parameter that you could adjust was the length of the window, the shape of the window was fixed. With the shape fixed, the attenuation provided by the window is also fixed. Adjusting the length of the window only allows us to control the width of the main lobe of the window. It would be desirable to have windows that allow us to adjust both the length and the shape of the window with one parameter that controls the shape of the window and the other that controls the length of the window, with both the parameters being adjusted independently. Several such windows have been proposed in the literature. Among these the Kaiser window is one of the most popular windows with two adjustable parameters. The Kaiser window has the β-shape or the ripple parameter that allows the designer to trade transition width for ripple. The window is defined as shown in Eq. (5.86), where $I_0(\cdot)$ is the zeroth-order Bessel function, β is the shape parameter, and M is half the length parameter.

$$w_k[n] = \frac{I_0\left\{\beta\sqrt{1-(n/M)^2}\right\}}{I_0(\beta)} \qquad -M \leq n \leq M \tag{5.86}$$

Very frequently the Bessel function is computed by its power series approximation as shown in Eq. (5.87).

$$I_0(x) = 1 + \sum_{k=1}^{\infty} \left[\frac{(x/2)^k}{k!}\right]^2 \tag{5.87}$$

To calculate the Bessel function using the approximation in Eq. (5.87), the summation need not be carried out till k becomes infinite. Generally, by the time $k = 20$, the terms are so small that they can be ignored. To calculate the two parameters of the Kaiser window, we generally follow the empirical relations that Kaiser himself developed. The steps for the window design are as follows:

Design Method 1: Begin with the specifications of the pass band edge ω_p, the stop band edge ω_s, the pass band ripple δ_p, and the stop band ripple δ_s.

Design Method 2: Evaluate the minimum ripple in dB by $A = -20\log_{10}(\min(\delta_s,\delta_p))$ and the transition width $\Delta\omega = (\omega_s - \omega_p)$. This is for the low-pass filter.

Design Method 3: Estimate the length of the required filter by

$$M = \begin{cases} \dfrac{A-7.95}{2.285\Delta\omega} & A > 21 \\[2mm] \dfrac{5.79}{\Delta\omega} & A < 21 \end{cases} \tag{5.88}$$

Design Method 4: Now determine β depending on the value of the attenuation A required by

$$\begin{aligned} \beta &= 0 & A < 21 \\ \beta &= 0.5842(A-21)^{0.4} + 0.07886(A-21) & 21 \leq A \leq 50 \\ \beta &= 0.1102(A-8.7) & 50 \leq A \end{aligned} \tag{5.89}$$

Design Method 5: Now compute the window coefficients using the formula given in Eq. (5.86).

In Figure 5.23, we see several plots of the Kaiser window. The two plots on the left-hand side are drawn with the length fixed at 35 samples and β is adjusted from 2.5 to 5.0 to 7.5. Notice that as the β increases, the attenuation in the stop band increases, and to get the

Kaiser Window Magnitude and Gain Plots

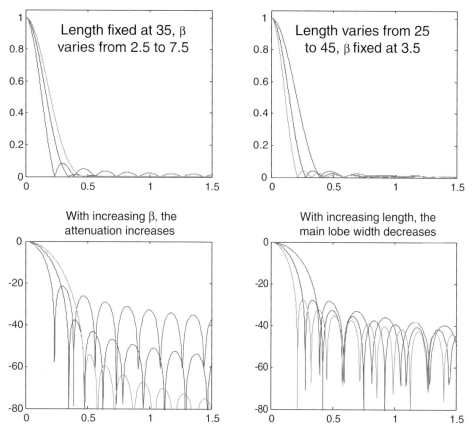

Figure 5.23 The effect of the ripple and the length parameters on the Kaiser window.

increased attenuation the width of the main lobe is also increasing. The plot on the right-hand side is drawn with β fixed at 3.5 with the length of the window changing from 25 to 35 to 45. This time since the shape (ripple) parameter β is fixed, the attenuation in the stop band is fixed, but as the length N of the window increases the width of the main lobe is shrinking.

Some of the other windows that are also used are Dolf Chebyshev window and the Kaiser–Bessel derived window. For the Dolf Chebyshev window, the window is of length $2N + 1$ and it is defined as

$$w_{DC}[n] = \frac{1}{2N+1} \left[\frac{1}{\gamma} + 2 \sum_{k=1}^{M} T\left(\beta \cos\left(\frac{k\pi}{2M+1} \right) \right) \cos\left(\frac{2nk\pi}{2M+1} \right) \right] \qquad -M \leq n \leq M$$

$$(5.90)$$

In Eq. (5.90), the parameter γ is the relative side lobe amplitude that is expressed as a fraction and β is expressed as shown in Eq. (5.91) and $T(\cdot)$ is the Chebyshev polynomial

$$\gamma = \frac{\text{side lobe amplitude}}{\text{main lobe amplitude}} \qquad \beta = \cosh\left(\frac{1}{2M}\cosh^{-1}\left(\frac{1}{\gamma}\right)\right) \tag{5.91}$$

This window can be designed with any specific relative side lobe value and as with other windows its main lobe width is adjusted by choosing the length of the window. The order of the window $N = 2M + 1$ is estimated by the formula in Eq. (5.92).

$$N = \frac{2.06A - 16.5}{2.29\Delta\omega} \tag{5.92}$$

Here also, A is the attenuation in dB as computed in step 2 of the Kaiser empirical method and $\Delta\omega = (\omega_s - \omega_p)$.

The third window, the Kaiser–Bessel window, is a derived window. To get this window, you first design the Kaiser window and from the coefficients of the Kaiser window you derive this window as shown in Eq. (5.93) where the coefficient $w_k(j)$ is the jth coefficient of the Kaiser window. The Kaiser window is of length $M + 1$ while the Kaiser–Bessel window is of length $2M + 1$. This window finds particular application in audio coding of the digital audio.

$$w_{kb}[n] = \begin{cases} \sqrt{\dfrac{\sum_{j=0}^{n} w_k(j)}{\sum_{j=0}^{M} w_k(j)}} & 0 \le n \le M \\[4mm] \sqrt{\dfrac{\sum_{j=0}^{2M-1-n} w_k(j)}{\sum_{j=0}^{M} w_k(j)}} & M \le n \le 2M \end{cases} \tag{5.93}$$

5.6
FIR Filter Design Using Optimization Techniques

In the designs of all the filters that we have come across, we have made no attempt at trying to control how the error in the pass band or in the stop band is distributed. In all the filters, the error in the pass band is maximum near the transition region and it gradually decreases as we move further away from the transition region. All the optimization techniques try in some way to control how the error in the pass band and in the stop band is distributed. The concept of how this is done is relatively simple, but its implementation requires a lot of computational power. So, instead of demonstrating the concept on an actual filter, we will demonstrate the concept on a function of a smaller order so that the math is tractable and the concept is made clear.

Let us state what we are attempting to do. We want to approximate a function of order N; this is the desired filter, with a function that is of an order less then N. Since we are trying to approximate a function of degree that is larger than the function that we are willing to use, there will be some error. We know that there will be some error and we are willing to accept this error, but we would like this error to be distributed equally at each of the sample points that we choose for our function. Also, we want this error as small as possible; however, we do not know how much this error will be. We explain the concept with the following example.

We have a quadratic function $D[x] = 1.5x^2 - 2x + 1$, which is a function of order 2, so we will try to approximate this function with a linear function, which is a function of order 1. We do not know this linear function, we only know it is of order 1, so we will make this function as $G[x] = a_0 + a_1 x$. In this function, we do not know the coefficients a_0 and a_1, so we will determine these to know the function $G[x]$. We want to approximate this function in the region $-1 \leq x \leq 1$, so the error is the minimum at each of the sampling points. We begin our approximation by defining our objective function as shown in Eq. (5.94).

$$\text{Minimize} \quad \underset{x \, \varepsilon [-1,1]}{\text{Max}} \left| 1.5x^2 - 2x + 1 - a_0 - a_1 x \right| \tag{5.94}$$

In this example, we have three unknowns, the error and the two coefficients of the function $G[x]$, so we begin by choosing three points (we can do this arbitrarily) in the range we want to minimize the error of the approximation in. Say, we choose the points $x = -0.5, 0,$ and 0.75 as our first guess where we think the error will be a minimum and equal at all the three sample points. Substitute each of the sample points in the approximation equation given in Eq. (5.95).

$$a_0 + a_1 x + \varepsilon = 1.5x^2 - 2x + 1 \tag{5.95}$$

Substituting each one of our three guesses where the error will be minimum, we get a set of three simultaneous equations as shown in Eq. (5.96).

$$\begin{bmatrix} 1 & -0.5 & 1 \\ 1 & 0 & -1 \\ 1 & 0.75 & 1 \end{bmatrix} \begin{bmatrix} a_0 \\ a_1 \\ \varepsilon \end{bmatrix} = \begin{bmatrix} 1.5(-0.5)^2 - 2(-0.5) + 1 = 2.375 \\ 1.5(0)^2 - 2(0) + 1 = 1.0 \\ 1.5(0.75)^2 - 2(0.75) + 1 = 0.3438 \end{bmatrix} \tag{5.96}$$

By solving the three simultaneous equations, we get $a_0 = 1.2813$, $a_1 = -1.627$, and $\varepsilon = 0.2813$. This tells us that using the a_0 and the a_1 values that we just got, the maximum error that we will encounter in the approximation is 0.2813, and this occurs at the sample points that we have chosen. The next question is, obviously, is this the best that we can do? To do this, let us determine the sample points where the error is maximum from Eq. (5.94). We do this in Eq. (5.97).

$$\frac{d(1.5x^2 - 2x + 1 - 1.2813 + 1.625x)}{dx} = 0 \Rightarrow x = 0.125 \tag{5.97}$$

So the maximum error with our approximation occurs at $x = 0.125$ and this error is -0.3047, which is more than the error approximation that we got using Eq. (5.96). Since we determined that the maximum error is more than the maximum error that we computed using Eq. (5.95), we have not reached the convergence point; the maximum error occurs at $x = 0.125$, and we should use this point as one of the points where we want to compute the error next time. All this result is shown in Figure 5.24. To include the point where we got the maximum error, we will replace the point that we had chosen, and that was nearest to this maximum error point, and repeat the process again. This iterative process will after a few iterations converge to the smallest error possible for the approximation that we have chosen.

Let us jump a little bit ahead and use the experience that we have gained and replace all the three points that we had chosen for the first iteration by three new points $-1, 0.125,$ and 1. Notice from the figure that the error is larger this time than in the previous estimate. This happens because the error in the previous estimate considered only a smaller region (from -0.5 to 0.75). Looking at the entire region from -1 to 1, the error at the end points is much larger. The three simultaneous equations that we get with the new sample points are given in Eq. (5.98). By solving these three equations, we get the three constants as $a_0 = 1.7612$,

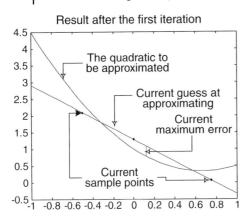

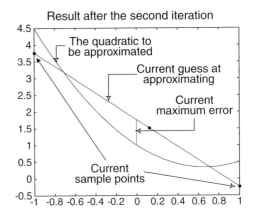

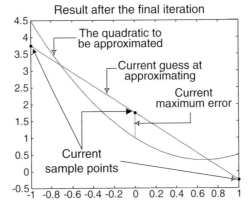

Figure 5.24 Demonstrating the Ramez exchange algorithm.

$a_1 = -2.0$, and $\varepsilon = 0.7383$.

$$\begin{bmatrix} 1 & -1 & 1 \\ 1 & 0.125 & 1 \\ 1 & 1 & 1 \end{bmatrix} \begin{bmatrix} a_0 \\ a_1 \\ \varepsilon \end{bmatrix} = \begin{bmatrix} 1.5(-1)^2 - 2(-1) + 1 = 4.5 \\ 1.5(0.125)^2 - 2(0.125) + 1 = 1.0 \\ 1.5(1)^2 - 2(1) + 1 = 0.5 \end{bmatrix} \tag{5.98}$$

By using these three values chosen for our sample points, we compute the location where the maximum error occurs using Eq. (5.97). By doing this, we find that the maximum error occurs at $x = 0$ and this error is $1.5(0)^2 - 2(0) + 1 = 1.7612 - 2(0) + \varepsilon \Rightarrow \varepsilon = -0.7612$. Since this error is larger than the error we have at our three sample points, we will once again replace the point closest to where the maximum error occurred with the point where the maximum error lies. This result is shown in Figure 5.24 also. So our new three points will be $-1, 0$, and 1. With these three points we get the three constants as $a_0 = 1.75$, $a_1 = -2.0$, and $\varepsilon = 0.75$. By using these values of the constants, we determine where the error is a maximum and what this error is. We find that the error is a maximum at $x = 0$ and it is 0.75. So, our iterative process has converged to the correct sampling points and the minimum error in the approximation. Using any other straight line to approximate the quadratic that we started with will give you a greater error somewhere in the range -1 to 1. The individual estimates and the error are also shown in Figure 5.24. This iterative process is based on the Ramez exchange algorithm.

MATLAB commands for designing optimum filters are as follows:

butter(N,Wn,'s')	This command designs a Butterworth filter of order N and cutoff frequency Wn.
Buttord(Wp,Ws,Rp,Rs,'s')	This command determines the order of the filter for given passband and stopband frequencies and ripples.
cheby1(N,Rp,Wn,'s')	This command designs a Chebyshev filter of order N, passband Ripple Rp and cutoff frequency Wn.
Cheby1ord(Wp,Ws,Rp,Rs,'s')	This command determines the order of the filter for given passband and stopband frequencies and ripples.
window('W_Name',N)	Designs the window specified by window name of length N.
BILINEAR(Z,P,K,Fs,Fp)	Designs a discrete filter from and analog filter given by Poles, Zeros, and gain. Fsis the sampling frequency and Fp is the pre-warping frequency.
impinvar(b,a,Fs)	Determines the impulse invariance design of an analog filter with numerator coefficients given by the vector b, denominator coefficients given by the vector a, and the sampling is done at frequency Fs.

5.7
Chapter Summary

In this chapter we studied several different methods of filter design.

We first began with the design of some classical analog filter designs. This review showed us how filters are specified both in the magnitude and in the gain format. In addition to the filter specifications, we also defined two special parameters that make the work of analog filter design a very simple procedure. These parameters are the discrimination parameter that measures in a special way the relative magnitudes in the pass band and the stop band and the transition parameter that measures the relative width of the transition band.

We also studied the details of how to design the Butterworth and the Chebyshev type I and type II filters. All the analog filters are invariably designed as prototype low-pass filters. Once you have the shape of the prototype filter correct, the filter is scaled in the frequency domain to the frequency range of interest and transformed to a different type filter if a filter other than the low-pass filter is desired.

Since our goal is discrete filters, we proceeded to transform the designed analog filters to be discrete filters by several different transform methods. The first method was the impulse invariance method. In the impulse invariance method, we map the poles of the analog filter to a corresponding location in the discrete domain; mapping the poles gives us the filter that we are looking for. This method has a drawback that there is aliasing in the high-frequency region of the transformed filter and for this reason this method is never used for high or band stop filters.

The other method that we learnt was the bilinear transform method. This is an algebraic substitution method in which we substitute a z-domain function for every s in the analog domain filter. The substitution method suffers from the warping of the frequency. To counter the frequency warping that occurs, we prewarp the frequency from the discrete domain so that the frequency specifications in the analog domain are based on the warped frequency that will be unwarped when we perform the algebraic substitution.

Both these methods map a stable filter to a stable filter, which is the basic requirement for any mapping method. The impulse invariance method pays no regard to the zeros of the filter. The zeros that are present in the analog filter show up in the discrete filter also, but they are not mapped like the poles are mapped. If the filter specifications are given in the discrete domain, as they usually are, and then transformed to the analog domain to design the filter, which is to be transformed back to the discrete domain, the sampling interval plays no part in the design of the filter and can be treated as any convenient value that you want to use. Generally for the bilinear transform method, we choose the sampling interval to be 2. Finally, we saw how a discrete filter is also scaled and transformed by the use of some simple all pass filters. This was given in Table 5.5.

The discrete filters that we obtained by transforming the analog filters have good magnitude response, but they suffer from nonlinear phase response, especially in and around the transition band. Finite impulse response filters can be and are designed for the express purpose that they exhibit linear phase response. The linear phase restriction on these filters required that the impulse response of the filter be either symmetric or antisymmetric. Also, the phase of the filter was not arbitrary but was strongly dependent on the length of the filter.

We studied two different methods to design the FIR filters. The window method used the IDTFT to transform an ideal filter from the frequency domain to the time domain. Since this transform gives us an infinite sequence, we had to limit the sequence by using one of the many window functions. This limited filter was a noncausal filter that we shifted by half the filter length to make it a causal filter. The frequency sampling method takes a finite number of samples around the unit circle of the frequency response. Since this sequence is a finite sequence, we determine its IDFT to get the impulse response of the filter in the time domain. With both the method, we are able to design filters that have arbitrary magnitude response, which was not true for the analog filters and the IIR filters.

The windows that we used in the FIR filter design methods to limit the length of the filter lead to the effect known as the Gibbs phenomenon. To ameliorate the effect seen when we use a rectangular window, several gently tapered windows were used. These windows have a fixed shape and variable length. Due to their fixed shape, they are able to provide a fixed amount of maximum attenuation, as increasing the length of the window only changes the width of the main lobe, it does not change the attenuation provided.

Two-parameter windows like the Kaiser window are able to trade ripple for length and provide windows in which we can control both the attenuation in the stop band and the ripple or the width of the transition band. Even in these two-parameter windows, keeping the shape constant by fixing the shape parameter and adjusting the length of the window only adjusts the width of the main lobe and hence the width of the transition band.

Finally, to close the chapter, we saw with a very simple example how the Ramez exchange algorithm works. It is an iterative process in which we begin with a guess of which frequencies will have the maximum error. We compute the error at these frequencies and check if the error is the maximum possible within the frequency range of interest. If the error is the maximum then we are done. If the error is greater at some other point, then we choose new frequency locations and recompute the error. Continuing this way the method converges after several iterations. It is not possible to design a filter of any size by hand with this method. Computing power is absolutely essential.

6
Computing the DFT

6.1
Introduction

The discrete Fourier transform (DFT) plays a very important role in the analysis, the design, and the implementation of the digital signal processing systems and algorithms. The DFT has been known for more than a 100 years, but it has gained widespread use and prominence only in the past 40 odd years since the publication of a very famous paper in 1965 by Cooley and Tukey. In this paper, the two authors describe an algorithm that is very efficient in computing the DFT. Now there are many such algorithms and they all together are known as the fast Fourier transform (FFT) algorithms.

In this chapter, we examine the issue of implementation of the DFT-based system in software. Hardware implementations of DFT-based systems are also possible, but they are not discussed here. We will demonstrate the algorithm using the simulation language MATLAB to demonstrate the main points of the algorithm and the required computations. All the algorithms are very similar to each other and they all rely on some simple concepts: first, W_N^{nk} is periodic with period N, so that $W_N^{(N-n)k} = \left(W_N^{nk}\right)^*$ and $W_N^{nk} = W_N^{(n+N)k} = W_N^{nk} W_N^{kN} = W_N^{nk}$, and second, there are several pairs of terms such as $x[n] W_N^{nk} + x[N-n] W_N^{(N-n)k}$ that can be grouped together so that they have computation efficiency as shown in Eq. (6.1).

$$
x[n] W_N^{nk} + x[N-n] W_N^{(N-n)k} = x[n]\left[\operatorname{Re}(W_N^{nk}) + j\operatorname{Im}\left(W_N^{nk}\right)\right] \\
+ x[N-n]\left[\operatorname{Re}\left(W_N^{(N-n)k}\right) + j\operatorname{Im}\left(W_N^{(N-n)k}\right)\right] \quad (6.1)
$$

On the left-hand side of Eq. (6.1), we need to perform two complex multiplications and one complex addition. This is equivalent to eight real multiplications and two real additions. Using the periodicity property of W_N^{nk}, we can rewrite Eq. (6.1) as shown in Eq. (6.2). In Eq. (6.2), we need to perform three real additions but only four real multiplications, which is an enormous saving in the required computations. The FFT algorithms make use of savings such as this and others to reduce the amount of computation required to compute the DFT.

$$
[x[n] + x[N-n]]\operatorname{Re}\left(W_N^{nk}\right) + [x[n] + x[N-n]]j\operatorname{Im}\left(W_N^{nk}\right) \quad (6.2)
$$

Digital Filters: Theory, Application and Design of Modern Filters, First Edition. Rajiv J. Kapadia.
© 2012 Wiley-VCH Verlag GmbH & Co. KGaA. Published 2012 by Wiley-VCH Verlag GmbH & Co. KGaA.

6.2
Direct Computation of the DFT

To really appreciate the various algorithms and the advances that have been made in the computations of the DFT, we must first understand what the difficulties themselves are in the computation of the DFT. To understand this, let us use MATLAB to first compute the DFT of a 16-point sequence using the definition of the DFT. (Here, we have used the term "point" to imply 16 sample points. We will use this notation in this and in subsequent chapters.) The MATLAB script is shown for you in MATLAB 6.1. We begin the script by first defining a complex sequence of length 16. Next, we set up a double loop to compute the DFT from the definition. In the outer or the "k" loop, we compute N values of the sequence $X[k]$. To compute each value of $X[k]$, we have to execute the inner or the "n" loop completely once. To completely execute the inner loop once, we need to go through different N values of the inner loop variable from $n=0$ to $n=N-1$. Thus, the inner loop executes N times for each computation of the outer loop, so there are N^2 computations to completely compute the DFT (Figure 6.1).

MATLAB 6.1. Computing the DFT from definition.

```
xr =[ 0 0 0 0.02 0.15 0.425 0.75 0.9 1.0 0.9 0.75 0.425 0.15 0.02 0 0];
xi =[ 1 1 1 1 0 0 0 0 0 0 0 0 0 1 1 1];
x = xr + j* xi;
N = length (x);
X = zeros (1,N);
%% For each X[k] execute the inner loop once.
for k = 0: (N-1)
%% Each pass through the loop requires four real computations
for n = 0: (N-1)
X (k+1) = X (k+1) + xr (n+1)* cos (2* pi* n* k/N) - j* xr (n+1)* sin
(2* pi* n* k/N)
                            + j* xi (n+1)* cos (2* pi* n* k/N) +
xi (n+1)* sin (2* pi* n* k/N);
end
end
subplot (321); stem (0: (N-1), xr);
subplot (322); stem (0: (N-1), xi);
subplot (323); stem (0: (N-1), abs (X));
subplot (324); stem (0: (N-1), angle (X)* 360/2/pi);
%% Compute the FFT directly using MATLAB to compare the results.
Xdir = fft (x, N);
subplot (325); stem (0: (N-1), abs (Xdir));
subplot (326); stem (0: (N-1), angle (Xdir)* 360/2/pi);
```

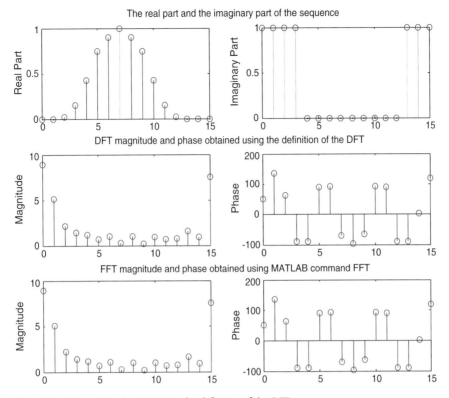

Figure 6.1 Computing the DFT using the definition of the DFT.

Now look at the computation that has to be performed. Since the two terms involved in the computations, the sample value x and the term W_N^{nk}, are both complex in general, we need to complete one complex multiplication every time through the inner loop. One complex multiplication equals four real multiplications as shown in the MATLAB script. Also, in each execution of the inner loop, we have to add the current result to the previous result that requires two real additions. So, to compute each value of $X[k]$, we need to perform 4^*N real multiplications and $2^*(N-1)$ real additions.

So, to compute an N point DFT, we need to perform N^*N complex multiplications and N $(N-1)$ complex additions. With $N=16$, the number of real multiplications are $4^*N^2 = 1024$ real multiplications. This is not a very large number and any computer will handle this amount of computation almost instantly. But a 16-point sequence is extremely rare and most of the times it exists only in textbooks. Sequences in real world are often of tens of thousands of sample points. A sequence that is of a modest length of 10 000 sample points will require 400 000 000 real multiplications. This will tax even the most modern computers. So any reduction in the computation required to get the DFT of a sequence is indeed very welcome. Our first attempt on this journey begins with the Goertzel algorithm, one of the early algorithms.

6.3
The Goertzel Algorithm

This method is an example of how we use periodicity of the term W_N^{nk} in our effort to compute the DFT. We first note that the term W_N^{nk} can be used in several ways and one of these is shown in Eq. (6.3) that is a direct result of the periodic nature of the factor W_N^{nk}.

$$W_N^{Nk} = 1 = W_N^{-Nk} \tag{6.3}$$

Using the result of Eq. (6.3), we can rewrite the definition of the DFT as shown in Eq. (6.4).

$$X[k] = \left[\sum_{n=0}^{N-1} x[n] W_N^{nk} \right] W_N^{-Nk} = \sum_{n=0}^{N-1} x[n] W_N^{-k(N-n)} \tag{6.4}$$

This last equation has the same form as a convolution, so we can say that Eq. (6.4) is a convolution of the sequence $x[n]$ with the impulse response W_N^{nk}. Since this is a convolution, we can represent the convolution as a system as shown in Figure 6.2 and given in Eq. (6.5).

$$y_k[n] = \sum_{r=0}^{N-1} x[r] W_N^{-k(n-r)} \tag{6.5}$$
$$X[k] = y_k[n]|_{n=N}$$

When we view Eq. (6.5) as a convolution, the result of the convolution can be thought of as an output from a system or a filter. This output of the system will be the same as the DFT every time $n = N$, as shown in Eq. (6.5) and in Figure 6.2. In Figure 6.2, since both $x[n]$ and W_N^{-k} are complex quantities, so the computation of each new $y_k[n]$ requires one complex multiplication. Since each output $X[k]$ appears when $n = N$, we have to compute N different $y_k[n]$ for each $X[k]$. So, there is no advantage in this arrangement as far as the computation effort is concerned compared to direct computation of the DFT. To compute N different DFT values $X[k]$, we need on the order of N^2 complex multiplications or $4N^2$ real multiplications. This is exactly how many computations are required for direct computations.

What we have done is that we have now removed the requirement of either computing or storing all the values of W_N^{nk}, these will be computed recursively by the system setup. From the second half of Figure 6.2, we can compute the system function as shown in Eq. (6.6).

$$H_k[z] = \frac{1}{1 - W_N^{-k} z^{-1}} \tag{6.6}$$

There is another modification we can make to this arrangement that we consider next. First, we realize that W_N^k and W_N^{-k} are complex conjugates of each other and we can separate

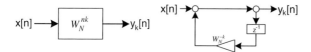

Figure 6.2 An alternative view of the DFT computation.

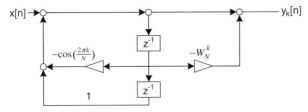

Figure 6.3 An alternative view of the DFT computation.

the real and the imaginary parts of the system transfer function $H_k[z]$ in Eq. (6.6) as shown in Eq. (6.7). The filter arrangement suggested in Eq. (6.7) is shown in Figure 6.3.

$$H_k[z] = \frac{1}{1 - W_N^{-k}z^{-1}} \frac{1 - W_N^k z^{-1}}{1 - W_N^k z^{-1}} = \frac{1 - W_N^k z^{-1}}{1 - 2\cos(2\pi l/N)z^{-1} + z^{-2}} \tag{6.7}$$

In the arrangement of Figure 6.3, we have built a system that represents two poles. To compute the output from this two-pole system, we need only real multiplications and not complex multiplications as all the terms in the feedback path, 1 and $-\cos(2\pi k/N)$, are real terms and hence require only real multiplications. The other coefficient $-W_N^k$ is a complex coefficient that is in the feedforward path. This coefficient builds a zero and being in the feedforward path needs to be computed only once when the Nth output is being computed. We do not need to complete this multiplication every time we compute $y_k[n]$ as we only want $y_k[n]$ when $n = N$. So not only we have effectively reduced the number of multiplications from $4N$ real multiplications to get one value of $X[k]$ down to $2N$ real multiplications (remember $x[n]$ is still complex) plus four more multiplications for the zero. Therefore, to get each required output value of $X[k]$, we need $2(N + 2)$ real multiplications.

In the arrangement of Figure 6.3, we have realized two poles. These two poles are the same poles that we need to realize the term $X[k]$ and the term $X[N - k]$. So at the end of N iterations, we are actually getting two coefficients. The two coefficients have different zero locations (the two zeros are also complex conjugate of each other); therefore, the last multiplication required to build the two coefficients are different, but the previous N multiplications are exactly the same for the two coefficients. So for a cost of $2(N + 2 + 2)$ real multiplications, we are getting two coefficients. This indeed is an achievement and a substantial reduction from the direct computation.

6.4
Decimation in Time Algorithm

This algorithm investigates if it is possible to first divide the sequence $x[n]$ of length N into two smaller sequences of length $N/2$ and then compute the DFT of two $N/2$ length sequences. The question "Are we able to get the DFT of a sequence of length N by combining the DFT of two length $N/2$ sequences?" is the main theme of this and the following algorithms. The reason why this helps is that direct computation of a length N sequence requires N^2 complex multiplications, but if the sequence is of length $N/2$, then the number of complex multiplications would be $2(N/2)^2 = N^2/2$ complex multiplications. This would be a saving of half as many multiplications. This algorithm is developed as follows.

We first divide the length N sequence into two length $N/2$ sequences (if N is odd then we can always pad it with one zero element to make it even) in the following way. The first length $N/2$ sequence will consist of all the samples of $x[n]$ that are in the even position of the original sequence and the second length $N/2$ sequence will consist of all the samples of $x[n]$ that are in the odd position of the original sequence. With this division of the sequence, the DFT of the length N sequence can be written as shown in Eq. (6.8).

$$X[k] = \sum_{n=0}^{N-1} x[n] W_N^{nk} = \underbrace{\sum_{n=0}^{\frac{(N/2)}{2}-1} x[2n] W_N^{2nk}}_{\text{Even position sequence}} + \underbrace{\sum_{n=0}^{(N/2)-1} x[2n+1] W_N^{(2n+1)k}}_{\text{Odd position sequence}} \tag{6.8}$$

Examine the even position sequence in Eq. (6.8), there we see a term W_N^{2nk}. This term can be rewritten as $W_{N/2}^{nk}$ as shown in Eq. (6.9).

$$W_N^{2k} = e^{-j2\pi nk/N} = e^{(-j\pi nk)/(N/2)} = W_{N/2}^{nk} \tag{6.9}$$

Similarly, in the odd position sequence in Eq. (6.8), we see a term $W_N^{(2n+1)k}$. This term can be rewritten as $W_N^k W_{N/2}^{nk}$ using the same development as that shown in Eq. (6.9). Using these two modifications, the computation of the DFT from Eq. (6.8) can be written as shown in Eq. (6.10).

$$X[k] = \underbrace{\sum_{n=0}^{(N/2)-1} x[2n] W_{N/2}^{nk}}_{\text{Even position sequence}} + W_N^k \underbrace{\sum_{n=0}^{(N/2)-1} x[2n+1] W_{N/2}^{nk}}_{\text{Odd position sequence}} \tag{6.10}$$

$$X[k] = \sum_{n=0}^{(N/2)-1} h[n] W_{N/2}^{nk} + W_N^k \sum_{n=0}^{(N/2)-1} g[n] W_{N/2}^{nk}$$

When we examine Eq. (6.10) carefully, we see that the even position sequence, $h[n]$, is a summation with the index of summation spanning $N/2$ terms, the sequence $h[n]$ is a sequence with $N/2$ terms, and the "twiddle" factor $W_{N/2}^{nk}$ samples only $N/2$ locations around the unit circle. So the first summation in Eq. (6.10) represents the DFT of a sequence of length $N/2$ that is the DFT of all the points in the even position of the original length N sequence. Similarly, the second summation in Eq. (6.10) represents the DFT of a sequence g $[n]$ that is of length $N/2$ and represents all the odd position terms in the sequence $x[n]$. This summation is, however, multiplied by a twiddle factor W_N^k. Equation (6.10) also tells us how to combine the even and the odd term DFT to get the overall DFT of length N sequence. Since the first summation in Eq. (6.10) is a DFT of a length $N/2$ sequence and since the second summation is the DFT of a length $N/2$ sequence, the DFT of the original length N sequence can be written as shown in Eq. (6.11).

$$X[k] = H[k] + W_N^k G[k] \tag{6.11}$$

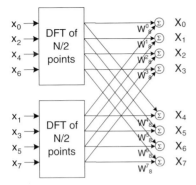

Figure 6.4 DFT of a length N sequence by combining two length $N/2$ sequences.

Examine Figure 6.4, there we see that to use the odd terms we first multiply each odd term with a twiddle factor, W_N^k, for the corresponding term and then add it to the appropriate term from the even sequence. So, for example, if we are interested in computing $X[4]$ from an eight-point sequence like the one shown in Figure 6.4, we first multiply the first term of the odd sequence DFT by W_8^4 and add this product to the first term from the even points DFT. This way all the terms of the DFT of the sequence $X[k]$ can be computed and we have achieved a gain of requiring reduced computations.

Now that we have reduced the large sequence DFT computation into a smaller sequence DFT computation, we can do it again if the smaller sequence of all the even points in Eq. (6.10) is an even sequence, and also for the sequence of all the odd points. If we are able to do this, we will have divided the original sequence of N points into four sequences of $N/4$ points each. We can determine the DFT of the entire length N sequence by combining the smaller DFTs of length $N/4$ sequences. We first compute four DFTs each of length $N/4$ and combine them to get two DFTs of length $N/2$ that are, in turn, combined together to get one length N DFT. This is shown in Figure 6.5.

There is of course no limit to how many times we can divide the sequence in its even and its odd position sequences. When we have sequences of length 1, the division process will stop. The DFT of length 1 sequence is computed as shown in Figure 6.6.

In Figure 6.6, the DFT term $X[0]$ is formed by adding the $x[0]$ term to the $x[1]$ term that is first multiplied by the twiddle factor W_2^0. The twiddle factor W_2^0 when computed by itself evaluates to plus 1. So the term $X[0]$ in a two-term DFT is formed by adding only the two terms of the sequence. Similarly, examine the computation for the term $X[1]$. It is formed by adding the $x[0]$ term to the $x[1]$ term that is first multiplied by the twiddle factor W_2^1. The twiddle factor W_2^1 when computed by itself evaluates to minus 1. So the term $X[1]$ in a two-term DFT is formed by only subtracting the $x[1]$ term from the $x[0]$ term of the sequence.

Using the algorithm just described, we need p stages where $p = \log_2(N)$ since at every stage we divide the sequence in half, so when $N = 2^p$ we can divide N till there are left with only two terms in every sequence and the computation of the DFT will be done as shown in Figure 6.6. Including these modifications, the twiddle factors can be modified as shown in Figure 6.7. In Figure 6.7a, we see the modifications to Figure 6.4 and Figure 6.7b we see the modifications to Figure 6.5.

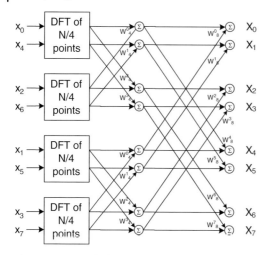

Figure 6.5 DFT of a length N sequence by combining two length $N/2$ sequences.

Another modification that we should consider in this computation deals with the periodicity of the twiddle factors. Notice that in Figures – there are two different twiddle factors that multiply each term from the odd sequence. Both the twiddle factors that multiply each odd term are related by the relation shown in Eq. (6.12).

$$W_N^n \quad \text{and} \quad W_N^{n+(N/2)}$$
$$W_N^{n+(N/2)} = W_N^{N/2} \, W_N^n = -W_N^n$$

$$(6.12)$$

Equation (6.12) tells us that the magnitudes of the two twiddle factors multiplying each odd term are the same, but the sign of the two is different. We can take advantage of this fact also to reduce the amount of computation required and this is shown in Figure 6.7. Now, if we look at any one stage of the computation of the DFT, we see that we need to multiply only $N/2$ terms and we need to add N terms together to get to the next stage. Consider one more modification: in Figures 6.4 and 6.5, we have twiddle factors that sample the unit circle at N equidistant points W_N^k in the last stage. In the stage prior to the last stage, we have twiddle factors that sample the unit circle at $N/2$ equidistant points $W_{N/2}^k$. We can of course replace the twiddle factor $W_{N/2}^k$ by its equivalent twiddle factor W_N^{2k}. So at each stage, the twiddle factors are computed as if they are sampled at N equidistant points, but we choose only the ones that we want. The saving here is that we need to compute the twiddle factors only once and use them at every stage.

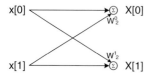

Figure 6.6 DFT of a length 2 sequence.

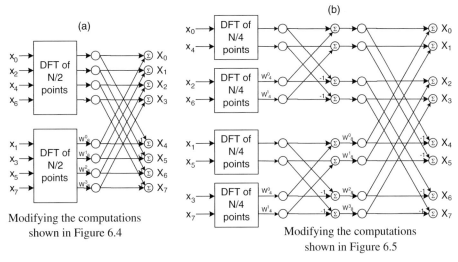

Figure 6.7 Recognizing the relation between W_N^n and $W_N^{(n+N/2)}$ and using it.

Another important point that we need to notice when we want to compute the DFT of an N point sequence is that the input sequence is not written in the order it was received. This of course does not matter as we can save the input sequence in any order that we want. It is, however, interesting to note how the samples are to be stored; they are stored in the "bit-reversed" order. This is clearly shown in Table 6.1 for a sequence of eight points.

In Table 6.1, we see that the binary representation of the sample numbers is listed in the first column. In the second column, we see that the binary representation is reversed as if we held a mirror at the edge of the column. This second column tells us where the input sample needs to be stored in order to execute the decimation in time algorithm. So, for example, if the input sample number is 3, which has a binary representation of (011), then this input sample is stored in location (110), which is 6 and the bit-reversed representation of the input sample number. It is also possible to enter and store the input in normal order and get the output in bit-reversed order. This is left for you to do as an exercise at the end of the chapter.

Table 6.1 Bit-reversed relation between sample number and position for computing the DFT.

Input sample number	Position of sample to start DFT computation
000 (0)	000 (0)
001 (1)	100 (4)
010 (2)	010 (2)
011 (3)	110 (6)
100 (4)	001 (1)
101 (5)	101 (5)
110 (6)	011 (3)
111 (7)	111 (7)

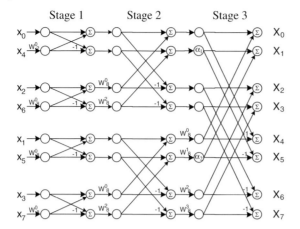

Figure 6.8 The complete sequence of computing the DFT using decimation in time algorithm.

Another important point to notice in the order of the computation recommended in Figure 6.5 and repeated in Figure 6.8 is that normally you would store the input sequence in one array, then perform the required computations, and store the result in a different array. Then, to perform the computation in the second stage, you would use the results stored in the second array and place the new results in the original or the first array. This way you need only two arrays to store and compute the results. This way by flipping the meaning of the arrays, you would continue to complete the computations at each stage. In the arrangement of Figure 6.8, the computations can, however, be performed in-place.

Performing computations in-place means that we start with the input samples stored in one array. Then, during the computation to be performed, we take two samples. For example, in stage 1 of the calculation in Figure 6.8, the first pair of samples are x[0] and x [4]. When we use these two samples to compute the corresponding two DFTs, we are done with these two samples. We do not need them for any other computation. So the space vacated in memory after using these two samples can now be used to store the two results computed. Thus, the DFT computed is placed back in the space that was just vacated. This paring of samples is true for all the stages. For example, look at the samples marked α_1 and α_2, these two samples enter together in a computation during stage 3. They are used to compute the DFT samples $X[1]$ and $X[5]$. Once we compute $X[1]$ and $X[5]$, we do not need the storage space marked α_1 and α_2, so this space is available to store the result $X[1]$ and $X[5]$. Thus, as soon as we use the input sample in a computation, its storage space becomes available to store the result of the computation. This algorithm is known as in-place computation.

Example 6.1

Using an algorithm similar to the decimation in time algorithm, show that if the input sequence is $N = 3^p$ long, then the DFT can be computed by dividing each stage into thirds.

Answer 6.1: Here, we are told that the sequence that we have can be repeatedly divided into thirds; so, from the definition of the DFT, we can write the computation for the DFT with three summations, each summation extending over a third of the points. We will divide the points so that the first summation will be all the points that can be divided by three, the second summation will be all the points that have a remainder of 1 after dividing the position by 3, and the third summation will consist of all the points that have a remainder of 2 after dividing the position by 3. The three grouped summations are shown in Eq. (6.13).

$$X[k] = \sum_{n=0}^{(N/3)-1} x[3n] W_N^{3nk} + \sum_{n=0}^{(N/3)-1} x[3n+1] W_N^{(3n+1)k} + \sum_{n=0}^{(N/3)-1} x[3n+2] W_N^{(3n+2)k}$$

(6.13)

Equation (6.13) is very similar to Eq. (6.8), so we can proceed just as we did in Eq. (6.8) and rewrite each of the three summations as shown in Eq. (6.14).

$$X[k] = \underbrace{\sum_{n=0}^{(N/3)-1} x[3n] W_{N/3}^{nk}}_{\substack{\text{DFT of terms with} \\ \text{no remainder}}} + W_N^k \underbrace{\sum_{n=0}^{(N/3)-1} x[3n+1] W_{N/3}^{nk}}_{\substack{\text{DFT of terms with} \\ \text{a remainder of 1}}} + W_N^{2k} \underbrace{\sum_{n=0}^{(N/3)-1} x[3n+2] W_{N/3}^{nk}}_{\substack{\text{DFT of terms with} \\ \text{a remainder of 2}}}$$

(6.14)

In Eq. (6.14), each summation is a DFT of length $N/3$. The two twiddle factors that multiply each term in the second and the third summation enable us to combine the three terms to obtain the overall DFT. If you wanted you could further divide the sequences in each of the summations in three more sequences each till you have only three terms left in a sequence.

6.5
Decimation in Frequency Algorithm

In Section 6.3, we developed the decimation in time algorithm to compute the DFT of a sequence that is $N = 2^p$ samples long. (Remember if the sequence is not exactly 2^p samples long, you can always make it sufficiently long by padding it with an appropriate number of zeros.) To develop the decimation in time algorithm, we divided the sequence into two sequences with the first sequence consisting of all the sample points in the even position and the second sequence with all the points in the odd position. There are other ways to divide the sequence into smaller sequences. One such a way was shown in Example 6.1. In this algorithm, we examine a different way of dividing the sequence.

This time we will divide the sequence so that the first smaller sequence consists of the first $N/2$ sample points and the second smaller sequence consists of the last $N/2$ sample points. Once again, we will assume that the entire sequence is of length $N = 2^p$; this division of the sequence and the corresponding required computation is shown in Eq. (6.15).

$$X[k] = \sum_{n=0}^{N-1} x[n]\, W_N^{nk} = \underbrace{\sum_{n=0}^{(N/2)-1} x[n]\, W_N^{nk}}_{\text{First half sequence}} + \underbrace{\sum_{n=N/2}^{N-1} x[n]\, W_N^{nk}}_{\text{Last half sequence}}$$

$$\quad (6.15)$$

$$X[k] = \underbrace{\sum_{n=0}^{(N/2)-1} x[n]\, W_N^{nk}}_{\text{First half sequence}} + \underbrace{\sum_{m=0}^{(N/2)-1} x\left[m + \frac{N}{2}\right] W_N^{(m + N/2)k}}_{\text{Last half sequence}}$$

In Eq. (6.15), the sequence of the first half of the points is of length $N/2$ and it has a factor W_N^{nk}, which samples the unit circle over N equally spaced points. So, the summation is not a DFT summation yet. In the summation of the second half of the sequence, we have a similar situation. The sequence is $N/2$ samples long, but the twiddle factor samples the unit circle at N equally spaced points. So, this summation is also not a DFT summation. Furthermore, in the second summation, the summation does not begin with $n=0$, but it begins at $n = N/2$. We can easily deal with this by substituting $n = m + N/2$. With this change in variable, we can rewrite Eq. (6.15) as shown in the second part of Eq. (6.15). Doing this and rewriting Eq. (6.15) can be written as shown in Eq. (6.16).

$$X[k] = \underbrace{\sum_{n=0}^{(N/2)-1} x[n]\, W_N^{nk}}_{\text{First half sequence}} + \underbrace{W_N^{k(N/2)} \sum_{m=0}^{(N/2)-1} x\left[m + \frac{N}{2}\right] W_N^{mk}}_{\text{Last half sequence}} \quad (6.16)$$

In Eq. (6.16), we have all the right conditions except for the twiddle factor. The twiddle factor samples the unit circle over N equally spaced points instead of sampling the unit circle over $N/2$ equally spaced points. To correct this let us look at only the even position terms of $X[k]$. We do this in Eq. (6.17).

$$X[2k] = \underbrace{\sum_{n=0}^{(N/2)-1} x[n]\, W_N^{n(2k)}}_{\text{First half sequence}} + \underbrace{W_N^{(2k)N/2} \sum_{m=0}^{(N/2)-1} x\left[m + \frac{N}{2}\right] W_N^{m(2k)}}_{\text{Last half sequence}}$$

$$X[2k] = \underbrace{\sum_{n=0}^{(N/2)-1} x[n]\, W_{N/2}^{nk}}_{\text{First half sequence}} + \underbrace{W_N^{kN} \sum_{m=0}^{(N/2)-1} x\left[m + \frac{N}{2}\right] W_{N/2}^{mk}}_{\text{Last half sequence}} \quad (6.17)$$

$$X[2k] = \sum_{n=0}^{(N/2)-1} \left(x[n] + x\left[n + \frac{N}{2}\right] \right) W_{N/2}^{nk}$$

In Eq. (6.17), the summations are over $N/2$ sample points, there are $N/2$ points in each sequence, and the twiddle factor samples the unit circle over only $N/2$ points. Thus, all the conditions for the computation of the DFT are satisfied and each of the two summations represents an $N/2$ point DFT. (Note that the factor $W_N^{kN} = 1$.) Now, we have all the even positions of the DFT computed. The method of computation is shown in the last equation in Eq. (6.17). Next, we will examine the odd positions of the DFT. For this we will examine the computation of the odd points in Eq. (6.16) as we do this in Eq. (6.18).

$$X[2k+1] = \underbrace{\sum_{n=0}^{(N/2)-1} x[n]\, W_N^{n(2k+1)}}_{\text{First half sequence}} + \underbrace{W_N^{(2k+1)N/2} \sum_{m=0}^{(N/2)-1} x\left[m + \frac{N}{2}\right] W_N^{m(2k+1)}}_{\text{Last half sequence}}$$

$$X[2k+1] = \underbrace{\sum_{n=0}^{(N/2)-1} x[n]\, W_N^n\, W_{N/2}^{nk}}_{\text{First half sequence}} + \underbrace{W_N^{N/2} \sum_{m=0}^{(N/2)-1} x\left[m + \frac{N}{2}\right] W_N^m\, W_{N/2}^{mk}}_{\text{Last half sequence}} \qquad (6.18)$$

$$X[2k+1] = \sum_{n=0}^{(N/2)-1} \left[\left(x[n] - x\left[n + \frac{N}{2}\right]\right) W_{N/2}^{nk}\, W_N^n\right]$$

The computation flow graph for the decimation in frequency DFT algorithm is shown in Figure 6.9. From Figure 6.9, we can see that the top half of the figure computes the even position terms of the DFT by adding the top half terms and the bottom half terms together and then computing the DFT. Similarly, the bottom half of the figure computes the odd position terms of the DFT by first subtracting the bottom half terms from the top half terms and then multiplying the result by the corresponding twiddle factor and finally computing the DFT. These computations are very similar to the computations involved in the decimation in time algorithm.

Just as we continued dividing the input sequence in half for every stage in the decimation in time algorithm until there were only two samples left, we can continue dividing the sequence in decimation in frequency algorithm also till we have a sequence of only two samples. Now, since each of the sequences is of two samples, the DFT can be computed directly as we showed in Figure 6.6. Thus, the complete computation sequence for a length 8 sequence using the decimation in frequency algorithm is shown in Figure 6.10.

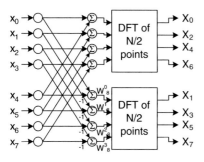

Figure 6.9 DFT of a length N sequence using decimation in frequency.

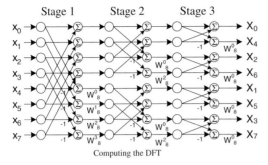

Computing the DFT

Figure 6.10 The complete sequence of computing the DFT using decimation in frequency algorithm.

In the decimation in frequency algorithm, we find that the input sequence is arranged in normal order and used for computation, but the DFT that we get as the output is in the bit-reversed order. Furthermore, this algorithm also performs the computation in-place so the storage requirements are the same for both the algorithms. Like the decimation in time algorithm, all the required twiddle factors are over N samples of the DFT, so this requirement is also the same for both algorithms. Both the algorithms are the same in computation requirement and in memory storage requirements.

Example 6.2

Using an algorithm similar to the decimation in frequency algorithm, show that if the input sequence is $N = 3^p$ long, then the DFT can be computed by dividing each stage into thirds.

Answer 6.2: Here, we are told that the sequence that we have can be repeatedly divided into thirds; so, from the definition of the DFT, we can write the computation for the DFT with three summations, each summation extending over a third of the points. We will divide the points so that the first summation will be over the first third of the points, the second summation will be over the middle third of the points, and the last summation will be over the last third of the points. The three grouped summations are shown in Eq. (6.19).

$$X[k] = \sum_{n=0}^{(N/3)-1} x[n]\, W_N^{nk} + \sum_{n=N/3}^{(2N/3)-1} x[n]\, W_N^{nk} + \sum_{n=2N/3}^{N-1} x[n]\, W_N^{nk} \tag{6.19}$$

Equation (6.19) is very similar to the first part of Eq. (6.15), so we can proceed just as we did in Eq. (6.15) and rewrite each of the summations in Eq. (6.19) as shown in Eq. (6.20).

$$X[k] = \underbrace{\sum_{n=0}^{(N/3)-1} x[n]\, W_N^{nk}}_{\substack{\text{Summation over} \\ \text{first } 1/3 \text{ terms}}} + \underbrace{\sum_{r=0}^{(N/3)-1} x\left[r + \frac{N}{3}\right] W_N^{(r+(N/3))k}}_{\text{Summation over second } 1/3 \text{ terms}}$$

$$+ \underbrace{\sum_{s=0}^{(N/3)-1} x\left[s + \frac{2N}{3}\right] W_N^{(s+(2N/3))k}}_{\text{Summation over last } 1/3 \text{ terms}} \tag{6.20}$$

Consider now looking at every third DFT value in Eq. (6.20), and to do this we replace the variable k by $3k$ to get Eq. (6.21).

$$X[3k] = \underbrace{\sum_{n=0}^{(N/3)-1} x[n] W_N^{n(3k)}}_{\substack{\text{Summation over}\\ \text{first } 1/3 \text{ terms}}} + \underbrace{\sum_{r=0}^{(N/3)-1} x\left[r + \frac{N}{3}\right] W_N^{(r+(N/3))(3k)}}_{\text{Summation over second } 1/3 \text{ terms}}$$

$$+ \underbrace{\sum_{s=0}^{(N/3)-1} x\left[s + \frac{2N}{3}\right] W_N^{(s+(2N/3))(3k)}}_{\text{Summation over last } 1/3 \text{ terms}}$$

$$X[3k] = \underbrace{\sum_{n=0}^{(N/3)-1} x[n] W_{N/3}^{nk}}_{\substack{\text{Summation over}\\ \text{first } 1/3 \text{ terms}}} + \underbrace{\sum_{r=0}^{(N/3)-1} x\left[r + \frac{N}{3}\right] W_{N/3}^{rk} W_N^{kN}}_{\text{Summation over second } 1/3 \text{ terms}}$$

$$+ \underbrace{\sum_{s=0}^{(N/3)-1} x\left[s + \frac{2N}{3}\right] W_{N/3}^{sk} W_N^{2kN}}_{\text{Summation over last } 1/3 \text{ terms}} \tag{6.21}$$

In Eq. (6.21), we recognize that the terms W_N^{kN} and W_N^{2kN} both evaluate to 1. So the last equation in Eq. (6.21) can be rewritten as shown in Eq. (6.22).

$$X[3k] = \sum_{n=0}^{(N/3)-1} \left(x[n] + x\left[n + \frac{N}{3}\right] + x\left[n + \frac{2N}{3}\right] \right) W_{N/3}^{nk} \tag{6.22}$$

In Eq. (6.22), we get every third term of the DFT. Similarly, we can evaluate the next two groups of terms as shown in Eq. (6.23).

$$X[3k+1] = \sum_{n=0}^{(N/3)-1} \left(x[n] + x\left[n + \frac{N}{3}\right] W^{1/3} + x\left[n + \frac{2N}{3}\right] W^{2/3} \right) W_N^n W_{N/3}^{nk}$$

$$X[3k+2] = \sum_{n=0}^{(N/3)-1} \left(x[n] + x\left[n + \frac{N}{3}\right] W^{2/3} + x\left[n + \frac{2N}{3}\right] W^{4/3} \right) W_N^{2n} W_{N/3}^{nk} \tag{6.23}$$

Equations (6.22) and (6.23) show how we can combine the output from the three groups of terms to get the DFT of a sequence that is of length divisible by three. Now that we have seen how we can use the FFT algorithm either for decimation in time or for decimation in frequency, the obvious question that you would ask is: Say, I have a length 18 sequence, can I use the decimation in time algorithm first by dividing the sequence in two equal sequences and then use the decimation in frequency algorithm by dividing the result into three length three sequences? The answer of course is yes and it is left for you to complete it as an exercise at the end of the chapter.

6.6
Algorithm when *N* is a Composite Number

The next question that would come to mind is: Can we use the FFT algorithm when *N* is a composite number? So, when *N* is a product of two numbers say $N = p^*q$. It was the affirmative answer to this very question that led to the explosive growth in the use of DFT as an analysis tool. This is the general form of the Cooley–Tukey algorithm.

Consider a length *N* sequence $x[n]$ where *N* can be factored into $N = p_1 p_2, \ldots, p_v$. This is exactly the condition that we had when we were examining the decimation in time and the decimation in frequency algorithms. In those two algorithms, all the factors of *N* were exactly the same and equal to 2. When the factors are 2, we can use the simple butterfly computation as shown in Figure 6.6 and we get a very efficient algorithm. We also know that we can extend the sequence so that the length of the resulting sequence is a power of two. There are, however, some special cases that require us to limit the length of the sequence. These special cases occur when the memory space is at a premium that is often the case in hand-held and embedded systems. When the memory space is at a premium, we will often use the algorithm of composite numbers. To explain this algorithm, we will use arguments that are similar to the decimation in time algorithm. To understand these algorithms, consider a sequence of length *N*, where *N* is a composite number with all its factors not equal to 2. So, we can write *N* as shown in Eq. (6.24).

$$N = p_1 p_2, \ldots, p_v = p_1 q_1 \qquad \text{where} \quad q_1 = p_2\, p_3, \ldots, p_v \qquad (6.24)$$

We begin the algorithm by dividing the input sequence into p_1 sequences each of length q_1 as suggested in Eq. (6.24) and shown in Figure 6.11. To do this, we will take every p_1th sample in the original sequence and place it in its own sequence. Thus, we will have p_1 sequences each of which is q_1 samples long. Figure 6.11 shows an example of a length 18 sequence that is divided into $p_1 = 3$ sequences each of which is $q_1 = 6$ samples long. As shown in Figure 6.11, this division takes a one-dimensional array and distributes it into a

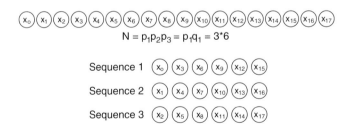

Figure 6.11   Mapping a one-dimensional array into a two-dimensional array.

two-dimensional array. Now, we can write the DFT of the array of length N as a sum of different p_1 arrays as shown in Eq. (6.25).

$$X[k] = \sum_{n=0}^{N-1} x[n] W_N^{nk}$$

$$X[k] = \sum_{r=0}^{q_1-1} x[p_1 r] W_N^{(p_1 r)k} + \sum_{r=0}^{q_1-1} x[p_1 r + 1] W_N^{(p_1 r)k} W_N^{(1)k} \tag{6.25}$$

$$+ \cdots + \sum_{r=0}^{q_1-1} x[p_1 r + q_1 - 1] W_N^{(p_1-1)k} W_N^{(p_1 r)k}$$

Look at the typical term in Eq. (6.25). It can be written as shown in Eq. (6.26). The typical term now has a summation that extends over q_1 terms; there are q_1 terms in the sequence being considered for this summation and the twiddle factor samples the unit circle at q_1 equally spaced points. Thus, the summation represents the DFT of the terms in the array formed by choosing every p_1 term from the original sequence. With $l = 0$, the DFT would be for sequence 1, with $l = 1$, the DFT would be for sequence 2, and with $l = 2$, the DFT would be for sequence 3 from Figure 6.11.

$$\sum_{r=0}^{q_1-1} x[p_1 r + l] W_N^{(p_1 r + l)k} = W_N^{lk} \sum_{r=0}^{q_1-1} x[p_1 r + l] W_N^{p_1 rk}$$

$$= W_N^{lk} \sum_{r=0}^{q_1-1} x[p_1 r + l] W_{N/p_1}^{rk} \tag{6.26}$$

$$= W_N^{lk} \sum_{r=0}^{q_1-1} x[p_1 r + l] W_{q_1}^{rk}$$

If we group all the terms from Eq. (6.25) together, then we can write the DFT as shown in Eq. (6.27). Equation (6.27) expresses the DFT of $x[n]$ as a sum of p_1 different DFTs each of length q_1 and each multiplied by its own twiddle factor to modify the DFT.

$$X[k] = \sum_{l=0}^{p_1-1} W_N^{lk} \sum_{r=0}^{q_1-1} x[p_1 + l] W_{q_1}^{rk} \tag{6.27}$$

Let us look at the computation requirement of the composite algorithm. Begin by looking at the last summation in Eq. (6.27). This summation represents a q_1 point DFT, so it requires q_1^2 complex multiplications. There are p_1 such summations so the total requirement is $p_1 q_1^2$ complex multiplications. In the outer summation in Eq. (6.27), we multiply each of the q_1 point DFT by its own twiddle factor W_N^{lk} and add all the results together. This requires $p_1 - 1$ complex multiplications. This will compute one value on the left-hand side of Eq. (6.27). Since there are N such terms that have to be computed to get the complete $X[k]$, we need N $(p_1 - 1)$ complex multiplications. With all these taken together, the complete number of complex multiplications that we need are given in Eq. (6.28).

$$N(p_1 - 1) + p_1 q_1^2 \tag{6.28}$$

Since we know that N is a composite number and we have just taken one factor out of N, we can definitely repeat the process on q_1 as looking at Eq. (6.24) we can see that q_1 is also a composite number that can be written as shown in Eq. (6.29).

$$q_1 = p_2 p_3, \ldots, p_v = p_2 q_2 \qquad \text{where} \qquad q_2 = p_3 p_4, \ldots, p_v \qquad (6.29)$$

Now that we know the inner summation in Eq. (6.27) represents a DFT of a sequence of length q_1 and that we know that q_1 is a composite number, we can break this DFT just as we did for the DFT of length N. When we do this, we will replace the last summation in Eq. (6.27) by an equation that is similar to Eq. (6.27). Now, the number of computations required to complete the DFT of q_1 points can be written as shown in Eq. (6.30).

$$q_1(p_2-1) + p_2 q_2^2 \qquad (6.30)$$

By substituting Eq. (6.30) into Eq. (6.28), where the entire Eq. (6.30) substitutes for the q_1^2 term, we get Eq. (6.31).

$$N(p_1-1) + p_1\left(q_1(p_2-1) + p_2 q_2^2\right)$$
$$N(p_1-1) + N(p_2-1) + p_1 p_2 q_2^2 \qquad (6.31)$$

This way if we keep on dividing the remaining sequence till there are no more composite factors left, the computation requirement for the algorithm can be written as shown in Eq. (6.32).

$$N(p_1 + p_2 + \cdots + p_v - v) \qquad (6.32)$$

In general, it can be seen from Eq. (6.32) that it is preferable to carry out the decomposition of the sequence on the basis of as many factors of N as possible, so each of the individual factors p_i are prime numbers. This can clearly be seen from Eq. (6.32) since if any of the p_i is not a prime number, then it is a composite number. Any composite number can be written as $p_i = r_i \cdot s_i$ and since a product is always (except for $2 \cdot 2$), the sum then $p_i \geq r_i + s_i$. So choosing any factor that is not a prime number will require us to perform more complex multiplications. Hence, we always choose prime factors to decimate N into. There is one exception to choosing prime numbers as suggested earlier. When we choose $p_i = 4$ rather than $p_i = 2$ and $p_{i+1} = 2$, there is no loss if we are careful since $2 + 2$ is equal to $2 \cdot 2$.

6.7
Computing the FFT of Only a Few Samples

Until now all the algorithms that we have seen compute the complete DFT; that is, the algorithms calculate all the N values equally spaced and on the unit circle. So, if we begin with a sequence of length N, we get the DFT that is also N samples long and all located on the unit circle. There are occasions when you are interested in the spectrum of the signal in only a limited frequency band. One such example of limited number of points from the entire spectrum is shown in Figure 6.12. The spectrum that relates to a narrow band of frequencies is a spectrum of K samples, where $K \leq N$ and N is the total number of samples in the signal; for now, we will assume that K is a factor of N so $N = K^* R$. This is always possible as we can pad N with zero values till K is a factor of N.

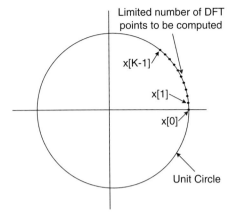

Figure 6.12 DFT to be computed for only a few points.

To compute the DFT of only K samples, we will first divide the original sequence of length N into R subsequences each of length K. We form the sequence $x_r[n]$ by taking every Rth sample of the original sequence, this way forming R sequences each of length K. We saw this done earlier in Figure 6.11 where we divided a sequence of length 18 into three sequences of length 6 each. To compute the DFT of only the required K points, we begin with Eq. (6.33).

$$X[k] = \sum_{n=0}^{N-1} x[n] \, W_N^{nk}$$

$$X[k] = \sum_{n=0}^{K-1} x[nR] \, W_N^{(nR)k} + \sum_{n=0}^{K-1} x[nR+1] \, W_N^{(nR+1)k} + \cdots + \sum_{n=0}^{K-1} x[nR+K-1] \, W_N^{(nR+K-1)k}$$

$$X[k] = \underbrace{\sum_{n=0}^{K-1} x[nR] \, W_K^{nk}}_{K\,\text{point DFT}} + \underbrace{W_N^{1k} \sum_{n=0}^{K-1} x[nR+1] \, W_K^{nk}}_{K\,\text{point DFT}} + \cdots + \underbrace{W_N^{(K-1)k} \sum_{n=0}^{K-1} x[nR+K-1] \, W_K^{nk}}_{K\,\text{point DFT}}$$

$$\underbrace{\hspace{8cm}}_{\text{Combine all the } R \text{ different } K \text{ point DFTs to determine the required spectrum}}$$

$$(6.33)$$

In Eq. (6.33), we begin with the definition of the DFT. In the second equation, we have separated the R different sequences of length K and have represented each as its own summation. Finally, noting that N is a product of R and K, we can change each of the summations as shown in the last equation in Eq. (6.33). This last equation tells us how to combine the R different DFTs so that we get a result that is only K points long and we do not have to compute the entire DFT of N points. This is represented in the MATLAB 6.2 script given below. There we begin with a 64-point sequence from which we want the DFT of the first 16 points.

Above we have determined the DFT of the first segment of the R segments. If the DFT required was for, say, the ith segment of K points where $i \leq K$, then we need to slightly modify Eq. (6.33) as shown in Eq. (6.34). Notice that this just requires that we multiply each of the

R individual DFTs that we computed by a different twiddle factor. In computing the required twiddle factor, we have used the fact that $N = R \cdot K$ and hence $W_N^{iK} = W_{N/K}^i = W_R^i$.

$$X[k+iK] = \underbrace{\sum_{n=0}^{K-1} x[nR] W_K^{n(k+iK)}}_{K \text{ point DFT}} + \underbrace{W_N^{(k+iK)} \sum_{n=0}^{K-1} x[nR+1] W_K^{n(k+iK)}}_{K \text{ point DFT}}$$

$$+ \cdots + \underbrace{W_N^{(K-1)(k+iK)} \sum_{n=0}^{K-1} x[nR+K-1] W_K^{n(k+iK)}}_{K \text{ point DFT}}$$

$$X[k+iK] = \underbrace{\sum_{n=0}^{K-1} x[nR] W_K^{nk}}_{K \text{ point DFT}} + \underbrace{W_N^k W_R^i \sum_{n=0}^{K-1} x[nR+1] W_K^{nk}}_{K \text{ point DFT}}$$

$$+ \cdots + \underbrace{W_N^{(K-1)} W_R^{(k+i)} \sum_{n=0}^{K-1} x[nR+K-1] W_K^{n(k+iK)}}_{K \text{ point DFT}}$$

$$(6.34)$$

MATLAB 6.2. Computing the spectrum of a limited number of points.

```
%% Compute the FFT of the entire sequence.
N = 64; %Define the length of the sequence.
x = randn(1,N)   ; %Define a sequence of 64 random values.
X = fft(x,64)    ; %Compute FFT of entire sequence.
%% Compute the FFT of four subsequences.
x1 = x(1:4:N)    ; %The 1st subsequence.
x2 = x(2:4:N)    ; %The 2nd subsequence.
x3 = x(3:4:N)    ; %The 3rd subsequence.
x4 = x(4:4:N)    ; %The 4th subsequence.
X1 = fft(x1,N/4); %FFT of the 1st subsequence.
X2 = fft(x2,N/4); %FFT of the 2nd subsequence.
X3 = fft(x3,N/4); %FFT of the 3rd subsequence.
X4 = fft(x4,N/4); %FFT of the 4th subsequence.
%% Combine the four FFTs after multiplying with respective twiddle
factor.
K = 0:1:(N/4-1) ; % multiply with the twiddle factors.
X16 = X1 + exp(-j*2*pi*k/N).*X2 + exp(-j*2*pi*2*k/N).*X3 + exp
(-j*2*pi*3*k/N).*X4
error = X(1:16) - X16 %Compare the two computations.
```

A question now arises why go through all this adjustment, why not just determine the DFT of the entire sequence of length N and then just choose the K points that you want from all the N points that we computed. The answer is simple, the number of computations required. Consider the computation done in the MATLAB 6.2 script. There $N=64$ and we were interested only in 16 points. The computations required for the entire sequence are of the order of $N^2 = 64^2 = 4096$ complex multiplications. The computations required for four 16-point DFTs are of the order of $4 \cdot K^2 = 4 \cdot (16)^2 = 1024$ complex multiplications, and finally to multiply the DFTs with the corresponding twiddle factor, we need additional $3 \cdot 16^2 = 768$ complex multiplications for a total of 1792 complex multiplications, which is a substantial saving in the required computation power.

6.8
The Chirp Z-Algorithm

In Section 6.6, we saw how we can compute the DFT of a select few points that lie on the unit circle. There are times when the points where you want to know the frequency spectrum do not lie on the unit circle. One such contour is shown in Figure 6.13. Notice that the contour is not coincident with the unit circle, it may or may not be circular and the first point of the Z-transform is not on the positive real axis. Here, we examine an algorithm that is very often used for evaluating the Z-transform of a finite sequence along an arbitrary contour in the Z-plane. This algorithm shows us that we can compute the required spectrum using convolution and hence it can be computed using techniques of fast convolution. This algorithm also removes many of the restrictions of the DFT. The most important of these being that to efficiently compute the FFT the number of points N must be a highly composite number and not a prime number.

We begin with the Z-transform pair $x[n] \Leftrightarrow X(z)$. With the Chirp Z-algorithm, we will compute this Z-transform as shown in Eq. (6.35).

$$X(z) = \sum_{n=0}^{N-1} x[n] z^{-n} \qquad \text{define} \qquad X[k] = X(z)\big|_{z=e^{j2\pi k/N}} \qquad (6.35)$$

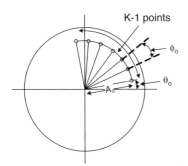

Figure 6.13 Chirp Z-transform. Spiral along which we determine the Z-transform.

The definition given in Eq. (6.35) will compute the Z-transform on the unit circle. To move the Z-transform to an arbitrary contour with an arbitrary starting point and a noncircular contour, we will define z in Eq. (6.35) as shown in Eq. (6.36).

$$z = A\Phi^{-k} \qquad k = 0,\ 1,\ \dots\ (K{-}1)$$
$$A = A_0 e^{j2\pi\theta_0} \quad \text{and} \quad \Phi = \Phi_0 e^{j2\pi\phi} \tag{6.36}$$

In Eq. (6.36), the index k ranges from 0 to $(K-1)$ where K is any arbitrary number of points where we are interested in determining the spectrum, K may be more or less than the N total number of points in the sequence. K represents the number of points where we want to evaluate the spectrum. The significance of the four parameters in Eq. (6.36) is explained here and in Figure 6.14.

A_0 represents the radius of the first point at which the spectrum is to be obtained. This parameter is always real.

θ_0 represents the angle that the first point makes with the positive real axis.

ϕ_0 represents the angular spacing between successive points where the spectrum is to be obtained

Φ_0 controls the spiral on which the points at which the spectrum is to be obtained are located. When $\Phi_0 = 1$ the spiral is a perfect circle, when $\Phi_0 > 1$ the spiral spirals inward to the origin, and when $\Phi_0 < 1$ the spiral spirals outward to infinity. This parameter is always real.

To understand how the Chirp Z-transform (CZT) is obtained, consider a typical point $z = z_k$, where we want to obtain the spectrum. Then from Eq. (6.35), we can write Eq. (6.37).

$$X[k] = \sum_{n=0}^{N-1} x[n] z_k^{-n} \qquad \text{where} \quad z_k = A\Phi^{-k}$$
$$X[k] = \sum_{n=0}^{N-1} x[n]\left(A\Phi^{-k}\right)^{-n} = \sum_{n=0}^{N-1} x[n] A^{-n} \Phi^{nk} \tag{6.37}$$

In Eq. (6.37), we can rewrite the term nk as $nk = \frac{1}{2}[n^2 + k^2 - (k-n)^2]$. By doing this we get Eq. (6.38).

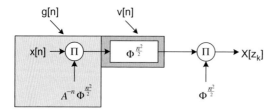

Figure 6.14 Computing the Chirp Z-transform.

$$X[k] = \sum_{n=0}^{N-1} x[n] A^{-n} \Phi^{nk} = \sum_{n=0}^{N-1} x[n] A^{-n} \Phi^{\frac{1}{2}\left(n^2 + k^2 - (k-n)^2\right)}$$

$$X[k] = \Phi^{k^2/2} \sum_{n=0}^{N-1} \underbrace{\left\{ x[n] A^{-n} \Phi^{\frac{n^2}{2}} \right\}}_{g[n]} \underbrace{\Phi^{-(k-n)^2/2}}_{v[k-n]}$$

$$X[k] = \Phi^{k^2/2} \sum_{n=0}^{N-1} g[n] v[k-n] \tag{6.38}$$

The last part of Eq. (6.38) has a summation that we can promptly recognize as a convolution between $g[n]$ and $v[n]$, so the computation can be performed as a filtering operation. This filtering operation is shown in Figure 6.13. In Figure 6.14, the signal $x[n]$ is multiplied by $A^{-n} \Phi^{n^2/2}$ to give us the input to the filter that is represented as the signal $g[n]$. This signal $g[n]$ is filtered by a filter with a transfer function $\Phi^{n^2/2}$ to get the output signal. The filtering operation is the convolution operation referred to in Eq. (6.38). The output of the filter is multiplied by $\Phi^{n^2/2}$ to get the CZT that we are after. All these operations are shown in Figure 6.14 and all the required operations are discussed in Figure 6.15.

Steps involved in determining the CZT are listed as follows:

1) First choose L to be the smallest integer that is greater than $(N + K - 1)$. Choose L so that it is compatible with the available FFT algorithm. This means that L is a highly composite number. This is required as we are going to use FFT to compute the CZT. The sequence N and the zero padded sequence of length L are shown in Figure 6.15a.

2) Form an L point sequence $g[n]$ as shown in Eq. (6.39). This sequence is shown in Figure 6.15b.

$$g[n] = \begin{cases} A^{-n} \Phi^{n^2/2} x[n] & n = 0, 1, \dots (N-1) \\ 0 & n = N, (N+1), \dots (L-1) \end{cases} \tag{6.39}$$

The frequency response of the filter $\Phi^{-n^2/2}$ is an infinite sequence as shown in Figure 6.15c. Form this infinite sequence, choose an L point sequence as shown in Eq. (6.40). This sequence $v[n]$ is shown in Figure 6.15d.

$$v[n] = \begin{cases} \Phi^{-n^2/2} & 0 \le n \le (K-1) \\ \Phi^{-(L-n)^2/2} & (L-N+1) \le n \le L \\ 0 & K \le n \le (L-1) \quad \text{These are any points not defined above} \end{cases} \tag{6.40}$$

Determine the L point DFT of the two sequences defined in step 2 (Figure 6.15b) and step 3 (Figure 6.15d) above. The DFT of the two sequences are shown in Figure 6.15e and f.

3) The two DFTs are multiplied together to get the output sequence from the filter. When this output sequence is multiplied by $\Phi^{n^2/2}$, we get the CZT transform that we are looking for; this is shown in Figure 6.15g.

4) Choose the K points that represent the CZT we are looking for. These are the first K points from the output. These required points are identified in Figure 6.15h.

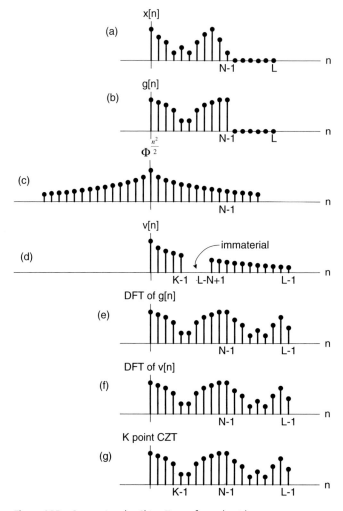

Figure 6.15 Computing the Chirp Z-transform algorithm.

The CZT is often used to sharpen the resonance peaks in a signal. The CZT is frequently used in radar systems as the radar system uses the chirp signals. A system similar to the one shown in Figure 6.12 is commonly used for spectrum analysis in radar systems. MATLAB has a special command to compute the CZT of a sequence and this command does all the steps outlined above directly for us. The comparison of the spectrum using the FFT and the CZT is shown below in MATLAB 6.3 script.

The CZT is often used to sharpen the resonance peaks in a signal. The CZT is frequently used in radar systems as the radar system uses the chirp signals. A system similar to one shown in Figure 6.12 is commonly used for spectrum analysis in radar systems. MATLAB has a special command to compute the CZT of a sequence and this command does all the steps outlined above directly for us. The comparison of the spectrum using the FFT and the CZT is shown below in MATLAB 6.3 script.

MATLAB 6.3. Computing the spectrum in a specified frequency range using the CZT.

```
%% Define a vector with two frequencies.
f1 = 30;
f2 = 45;
fs = 100;
N = 265;
t = 0:1/(fs):1;
x = cos(2*pi*(f1 + 5)*t) + 2*cos(2*pi*(f2-5)*t);
%% Define the coefficients needed for the Chirp Z Transform.
m = N;
A = exp(1i*2*pi*(f1)/fs);
W = exp(-1i*2*pi*((f2)-(f1))/(m*fs));
%% Compute the FFT and the CZT, build the frequency vectors.
Y = fft(x,N);
Z = czt(x,m,W,A);
fy = (0:N/2-1)*fs/N;
fz = (0:N-1)*((f2)-(f1))/N + f1;
%% Plot the two responses
subplot(311);plot(fy,abs(Y(1:N/2)));
title(['FFT plot of the entire signal ',num2str(N),' points'])
subplot(312);plot(fz,abs(Z));
title(['CZT plot of selected spectrum ',num2str(N),' points'])
%% Plot from the FFT in the frequency range from f1 to f2.
fy1 = zeros(1,N/2);j = 1;
  for i = 1:1:N/2
     if fy(i)>= f1 && fy(i)<= f2
        if j ==1, first = i; end
        last = i;
        fy1(j) = fy(i);
        j = j + 1;
     end
  end
subplot(313);plot(fy1(1:(last-first + 1)),abs(Y(first:last)));
title([ 'FFT plot in the range ',num2str(last-first + 1),' points'])
```

The plots indicated are shown in Figure 6.16, where we see the comparison of the CZT and the FFT. For this MATLAB script, we have chosen $A_0 = 1$ and Φ_0 is also 1, so the CZT computes the FFT. From the plot, we see that the FFT spectrum extends from 0 to 50 Hz and we see the peaks at 33 and 40 Hz. In this plot, we also see the spectrum that covers the

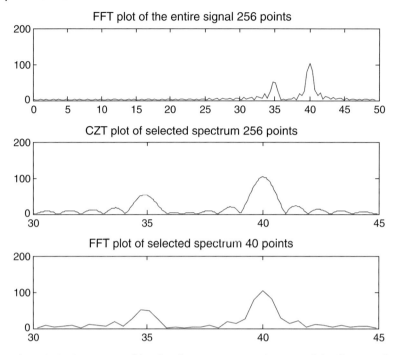

Figure 6.16 Comparison of the plot of a spectrum using the FFT and the Chirp transform.

frequency range that is of no interest to us, which is from 0 to 30. When we look at the spectrum computed using the CZT, we see the spectrum covers the frequency range from 30 to 45, but in this frequency range we have the plot that covers 256 points that are all the points we used to compute the spectrum. This is how we sharpen the resonant peaks of the signal, by using more points to represent the spectrum of a narrow band of frequencies. Finally, the third spectrum is drawn by choosing the points in the frequency range of 30–45 from the FFT that we computed earlier. From the plot, we see that the FFT has only 40 points in this range and the plot shows that the resonance peaks are rather rough. This clearly demonstrates that the CZT is effectively used to sharpen the resonance peaks of a signal in a narrow frequency range.

The CZT has several advantages over the conventional DFT. These advantages are as follows:

1) The length of the sequence N or the number of points of the spectrum K need not be the same as is required by the FFT.
2) Both the numbers N and K can be prime numbers. There is no need to pad the sequence by zeros.
3) The angular spacing between the K points of the spectrum can be arbitrary. With the FFT the angular spacing is always $2\pi/N$.
4) The contour over which the DFT is computed does not have to be circular or be restricted to the unit circle, which is the restriction for the DFT.

5) The starting point and the ending point over which the spectrum is computer are arbitrary. In the DFT, the starting and the ending points are predefined.

6) If we set $A = 1$, $\Phi = 1$, and $N = K$, the CZT can be used to evaluate the DFT efficiently even when N is a prime number.

MATLAB Commands:

fft(x,N)	This command computes the Fast Fourier transform of the sequence x. The FFT is computed as an N point sequence.
ifft(x,N)	This command computes the Inverse Fast Fourier transform of the sequence x. The Inverse FFT is computed as an N point sequence.
Fftshift(x)	This command shifts the zero frequency component to the center. It swaps the left half with the right half. Used to display results in a nice format.
CZT(X, M, W, A)	This command computes an M element Z transform of the sequence X where M is the length of the transform, W is a complex ratio between points on the contour, and A is the complex starting point.

6.9
Chapter Summary

In this chapter, we investigated several techniques for the computation of the DFT. In all the techniques, we have been careful to show how periodicity and symmetry of the complex factor W_N^{nk} can be used to increase the efficiency and to decrease the number of computations required to calculate the DFT.

This chapter began by looking at the Goertzel algorithm. This algorithm arranged the computation as a feedback loop, so there was no need to compute all the individual W_N^{nk} as they were computed automatically as the signal went through the loop. The algorithm by itself did not reduce the computation requirement. Next, when we extended the Goertzel algorithm to compute two poles simultaneously, we achieved significant savings. In both the algorithms, we got the required output every Nth time through the loop.

Next, we described two of the most popular DFT algorithms; they are decimation in time and decimation in frequency. These algorithms operate at their peak efficiency when N is a power of 2 because they both operate by dividing the sequence in half repeatedly till there are only two terms left in the sequence and then computing the DFT.

The decimation in time algorithm divides the sequence in the even position terms and the odd position terms. So, for this algorithm the sequence is first stored in the bit-reversed order and the FFT computed to get the output in the sequential order. The decimation in frequency algorithm divides the sequence in the first half and the second half sequences. So, for this algorithm the sequence is first stored in sequential order and the output that we get is in the bit-reversed format.

We drew the diagram that showed how the computation is to be done for a small number of terms ($N = 8$), but looking at the computation diagram makes it very clear how the diagram can be extended so the FFT computed is for more than eight points. All the computations in these algorithms are "butterfly" computations. The butterfly computations use two input points to compute two output points. As a result of this, the output points computed can be stored in the storage location vacated by the input points. Hence, both these algorithms are known as "in-place" algorithms as they do not need any extra storage locations.

Not all sequences are a power of 2 or can be extended to be a power of 2. So, we examined the original Cooley–Tukey algorithm. This algorithm showed us how the algorithm would work if the sequence length N was a composite number. Once again, we used the basic decimation principles and developed an algorithm when N is a composite number made of a sequence of products of prime numbers. Again, we found that if the prime numbers were small, then the total computation requirement would be reduced.

Next, we examined how we can determine the DFT of a limited spectrum. This would allow us to examine a small portion of the spectrum without computing the entire spectrum of N points. For this, we had to have the N length of the sequence to be a product of $K \cdot R$ where we were interested in the DFT of K points. This was an extension of the Cooley–Tukey algorithm.

Finally, we examined the Chirp Z-transform. This is an efficient way of computing the DFT of a sequence N when N is not a composite number, but a prime number, and it cannot be extended by padding it with zeros to make it a composite number. We can, of course, compute the DFT using the definition but the CZT is more efficient. On examining the CZT further, we found that it was possible to use this transform to sharpen the resonant peaks in the signal as the CZT can zoom into a specific frequency band and compute the spectrum in that frequency band.

7
Multirate Signal Processing and Devices

7.1
Introduction

All the discrete structures and algorithms that we have studied in this book so far have one particular characteristic in common. They all operate on just one sampling rate. The sampling rate at the input is the same as the sampling rate at the output and everywhere else within the system. This works fine most of the time as long as you do not want to import data from one system to another that has a different sampling rate. Consider, for example, the sampling rate used in your CD player; this sampling rate is 44.1 kHz, while the sampling rate used in a digital audio tape is 48 kHz. If you were interested in converting the signal from one to the other, you would not have the same experience because the sampling rate is different and hence you will experience a shift in the frequency. In today's discrete world, there are many such examples.

Sometimes, even if you are working in the same system and do not need to change the sampling rate there are efficiencies that are available if we are able to change the sampling rate. So in these systems we will want to increase and decrease the sampling rate at various internal points for efficient processing of signals. To change the sampling rate of a discrete signal, our goal is to stay in the discrete domain entirely. It is definitely possible to use a D/A converter to convert the discrete signal to its analog equivalent. Now resample the signal at the new rate of the target system. But this would defeat the purpose of converting the signal first to its discrete equivalent. Here, we plan to show that it is possible to work only in the discrete domain and convert the sampling rate from one to another.

As we study the sampling rate alteration devices and techniques, we will find efficiencies in using multirate structures not only for changing the sampling rate from one to another but also for simple applications such as linear filtering using FIR filters. In fact, the study and the use of FIR filters is the basis of multirate systems. To study the multirate structures, we will examine the polyphase decomposition of the discrete signal and see how the two go hand in hand.

Digital Filters: Theory, Application and Design of Modern Filters, First Edition. Rajiv J. Kapadia.
© 2012 Wiley-VCH Verlag GmbH & Co. KGaA. Published 2012 by Wiley-VCH Verlag GmbH & Co. KGaA.

7.2
Time Domain Characteristics of the Sampling Rate Alteration Devices

There are basically two sampling rate alteration devices. One of the devices is known as the upsampler or the expander and the other is known as the downsampler or the compressor. In this chapter, we will study the characteristics of these devices and see how they function in a multirate system.

7.2.1
The Upsampler

This device takes the discrete sequence as its input and outputs a different sequence that has $L-1$ zeros inserted between two consecutive samples of the input sequence. So this device takes in N samples and outputs L^*N samples. Of these L^*N output samples $(L-1)^*N$ are zeros. The block representation of an upsampler is shown in Figure 7.1. The time domain representation of the device is given in Eq. (7.1).

$$x_u[n] = \begin{cases} x\left[\dfrac{n}{L}\right] & L = 0, \pm 1L, \ \pm 2L, \ldots \\ 0 & \text{otherwise} \end{cases} \tag{7.1}$$

With the upsampler, we are increasing the number of samples within the same time interval, so we are in effect increasing the sampling rate. This is shown in Eq. (7.2). In Eq. (7.2), F_T is the sampling rate of the input signal, while F_{Tu} is the sampling rate of the upsampled signal. An example of upsampled signal is shown in Figure 7.2.

$$\begin{aligned} \text{Input frequency} \quad & F_T = \frac{1}{T} \\ \text{Output frequency} \quad & F_{Tu} = \frac{L}{T} = LF_T \end{aligned} \tag{7.2}$$

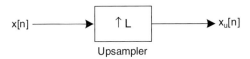

Figure 7.1 Block diagram of an upsampler.

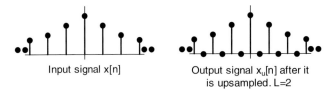

Input signal x[n]　　　Output signal x_u[n] after it
　　　　　　　　　　　is upsampled. L=2

Figure 7.2 Relation between input and output signal after it is upsampled by $L=2$.

The upsampler is a linear system, but it is a time varying system. It is very easy to show that the upsampler is a linear system. To show that the system is time varying, consider the input at time 0 and the output at time 0 of an upsampler that upsamples by L; this is the first sample of the input and the output. Now shift the output by any number of samples that is not a multiple of L. The output after the same amount of delay is not the first sample, but it is one of the zero samples. Since the output before and after the shift is different, the system is a time varying system.

7.2.2
The Downsampler

This device takes the discrete sequence as its input and outputs a different sequence. The output sequence is every Mth sample of the input sequence. So this device throws away $M-1$ consecutive samples from the input sequence and outputs the Mth sample. This device is shown in Figure 7.3 and the time domain representation of the device is given in Eq. (7.3).

$$x_d[n] = x[Mn] \tag{7.3}$$

With the downsampler, we are decreasing the number of samples at the output within the same time interval, so we are in effect decreasing the sampling rate. This is shown in Eq. (7.4). In Eq. (7.4), F_T is the sampling rate of the input signal, while F_{Td} is the sampling rate of the downsampled signal. An example of downsampled signal is shown in Figure 7.4.

$$\text{Input frequency} \quad F_T = \frac{1}{T}$$

$$\text{Output frequency} \quad F_{Td} = \frac{1}{MT} = \frac{F_T}{M} \tag{7.4}$$

The downsampler is a linear system, but it is a time varying system. It is very easy to show that the downsampler is a linear system. To show that the system is time varying, consider the input at time 0 and the output at time 0 of a downsampler that downsamples by M; this is the first sample of the input and the output. Now shift the input by any number of

x[n] ⟶ ↓ M ⟶ x_d[n]

Downsampler

Figure 7.3 Block diagram of an upsampler.

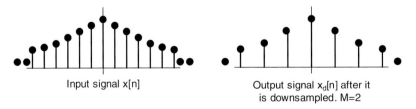

Input signal x[n]   Output signal x_d[n] after it is downsampled. M=2

Figure 7.4 Relation between input and output signal after it is upsampled by $L = 2$.

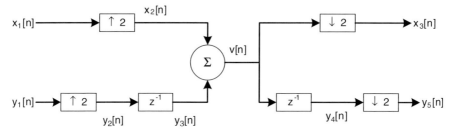

Figure 7.5 A multirate system.

samples that is not a multiple of *M*. The output after the same amount of delay is not the first sample any more. Since the output before and after the shift is different, the system is a time varying system.

As an example of a simple multirate system, consider the system shown in Figure 7.5. In Figure 7.5, we have two different signals that we want transmitted through a single channel. To accomplish this, we upsample both the signals first and delay one of the signals by one sample period. Next, the two signals are merged into one channel and transmitted. When the merged signal is received, it is downsampled to get back one of the original signals. To get back the other signal, we first delay the signal by one sample period and then downsample the signal. This gives us the other signal that we had. In what follows, we examine the signals after each block in the signal flow graph in Figure 7.5.

Examining Table 7.1 we see first that the signal $x_1[n]$ has a sample at every time instant. After this signal is upsampled by 2, we get the signal $x_2[n]$, and the signal $x_2[n]$ represents the signal $x_1[n]$ after it has been upsampled by factor 2. Thus, we have a zero sample between each original sample.

Next, we examine the signal $y_1[n]$; it is an entirely different signal with a sample at each sampling instant. The signal $y_2[n]$ is the upsampled version of signal $y_1[n]$, which is then delayed, so we get the shifted version of the signal $y_2[n]$ in signal $y_3[n]$.

The two signals $x_2[n]$ and $y_3[n]$ are added together and because of the upsampling and the delay, the two signals are interleaved. This signal is at a higher sampling rate than either $x_1[n]$ or $y_1[n]$. This signal can now be transmitted on a single channel and received at a remote station where the two signals can be separated. This is the signal $v[n]$.

Table 7.1 Signal progression through system in Figure 7.5.

N	0	1	2	3	4	5	6	7	8
$x_1[n]$	x_0	x_1	x_2	x_3	x_4	x_5	x_6	x_7	x_8
$x_2[n]$	x_0	0	x_1	0	x_2	0	x_3	0	x_4
$y_1[n]$	$y0$	$y1$	$y2$	$y3$	$y4$	$y5$	$y6$	$y7$	$y8$
$y_2[n]$	$y0$	0	$y1$	0	$y2$	0	$y3$	0	$y4$
$y_3[n]$	0	$y0$	0	$y1$	0	$y2$	0	$y3$	0
$v[n]$	x_0	y_0	x_1	y_1	x_2	y_2	x_3	y_3	x_4
$x_3[n]$	x_0	x_1	x_2	x_3	x_4	x_5	x_6	x_7	x_8
$y_4[n]$	$y?$	x_0	y_0	x_1	y_1	x_2	y_2	x_3	y_3
$y_5[n]$	$y?$	$y0$	$y1$	$y2$	$y3$	$y4$	$y5$	$y6$	$y7$

First, to get the signal $x_3[n]$ we downsample the signal $v[n]$ to get back the original signal $x_1[n]$ in the identical form and sampling rate. In the lower channel, we first delay the signal $v[n]$ to get the signal $y_4[n]$. Then, when we downsample the signal $y_4[n]$ by a factor of 2, we get back the original signal $y_1[n]$; it is in identical form to the original signal except it is delayed by only one sampling instant. Thus, this system represents a time division multiplexed transmission of two signals to a remote station. Notice that there are two delay elements in the path of the signal $y[n]$, yet the delay observed at the output of the system is a unit delay. This is because the two delays take place at a higher sampling rate where each delay is only half the sampling time interval compared to the system input and output.

7.3
Frequency Domain Characteristics of the Sampling Rate Alteration Devices

We have seen that both the upsampler and the downsampler change the sampling rate of the input signal. The upsampler increases the sampling rate, while the downsampler reduces the sampling rate. Changing the sampling rate should have the same effect in the frequency domain as we have seen in Chapter 2. Decreasing the sampling rate could introduce aliasing in the output signal. Increasing the sampling rate will represent the signal along with its image spectrums.

7.3.1
The Upsampler

The diagram of the upsampler by a factor of L is shown in Figure 7.1. The input–output relations of the upsampler are given in Eq. (7.1). If we were to obtain the Z-transform of the upsampler, we can write Eq. (7.5) as follows:

$$X_u(z) = \sum_{n=-\infty}^{\infty} x_u[n]z^{-n} = \sum_{n=-\infty}^{\infty} x[n/L]z^{-n}$$

$$X_u(z) = \sum_{m=-\infty}^{\infty} x_u[m]z^{-Ln} = \sum_{n=-\infty}^{\infty} x[m]\left(z^L\right)^{-n} = X\left(z^L\right)$$

(7.5)

If we let $z = e^{j\omega}$ in Eq. (7.5), we can evaluate the implications of the above relation on the unit circle. In this case $X_u(e^{j\omega})$ equals $X(e^{jL\omega})$; that is, the spectrum of $X(e^{j\omega})$ is repeated L times to become the spectrum of $X_u(e^{j\omega})$. This is shown in Figure 7.6. For a factor of L sampling rate increase, we see $L-1$ additional spectrum replica in the range 0–2π. These additional spectrums are called as image spectra. The image spectra are replica of the original spectrum and arise because we inserted the zero values between the samples of the sequence. From Figure 7.6, we see that if we pass the upsampled signal through an appropriately designed low-pass filter, we can eliminate the image spectra.

This is done for you in Figure 7.7. In Figure 7.7, we take the upsampled signal and pass it through an appropriate low-pass filter. Now, when we look at the upsampled signal in the time domain after low-pass filtering, we will find that all the zeros that we inserted have been

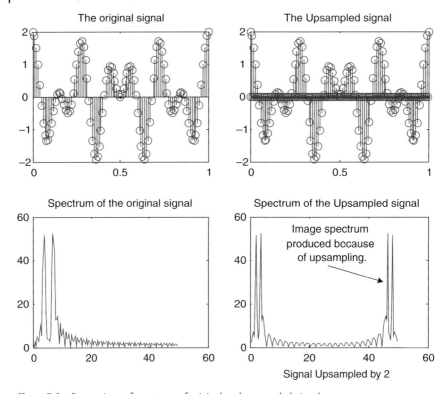

Figure 7.6 Comparison of spectrum of original and upsampled signal.

"filled" out; the low-pass filter "estimates" what the values at the in-between points should be and fills them in its estimate in place of the zero values and hence we no longer see the image spectra.

7.3.2
The Downsampler

The downsampler by a factor of M is shown in Figure 7.3. The input–output relations of the downsampler are given in Eq. (7.3). Determining the Z-transform of the downsampler is a bit tricky as we are eliminating some of the sample points, so we have to take a circuitous route rather than the direct route that we took for the upsampler; the following development shows the steps required. We begin with Eq. (7.6).

$$Y(z) = \sum_{n=-\infty}^{\infty} x[Mn]z^{-n} \tag{7.6}$$

Our goal is to represent $Y(z)$ in terms of $X(z)$, but as we mentioned we cannot do that here directly because we have removed some of the samples of $x[n]$. To get around this problem, let us define an intermediate sequence that is the same length as $x[n]$ and every Mth sample

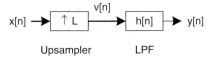

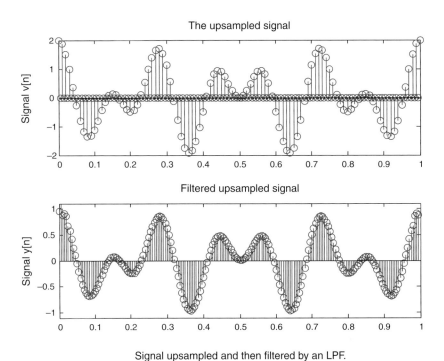

Signal upsampled and then filtered by an LPF.

Figure 7.7 Signal after it is filtered by an appropriate low-pass filter.

of this sequence is the same as the sequence $x[n]$, and the in-between $M - 1$ samples are zero samples, so the intermediate sequence can be written as shown in Eq. (7.7).

$$x_{\text{int}}[n] = \begin{cases} x[n] & n = 0, \pm M, \pm 2M, \ldots \\ 0 & \text{otherwise} \end{cases} \tag{7.7}$$

We can substitute Eq. (7.7) in Eq. (7.6) to get Eq. (7.8).

$$Y(z) = \sum_{n=-\infty}^{\infty} x[Mn]z^{-n} = \sum_{n=-\infty}^{\infty} x_{\text{int}}[Mn]z^{-n}$$

$$Y(z) = \sum_{k=-\infty}^{\infty} x_{\text{int}}[k]z^{-k/M} = X_{\text{int}}\left(z^{1/M}\right) \tag{7.8}$$

Equation (7.8) gives us the transform of the sequence $x_{\text{int}}[n]$ that has the same sample values as the downsampled signal except it has zeros inserted between the sample values.

Our next task is to relate $x_{int}[n]$ to $x[n]$. To do this let us define another sequence $\gamma[n]$ as shown in Eq. (7.9).

$$\gamma[n] = \begin{cases} 1 & n = 0, \pm M, \pm 2M, \ldots \\ 0 & \text{otherwise} \end{cases} \tag{7.9}$$

With this definition, the relation between $x_{int}[n]$ and $x[n]$ can be written as $x_{int}[n] = \gamma[n]x[n]$. We will write $\gamma[n]$ in a little more convenient way as shown in Eq. (7.10).

$$\gamma[n] = \frac{1}{M}\sum_{k=0}^{M-1} W_M^{kn} = \frac{1}{M}\sum_{k=0}^{M-1} e^{-j2\pi nk/M} \tag{7.10}$$

With this value for $\gamma[n]$, we can now write $x_{int}[n] = \frac{1}{M}\sum_{k=0}^{M-1} x[n]W_M^{nk}$. Now, we can relate the Z-transform of $x_{int}[n]$ as shown in Eq. (7.11).

$$X_{int}[z] = \sum_{n=-\infty}^{\infty} x_{int}[n]z^{-n} = \frac{1}{M}\sum_{n=-\infty}^{\infty}\left(\sum_{k=0}^{M-1} W_M^{nk}\right)x[n]z^{-n}$$

$$X_{int}[z] = \frac{1}{M}\sum_{k=0}^{M-1}\left(\underbrace{\sum_{n=-\infty}^{\infty} x[n]W_M^{nk}z^{-n}}_{\text{Replace }(W_M^{-k}z)\text{ with }\xi}\right) = \frac{1}{M}\sum_{k=0}^{M-1}\left(\underbrace{\sum_{n=-\infty}^{\infty} x[n]\xi^{-n}}_{\text{Z-transform}}\right) \tag{7.11}$$

Equation (7.11) gives us the Z-transform of $x_{int}[n]$ in terms of the sequence $x[n]$. When we substitute this in Eq. (7.8), we get the transform of the downsampled signal as shown in Eq. (7.12).

$$Y(z) = \frac{1}{M}\sum_{k=0}^{M-1} X\left(z^{1/M}W_M^{-k}\right) \tag{7.12}$$

Equation (7.12) is an interesting equation. To understand what the equation represents, let us take a specific value for M and see what the spectrum looks like. We will consider the specific example of a downsampler by 3. By setting $M = 3$ and replacing z by $e^{j\omega}$, we will be able to see the effect downsampling has on the frequency domain. This is shown in Eq. (7.13).

$$Y(e^{j\omega}) = \frac{1}{3}\left[X\left(e^{j\omega/3}e^{2j\pi\cdot0/3}\right) + X\left(e^{j\omega/3}e^{2j\pi\cdot1/3}\right) + X\left(e^{j\omega/3}e^{2j\pi\cdot2/3}\right)\right]$$

$$Y(e^{j\omega}) = \frac{1}{3}\left[X\left(e^{j\omega/3}\right) + X\left(e^{j(\omega-2\pi\cdot2)/3}\right) + X\left(e^{j(\omega-2\pi\cdot1)/3}\right)\right] \tag{7.13}$$

To understand the relations of the three terms with each other, we have rewritten the equation in a special format in the second equation in Eq. (7.13), where we have made use of the fact that $+2\pi/3 = -4\pi/3$ and $+4\pi/3 = -2\pi/3$. From this we can see that the spectrum of $Y(e^{j\omega})$ consists of a sum of M spectra (three in the present case since $M = 3$) with each spectra shifted from the previous one in frequency by $2\pi/M$. We can now rewrite the three terms to represent the general expression for the spectra as shown in Eq. (7.14).

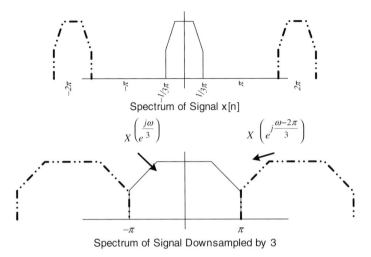

Spectrum of Signal x[n]

$$X\left(e^{\frac{j\omega}{3}}\right) \qquad\qquad X\left(e^{j\frac{\omega-2\pi}{3}}\right)$$

Spectrum of Signal Downsampled by 3

Figure 7.8 Spectrum of a downsampled signal when there is no aliasing.

$$Y\left(e^{j\omega}\right) = \frac{1}{M}\sum_{k=0}^{M-1} X\left(e^{j(\omega-2\pi k)/M}\right) \tag{7.14}$$

Equation (7.14) says that the spectra of the downsampled signal is a sum of M spectra that are uniformly shifted and stretched versions of the spectrum of $X(e^{j\omega})$ and scaled by a factor of M. There is a very good chance that the downsampled signal will be aliased unless the original signal is band limited to $\pm\pi/M$. In our example, $M=3$. Figure 7.8 shows a signal that is band limited to $\pi/3$; this signal is then downsampled by 3 and the spectrum of the downsampled signal is drawn. The figure shows that after downsampling there is no aliasing. A different situation is shown in Figure 7.9. In Figure 7.9, the signal is band limited to $\pi/2$, and it is downsampled by $M=3$ and the resulting spectrum shows that there is aliasing.

7.3.3
The Upsampler–Downsampler Cascade Arrangement

Until now we have examined the upsampler and the downsampler individually. In Figure 7.5, we saw that the upsampler and the downsampler are often used in a cascade arrangement. Toward the end of this chapter, we will have occasions where we will want to interchange the positions of the upsampler and the downsampler as it will lead to efficiencies in computation. Here, we examine certain cascade arrangements, as shown in Figure 7.10, and their equivalences that leave the input–output relations unchanged.

Consider the structure where we have the upsampler first and it is followed by the downsampler. The output from the two sampling rate alteration devices can be written individually in terms of the input to that device as shown in Eq. (7.15).

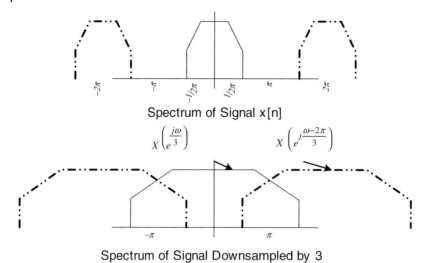

$$X\left(e^{\frac{j\omega}{3}}\right) \qquad\qquad X\left(e^{j\frac{\omega-2\pi}{3}}\right)$$

Spectrum of Signal Downsampled by 3

Figure 7.9 Spectrum of a downsampled signal when there is aliasing.

$$V_1(z) = X(z^L)$$

$$Y_1(z) = \frac{1}{M}\sum_{k=0}^{M-1} V_1\left(z^{1/M} W_M^{-k}\right) \tag{7.15}$$

Combining the two equations together by substituting the first in the second, we get one of the relations we are looking for, as shown in Eq. (7.16).

$$Y_1(z) = \frac{1}{M}\sum_{k=0}^{M-1} X\left(z^{L/M} W_M^{-kL}\right) \tag{7.16}$$

Next, we consider the structure where we have the downsampler first and then the upsampler. The output from the two sampling rate alteration devices can be written individually in terms of the input to that device as shown in Eq. (7.17).

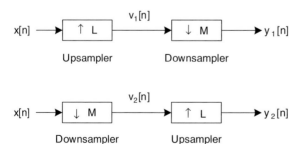

Figure 7.10 Up–down sampler cascade arrangement.

$$V_2(z) = \frac{1}{M}\sum_{k=0}^{M-1} X\left(z^{1/M} W_M^{-k}\right)$$

(7.17)

$$Y_2(z) = V_2(z^L)$$

Combining the two equations together by substituting the first in the second, we get the other relation we are looking for, as shown in Eq. (7.18)

$$Y_2(z) = \frac{1}{M}\sum_{k=0}^{M-1} X\left(z^{L/M} W_M^{-k}\right)$$

(7.18)

For the output Y_1 to be equal to the relation Y_2, the relations in Eq. (7.16) must be equal to the relation in Eq. (7.18) as shown in Eq. (7.19).

$$\frac{1}{M}\sum_{k=0}^{M-1} X\left(z^{L/M} W_M^{-kL}\right) = \frac{1}{M}\sum_{k=0}^{M-1} X\left(z^{L/M} W_M^{-k}\right)$$

(7.19)

When we examine Eq. (7.19), we see that the two sides will be equal to each other only when the two alteration rates, L and M, are relatively prime to each other; that is, they do not have any common factors between them. This is because the two factors W_M^{-k} and W_M^{-kL} will take on the same values only when L and M are relatively prime otherwise that is not true.

7.3.4
The Noble Identities

In Section 7.2.3, we saw the cascade of an upsampler and a downsampler. Here, we examine what effect there is of moving the filter to the other side of the sampler. Our goal is to identify the relation that would exist if we interchanged the cascade arrangement. Consider the downsampler followed by the filter. This is shown in Figure 7.11 identity 1. We first determine the output $Y_1(z)$ as shown in Eq. (7.20).

$$V_1(z) = \frac{1}{M}\sum_{k=0}^{M-1} X\left(z^{1/M} W_M^{-k}\right)$$

(7.20)

$$Y_1(z) = V_1(z)H(z)$$

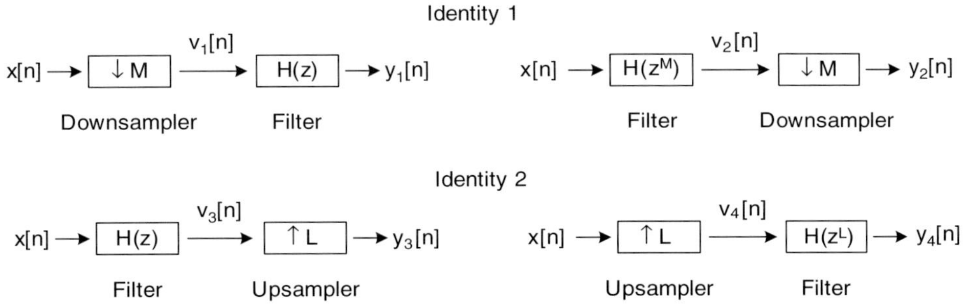

Figure 7.11 Noble identities for multirate systems.

When we substitute the value for $V_1(z)$ from the first equation into the second equation, we get the result that we want, as shown in Eq. (7.21).

$$Y_1(z) = \frac{1}{M} \sum_{k=0}^{M-1} X\left(z^{1/M} W_M^{-k}\right) H(z) \tag{7.21}$$

Next, we determine $Y_2(z)$ that has the filter and the downsampler with interchanged places. This is shown in Eq. (7.22).

$$V_2(z) = H(z^M) X(z)$$

$$Y_2(z) = \frac{1}{M} \sum_{k=0}^{M-1} V_2\left(z^{1/M} W_M^{-k}\right) \tag{7.22}$$

Again, we substitute the value of $V_2(z)$ from the first equation into the second equation to get the result we want, as shown in Eq. (7.23).

$$Y_2(z) = \frac{1}{M} \sum_{k=0}^{M-1} H\left(z^{M/M} W_M^{-kM}\right) X\left(z^{1/M} W_M^{-k}\right)$$

$$Y_2(z) = \frac{1}{M} \sum_{k=0}^{M-1} H(z) X\left(z^{1/M} W_M^{-k}\right) \tag{7.23}$$

When we compare Eqs. (7.21) and (7.23), we see that they are both identical and hence the two arrangements are identical to each other. This Noble identity tells how to modify the filter when we move the filter on the input side of the downsampler. The second Noble identity involves the upsampler. It is proved in the same way as we proved the first Noble identity. First, consider the cascade of the upsampler and then the filter. This combination is evaluated in Eq. (7.24).

$$V_3(z) = X(z^L)$$

$$Y_3(z) = H(z^L) V_3(z^L) = X(z^L) H(z^L) \tag{7.24}$$

By interchanging the positions of the filter and the upsampler, we get Eq. (7.25).

$$V_4(z) = H(z) X(z)$$

$$Y_4(z) = V_4(z^L) = H(z^L) X(z^L) \tag{7.25}$$

Once again, we see that the two arrangements are indeed equal. This Noble identity tells how to modify the filter when we move the filter on the output side of the upsampler. There is, however, one condition that must be satisfied when we are using the Noble identities. In both the examples, we had the filter and the sampler power match each other. So if the signal was to be upsampled by L, the filter was also z^L. It would be wrong if the filter turned out to be a fraction power after moving it past the sampler in either case. Take a look at Figure 7.12. This use of the Noble identities would be wrong. The error occurs because we cannot have a filter $H(z^{1/2})$. These Noble identities are extremely useful in almost all the applications in multirate systems; however, the identities require to be used in a correct way. In Figure 7.12,

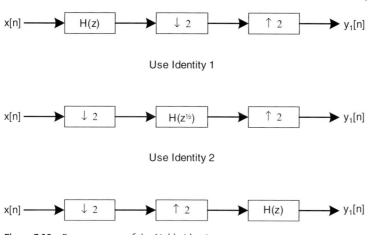

Figure 7.12 Erroneous use of the Noble identity.

if the filter $H(z)$ were instead $H(z^2)$ to begin with, then the use of identity as shown in Figure 7.12 would be absolutely correct.

7.4
Basic Sampling Rate Converters

Until now we have used either the upsampler or the downsampler by themselves and this can give us a change in the sampling rate by an integer value. A fractional change in sampling rate, when the required rate change is a rational fraction, can be achieved by a cascade combination of an upsampler followed by a downsampler as shown in Figure 7.13. In Figure 7.13, we have two filters, the antiimage filter and the antialiasing filter. We need these filters as we have seen in Figure 7.6 that the upsampler produces image spectrums we do not want. To get rid of these spectrums, we always follow the upsampler with a low-pass filter that is used to remove the image spectrums. Similarly, we have seen in Figure 7.7 that the signal output from the downsampler could be an aliased spectrum, so the downsampler is almost always preceded by an antialiasing filter. In the cascade arrangement of the upsampler and

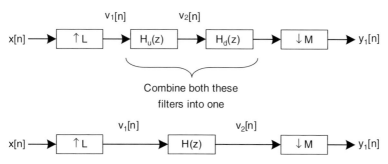

Figure 7.13 Fractional sampling rate converter.

downsampler, to change the sampling rate by a rational fraction, we will have an antiimaging and an antialiasing filter. Of course, we can combine the two filters into one filter. This one filter will have specifications that are equal to the filter that has the more stringent specifications. We can do this primarily because both the filters operate at the same sampling rates. The input–output relations for such an arrangement are the subject of study in the next section.

In the above discussion and shown in Figure 7.13, we have used a filter as both an antialiasing filter and an antiimaging filter. The same filter can serve both the purposes if the requirement for the filter is the stringent of the two requirements. If the sampled signal $x[n]$ is band limited to π radians, then there will be no aliasing in the sampled spectra of the signal. So we want to preserve the information up to π radians as the signal goes through the fraction rate changing structure. Assuming that the signal $x[n]$ is band limited to π radians, we first upsample the signal by a factor L. This will cause the signal to produce image spectra in the range 0–π and the original signal will occupy frequency range from 0–π/L. So after the signal is upsampled, we need to eliminate all the images that occupy the frequency range from π/L to π and hence the antiimage filter should have a cutoff frequency of π/L.

When we look at the decimation and the antialiasing filter, we look at the signal after the downsampler. The downsampler expands the signal; since we want to avoid aliasing, we can allow the downsampler to expand the signal only up to π radians. If the signal that is downsampled by factor of M is expanded to π radians, then the signal before downsampling can only be allowed to be π/M radians. So the antialiasing filter has to have a cutoff frequency that is no more than π/M radians. Combining the two requirements, we can say that the cutoff frequency of the combined antialiasing and antiimaging filter has to be as shown in Eq. (7.26).

$$\omega_c = \min\left\{\frac{\pi}{L}, \frac{\pi}{M}\right\} \tag{7.26}$$

7.4.1
Input–Output Relations of a Fraction Rate Structure

We look at the general arrangement to change the sampling rate by a rational fraction in Figure 7.13. We examine the structure in the time domain first. We begin with the upsampler and then the filter as shown in Eq. (7.27).

$$V_1[nL] = x[n]$$

$$V_2[n] = \sum_{k=-\infty}^{\infty} V_1[k]h[n-k] \tag{7.27}$$

$$V_2[n] = \sum_{k=-\infty}^{\infty} x[k]h[n-Lk]$$

Now, when this filtered signal goes through the downsampler, we get the relation as shown in Eq. (7.28).

$$Y[n] = V_2[nM] = \sum_{k=-\infty}^{\infty} x[k]h[nM-kL] \tag{7.28}$$

Next, we see the relation in the frequency domain using the same sequence of upsampler followed by the filter and then the downsampler as shown in Eq. (7.29). We first look at the signal as it goes through the upsampler and the filter.

$$V_1(z) = X(z^L)$$
$$V_2(z) = H(z)V_1(z) = H(z)X(z^L) \tag{7.29}$$

Now, when this filtered signal goes through the downsampler, we get the relation as shown in Eq. (7.30).

$$Y(z) = \frac{1}{M} \sum_{k=0}^{M-1} H\left(z^{1/M} W_M^{-k}\right) X\left(z^{L/M} W_M^{-Lk}\right) \tag{7.30}$$

7.4.2
Multistage Design of Fraction Rate Converter

The fraction rate converter of Figure 7.13 is shown completing the upsampling and the downsampling in a single stage. This single stage conversion requires only one filter between the upsampler and the downsampler. This works fine most of the time, but there are efficiencies that can be gained if the process of upsampling and downsampling is performed in stages; this is especially so when the factor of upsampling and/or downsampling is a large number or the required filter is of a very large order. Dividing the sampling process is possible when the factor of upsampling can be factored into two or more factors as $L = L_1{}^*L_2$ and/or the factor of downsampling can be factored into two or more factors as $M = M_1{}^*M_2$. This is shown in Figure 7.14.

There is computational efficiency to be gained when a sampling rate alteration arrangement is organized as a cascade of several stages. This is demonstrated by the following computational representation. Consider the interpolation for increasing the sampling rate of the signal from a signal sampled at 500 Hz–20 kHz. The signal that is to be preserved is from 0 to 200 Hz. The change in the sampling rate requires us to interpolate by $L = 20\,000/500 = 40$. We can interpolate the signal as required using one stage and an antiimaging filter as shown in Figure 7.15 or we can interpolate using two stages as shown in Figure 7.16. There are many possible choices for the two interpolation factors; in this computation, we have chosen the factors 5 and 8, so the product equals 40, which is the required upsampling factor. Below we examine the computational requirements for both the options. To complete this comparison, let us define the requirement for the filter as $f_p = 200$, $\delta_p = 0.002$, and $\delta_s = 0.001$.

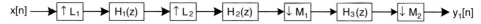

Figure 7.14   Multistage fractional sampling rate converter.

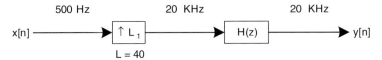

Figure 7.15 Single-stage interpolator that is to be compared with the multistage interpolator in Figure 7.17.

7.4.2.1 Single-Stage Design

First, we determine the required size of the filter. This filter is required to preserve the spectrum up to 200 Hz and remove all the images that are present from the first image to 20 000 Hz. To do this we must determine where the first image spectrum begins. Then, we can design the filter so that the transition band extends from the required pass band to the beginning of the first image spectrum. The pass band extends up to 200 Hz and the first image spectrum begins at 300 Hz as calculated in Eq. (7.31).

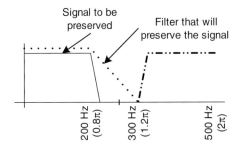

Filter ω_p = signal ω_P
Filter ω_s = ω_{Samp} - signal ω_s

Original signal spectrum and filter requirement

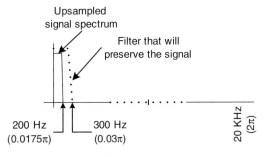

Filter ω_p = signal ω_P

Filter cutoff freq. $\omega = \dfrac{\omega_{samp} - L * \omega_s}{L} = \dfrac{2\pi - 40 * 0.02\pi}{40} = 0.03\pi$

Upsampled signal spectrum and filter requirement

Figure 7.16 Filter specifications for upsampling in one stage.

$$\text{Filter cutoff frequency} = \frac{f_{\text{sampling}} - L^* f_{\text{pass band}}}{L} = \frac{20000 - 40^*200}{40} = 300\ \text{Hz} \qquad (7.31)$$

The spectrum to be preserved is shown in Figure 7.16. By using Kaiser's method to determine the filter order, we get the filter order as shown in Eq. (7.32). Equation (7.32) shows that the filter size required is 659.

$$N_{\text{single stage}} = \frac{-20\log\left(\sqrt{\delta_p\delta_s}\right) - 13}{14\left(\omega_s - \omega_p\right)/(2\pi)} = \frac{-20\log\left(\sqrt{0.002^*0.001}\right) - 13}{14(300 - 200)/(20000)} = 659 \qquad (7.32)$$

7.4.2.2 Multistage Design

In this design, we will divide the upsampling process into two stages, with the first stage upsampling by 5 and the second stage upsampling by 8. When we divide the upsampling this way, we need to design two antiimaging filters, one filter after each upsampler. This arrangement is shown in Figure 7.17. With the two filters, we need to make sure that the overall requirements of the signal remain the same. For this reason, we will take the allowed pass band ripple ($\delta_p = 0.002$) and divide it equally among the two filters, so each of the filters will be allowed the pass band ripple equal to $\delta_p = 0.001$. This will keep the pass band ripple the same in the two designs. We do not need to adjust the stop band ripple and we will leave it as $\delta_s = 0.001$ for both the filters. This will make the stop band ripple even less in the multistage design compared to the single-stage design. For the first filter, we will have a transition band that begins at 200 Hz and extends to the first image that begins at 300 Hz, which can be computed using Eq. (7.31); this time the sampling frequency is 2500 Hz and the upsampling factor is 5. With these specifications, the first antiimaging filter size can be computed as shown in Eq. (7.33) that says that the filter size required is 84. The expected signal spectrum is shown in Figure 7.18.

$$N_1 = \frac{-20\log\left(\sqrt{\delta_p\delta_s}\right) - 13}{14\left(\omega_s - \omega_p\right)/(2\pi)} = \frac{-20\log\left(\sqrt{0.001^*0.001}\right) - 13}{14(300 - 200)/(2500)} = 84 \qquad (7.33)$$

The signal coming out of this filter occupies a frequency band from 0 to 200 Hz and there is no signal from 200 to 1250 Hz. We use this fact to design the next filter. The second filter in the multistage design will have its specifications as follows: the pass band error as $\delta_p = 0.001$ and the stop band error as $\delta_s = 0.001$. For the second filter also, we will have a transition band that begins at 200 Hz and extends to the first image that begins at 2300 Hz, which can be computed using Eq. (7.31); this time, the sampling frequency is 20 kHz and the upsampling factor is 8. With these specifications, the second antiimaging filter size can be computed as

Multistage operation

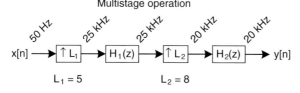

$$L_1 = 5 \qquad\qquad L_2 = 8$$

Figure 7.17 Multistage interpolator to match the single-stage interpolator.

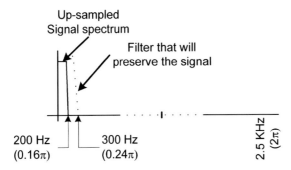

Up-sampled
Signal spectrum

Filter that will
preserve the signal

200 Hz   300 Hz   2.5 KHz
(0.16π)   (0.24π)   (2π)

Filter ω$_p$ = signal ωP

$$\text{Filter cutoff freq } \omega = \frac{\omega_{samp} - L*\omega_s}{L} = \frac{2\pi - 5*0.16\pi}{5} = 0.24\pi$$

Up-sampled Signal Spectrum after up-sampling by 5

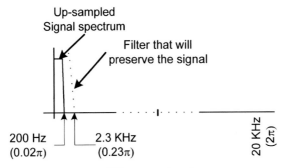

Up-sampled
Signal spectrum

Filter that will
preserve the signal

200 Hz   2.3 KHz   20 KHz
(0.02π)   (0.23π)   (2π)

Filter ω$_p$ = signal ωP

$$\text{Filter cutoff freq } \omega = \frac{\omega_{samp} - L*\omega_s}{L} = \frac{2\pi - 8*0.02\pi}{8} = 0.23\pi$$

Up-sampled Signal Spectrum after up-sampling by 8

Figure 7.18 Filter specifications for up-sampling in two stages.

shown in Eq. (7.34), which says that the filter size required is 32. The expected signal spectrum is shown in Figure 7.18.

$$N_2 = \frac{-20\log\left(\sqrt{\delta_p\delta_s}\right)-13}{14\left(\omega_s-\omega_p\right)/(2\pi)} = \frac{-20\log\left(\sqrt{0.001*0.001}\right)-13}{14(2300-200)/(20000)} = 32 \tag{7.34}$$

With the specifications for the two filters defined, we can see that the two separate filters taken together are much smaller than the single filter. Smaller filters lead directly to a fewer number of required computations to accomplish the upsampling and the removal of the image spectrum.

7.4.3
Application of Sampling Rate Converter

In the previous section, we saw how we can reduce the required computation by decimating or interpolating a signal in two or more stages as shown in Figure 7.16. There are other places we can use the sampling rate converters also. For this consider the simple CD player that plays audio music from the CD. The CD player uses a sampling rate of 44.1 kHz. This implies that the audio signal up to 22.05 kHz can be recovered. Trying to recover a signal up to 22.05 kHz would require a brick wall filter to eliminate the entire signal outside the pass band frequency. Even assuming that there is a guard band of 0.1 kHz, we would require a filter that has a very large size as shown in Eq. (7.35). Consider now an alternative arrangement where the sampled audio signal is first upsampled by an upsampling factor of 3. This arrangement is shown in Figure 7.19a.

$$N_{\text{CD player}} = \frac{-20 \log\left(\sqrt{\delta_p \delta_s}\right) - 13}{14\left(f_s - f_p\right)/f_{\text{sampling}}} = \frac{-20 \log\left(\sqrt{0.001 \cdot 0.001}\right) - 13}{14(22.1 - 22.0)/44.1} = 1481 \quad (7.35)$$

Consider now an alternative arrangement where the sampled audio signal is first upsampled with upsampling factor of 3. Next, consider an arrangement where the input

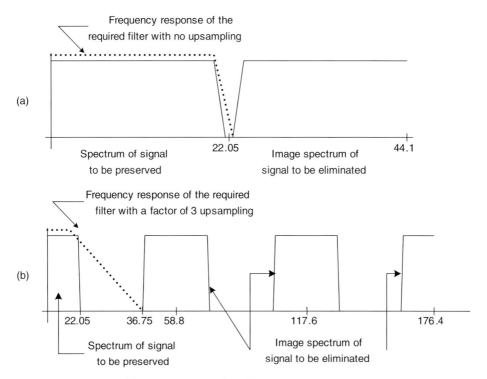

Figure 7.19 (a) Required frequency response of the filter with no upsampling. (b) Required frequency response of the filter with factor of 3 upsampling.

signal sampled at the rate of 44.1 kHz is upsampled by a factor of 3. In the new signal, the sampling rate is 176.4 kHz, but the signal that we want to retain and recreate is still 22.0 kHz. Now, we also have image spectra located between 36.75 and 80.85 kHz, between 95.55 and 139.65 kHz, and the final one between 154.35 and 176.4 kHz. These image spectra have to be eliminated by the filter to be designed. This filter now will have a pass band edge at 22.05 kHz, as before, but now the stop band edge is at 36.75 kHz and the sampling rate is also changed to 176.4 kHz. Now, if we use the same ripple limits for the filter, the size of the required filter is much smaller than the original filter as shown in Eq. (7.36).

$$
N_{\substack{\text{CD player} \\ \text{Upsampled}}} = \frac{-20\log\left(\sqrt{0.001^*0.001}\right) - 13}{14(36.75 - 22.05)/176.4} = 41 \tag{7.36}
$$

The difference in the two filter sizes is because the transition band of the filter before upsampling is an extremely narrow band. Upsampling separates the required frequency spectrum from the image spectrum, so the transition band does not have to be so steep and hence the filter is of a much smaller size. This is a significant advantage in the computation power required for the CD player.

7.5
Polyphase Decomposition

An interesting realization of an FIR filter is based on the polyphase decomposition of the filter impulse response. This results in implementation of the filter as a parallel structure. First, consider the polyphase decomposition of an arbitrary sequence. The sequence $x[n]$ that has a z-transform of $X(z)$ is written as shown in Eq. (7.37).

$$
X(z) = \sum_{n=-\infty}^{\infty} x[n]z^{-n} \tag{7.37}
$$

To rewrite the sequence in its polyphase form, we first define a new subsequence X_k that consists of every kth term from the sequence X as shown in Eq. (7.38).

$$
\text{Define } X_k = \sum_{n=-\infty}^{\infty} x_k[n]z^{-n} = \sum_{n=-\infty}^{\infty} x[nM+k]z^{-n} \tag{7.38}
$$

With the definition of Eq. (7.38), we can rewrite Eq. (7.37) as shown in Eq. (7.39). The breakdown of the sequence $x[n]$ is also shown in Figure 7.20.

$$
X(z) = \sum_{k=0}^{M-1} X_k(z)z^{-k} = \sum_{k=0}^{M-1}\left(\sum_{n=-\infty}^{\infty} x[nM+k]z^{-n}\right)z^{-k} \tag{7.39}
$$

Just as we split up the sequence $x[n]$ into its polyphase equivalent branches, we can divide an FIR filter into an M branch polyphase decomposition. Consider a causal FIR filter of order N as shown in Eq. (7.40).

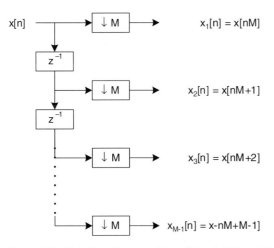

Figure 7.20 Polyphase decomposition of signal $x[n]$ into M bands.

$$H(z) = \sum_{n=0}^{N-1} h[n]z^{-n} = h[0] + h[1]z^{-1} + h[2]z^{-2} + \cdots + h[N-1]z^{-(N-1)} \qquad (7.40)$$

This equation can be written as a sum of M terms as shown in Eq. (7.42) where we have used the term $E_k(z)$ that we define as shown in Eq. (7.41); notice that this definition is the same as the definition in Eq. (7.38).

$$E_k(z) = \sum_{k=0}^{\text{int}[(N+1/M)]} h[nM+k]z^{-k} \quad 0 \le k \le M-1 \qquad (7.41)$$

Using this definition, we can write the transfer function of the FIR filter $H(z)$ as shown in Eq. (7.42). This filter realization is shown in Figure 7.21.

$$H(z) = \sum_{k=0}^{M-1} E_k\left(z^M\right)z^{-k} = \sum_{k=0}^{M-1}\left(\sum_{n=0}^{\text{int}[(N+1/M)]} h[nM+k]z^{-n}\right)z^{-k} \qquad (7.42)$$

The realization shown on the left-hand side of Figure 7.21 is described in Eq. (7.42). Transposing that realization gives us the structure on the right-hand side of Figure 7.21. The relation between the individual filters in the two realizations is given in Eq. (7.43).

$$R_k\left(z^M\right) = E_{M-1-k}\left(z^M\right) \quad 0 \le k \le M-1 \qquad (7.43)$$

With the equality shown in Eq. (7.43), the realization of filter $H(z)$ can be written as shown in Eq. (7.44).

$$H(z) = \sum_{k=0}^{M-1} z^{-(M-1-k)} R_k\left(z^M\right) \qquad (7.44)$$

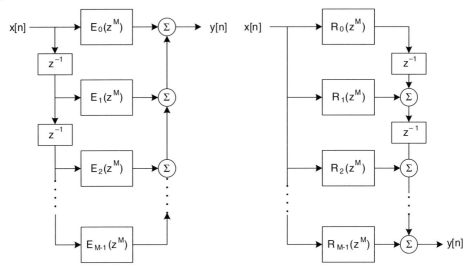

Figure 7.21 Realization of filter $H(z)$ using two different polyphase decomposition.

7.6
Computationally Efficient Interpolator and Decimator

The polyphase structures seen in Figure 7.21 can be used to efficiently decimate or interpolate an input signal. Using these structures, we will show how to implement an arbitrary sampling rate converter efficiently. To see how this is done, first consider the typical downsampling arrangement shown on the top of Figure 7.22. When we arrange the decimator in Figure 7.22 using polyphase decomposition for the filter, we get the filter arrangement on the left-hand side of Figure 7.23.

On the left-hand side, we have a filter $H(z)$ implemented as a polyphase filter with M polyphase branches. Following the filter we have a downsampler of order M. In this arrangement, the input signal $x[n]$ is sampled at frequency F_1. This signal goes through each of the polyphase branches at sampling frequency F_1. So each of the polyphase filter arrangement made of individual components $E_k(z)$ is operating at the sampling frequency F_1. After the signal is filtered, it is downsampled by a downsampler of order M. So the sampling frequency after downsampling is F_1/M. This means that ever Mth sample is preserved and all the other computations are thrown away. The computation effort by the filter for the $M-1$ samples that are to be thrown away is wasted. In this arrangement, to compute every sample that is output, the computation requirement is N multiplications. These N multiplications have to be completed in one sample time that is at the rate of sampling frequency F_1.

Consider now the arrangement on the right-hand side of Figure 7.23. In this arrangement, we have taken advantage of the Noble identity to move the downsampler in front of the filter rather than after the filter. Now, we throw away the $M-1$ samples, which were going to be thrown away anyway, before we waste our time using them in the computations. In this arrangement also, we need to perform N multiplications to get every output, but in this

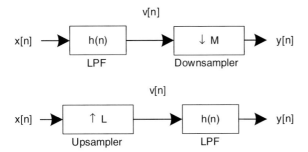

Figure 7.22 Typical up- or downsampling arrangement to be converted to polyphase.

arrangement the filter is operating at the frequency rate of F_1/M. So to perform the N multiplications, the filter now has M times more time to complete the required N multiplications.

Similar savings are also obtained when we consider the upsampler; this is shown in Figure 7.24. Here, we have arranged the filter in L polyphase branches. In Figure 7.24, we are inserting $L-1$ zero values before the filter and hence the filter has to operate at the frequency of L^*F_1. But since these are zero values, they do not add to the information of the signal. Filtering the signal first and then using the upsampler reduces the frequency at which the filter is operating to the frequency F_1 instead of frequency L^*F_1. In Figures 7.23 and 7.24, we have seen the upsampler and the downsampler with a polyphase filter arrangement. It is possible to combine the two together and have an efficient rational fraction rate converter.

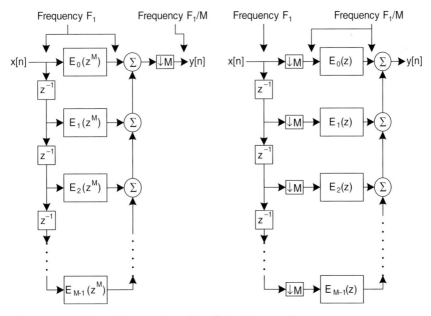

Figure 7.23 Polyphase decomposition of an efficient downsampler.

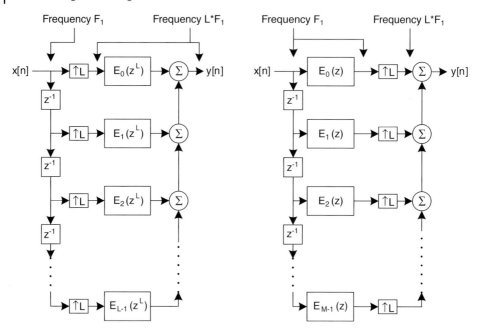

Figure 7.24 Polyphase decomposition of an efficient upsampler.

Here though, we need to be careful and make sure that we avoid the problem like the one that we saw in Figure 7.12.

7.6.1
Computationally Efficient Fraction Rate Converters

The design of the fractional sampling rate converter depends on the ratio of the input sampling rate and the output sampling rate. The required sampling rate is written as a ratio L/M with L and M being mutually prime to each other. To avoid the difficulty discussed in Figure 7.12 that comes about because there is increasing amount of delays in going from one polyphase branch to another as seen in each branch of two diagrams in Figure 7.25, we make use of a result from number theory. This result states that two mutually prime integers L and M satisfy the relation $\mu M - \lambda L = 1$ for a set of unique positive integers μ and λ. That is, given two mutually prime integers, you can always find two other mutually prime integers so that the result of the subtraction is always $\mu M - \lambda L = 1$. With this understanding, we can begin to draw the general fraction sampling rate converter as shown in Figure 7.25.

   The top of Figure 7.25 shows the fractional sampling rate converter that we wish to implement. We begin by using the polyphase structure of Figure 7.23 to first replace the filter $H(z)$ into L polyphase branches as shown on the left-hand side of Figure 7.25. Now that we have broken up the filter in its polyphase branch representation with L branches, we can move the upsampler through the various polyphase branches. By doing this we get the figure on the right-hand side of Figure 7.25. When we examine a typical branch on the right-hand side of Figure 7.25, we get the top row of Figure 7.26. In Figure 7.26, we have

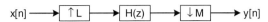

An arbitrary rational fraction rate converter

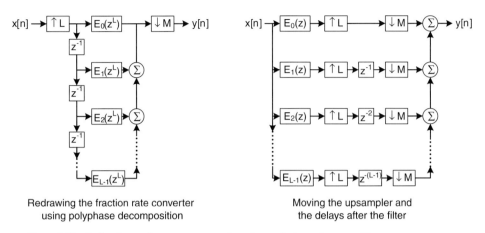

Redrawing the fraction rate converter
using polyphase decomposition

Moving the upsampler and
the delays after the filter

Figure 7.25   Rational sampling rate converter based on polyphase decomposition.

made successive modifications that will allow us to move the delays on the other side of the upsampler and the downsampler as shown in the bottom row of Figure 7.26. When we redraw the entire structure taking in account the modification shown in Figure 7.26, we get Figure 7.27.

The left-hand side of Figure 7.27 is a direct extension of a typical branch shown in Figure 7.26. In this figure, we see several blocks that represent time advances. These blocks are labeled z^{λ}. Advancing samples is not possible as this would require us to predict what

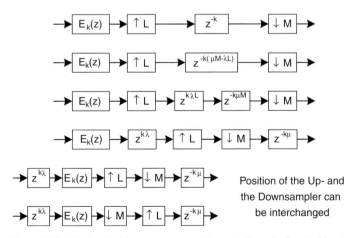

Position of the Up- and
the Downsampler can
be interchanged

Figure 7.26   Moving the delay operator around in a typical branch of a typical fractional sampling rate converter.

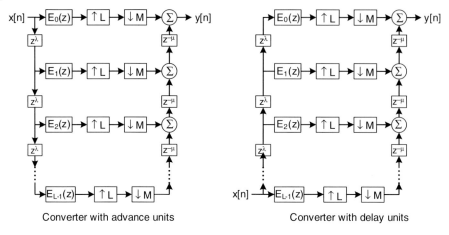

Converter with advance units   Converter with delay units

Two equivalent realizations for a fractional sampling rate converte

Figure 7.27   Using the typical branch from Figure 7.26 to draw the entire sampling rate converter.

the samples would be in the future. Therefore, we need to rearrange the blocks so that we have only delay elements. This adjustment is shown on the right-hand side of Figure 7.26. The arrangement on the right-hand side of Figure 7.26 suggests that we connect the input at the bottom and let it delay its way to the top. Also, now that the upsampler and the downsampler are adjacent to each other, they may be interchanged if we so desire.

There is one more possible simplification that we can add to the right-hand side of Figure 7.27. Look at the bottom branch in Figure 7.26. In this branch we have interchanged the upsampler and the downsampler. With the interchange, we now have a filter followed by a downsampler. This is exactly the same arrangement as we saw on the top of Figure 7.22. We were able to modify the downsampling arrangement on the top of Figure 7.22 as shown in Figure 7.23. We plan to use the same modifications we saw in Figure 7.23 in each branch of Figure 7.27. Since each branch of Figure 7.27 has the exact same arrangement as we saw on the left-hand side in Figure 7.22, this modification is obvious. Using this modification, we can now rearrange each branch in Figure 7.27 as shown in Figure 7.23 and move the downsampler before the filter instead of having it after the filter. This way we gain all the advantages of polyphase redesign of the arbitrary sampling rate converters. The next section shows us the application of one such an arrangement.

7.6.2
Building an Efficient Fraction Rate Converter

A complete efficient fraction rate converter is shown in Figure 7.28. In Figure 7.28, we want to increase the sampling rate by 3/2. The simplest arrangement is shown in Figure 7.28a. Here, we have taken no advantage of polyphase arrangement and hence the filter is processing zero samples inserted by the upsampler since it is located after the upsampler and it is processing samples that will be thrown away by the downsampler as the downsampler is after the filter. So, the first modification that we will make is to arrange the filter in between the upsampler

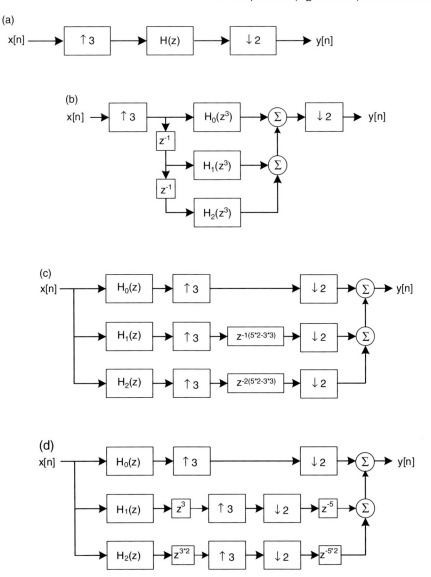

Figure 7.28 (a) The required sampling rate converter. (b) Arrangement after breaking the filter into its polyphase components. (c) Arrangement after moving the upsampler past the polyphase filters. (d) Moving the delays past the up- and the downsamplers. (e) Interchanging the location of the up- and the downsampler. Moving the delays. (f) The final arrangement before we make the advance into a delay. (g) The final arrangement after the input location is changed so that we have all delays.

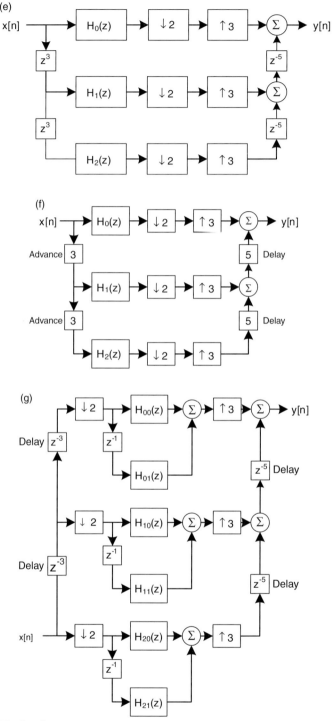

Figure 7.28 *(Continued)*

and the downsampler as a polyphase filter with L polyphase branches, which is shown in Figure 7.28b. Once this is done, the upsampler can be moved first into each of the polyphase branches and then moved after each filter with an appropriate adjustment in the filter, using the Noble identity as shown in Figure 7.28c. In Figure 7.28c, we have also moved the downsampler into each of the polyphase branches.

Next, we want to interchange the location of the upsampler and the downsampler. To do this, we need to adjust the delays so that we do not violate the principle of Figure 7.12. To do this, we use the result from mathematics regarding prime numbers. With $L = 3$ and $M = 2$, we can choose $\lambda = 3$ and $\mu = 5$, so $\mu M - \lambda L = = 5 \cdot 2 - 3 \cdot 3 = 1$. With the inclusion of this, the delays in each of the branches in Figure 7.28c are written so that we can proceed to the next step shown in Figure 7.28d.

In each of the branches of Figure 7.28d, we have moved the delay and the advance past the upsampler and the downsampler and made the appropriate change in the delay to account for the effect of the upsampler and the downsampler. This allows us to now interchange the location of the up- and the downsampler. Finally, we move the delay and the advance, so we get the individual polyphase branches as shown in Figure 7.28e. In this arrangement, the filter is still processing samples that we will be throwing away in the downsampler. We can modify the filter and the downsampler arrangement in each branch of Figure 7.28e as we did in Figure 7.23. This will allow us to throw away the unwanted samples before they are processed in the filter. In Figure 7.28e, we have also moved the delay and the advance to the two ends of each branch. Figure 7.28f shows where the advance and the delay occur.

In Figure 7.28f, we have the downsampler in each branch with the filter preceding the downsampler. This is the same arrangement as we saw on the top in Figure 7.22, which we arranged in polyphase form in Figure 7.23. We can do the same modification in each of the branches of Figure 7.28f. This is done in Figure 7.28g, where we have also decided that the input has to be connected to the last filter and all the advances made into delays. With this arrangement, neither the filter is processing any zero samples nor are we throwing away any of the samples processed by the filter. This arrangement is indeed very much efficient compared to the arrangement in Figure 7.28a.

7.7
Half Band and Nyquist Filters

In all the multirate systems that employ an upsampler, we always need an image elimination filter; this is a low-pass filter whose job is to estimate the values for the zero samples while keeping the input sample as it was with no modification. Here, we examine a specific type of low-pass filter that satisfies the requirements of the image elimination filter.

Consider such an L-fold interpolation filter shown on the top of Figure 7.29, the impulse response of one such filter with $L = 3$ is shown in Figure 7.29 (there are many different filters possible but this one is special as we will see); the output of this filter $y[n]$ has a Z-transform given by Eq. (7.45).

$$Y(z) = X(z^L) H(z) \tag{7.45}$$

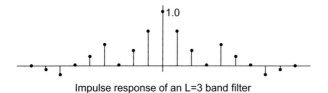

Filtering arrangement

Impulse response of an L=3 band filter

Figure 7.29 Impulse response of a filter with every third sample zero.

Consider now implementing the filter $H(z)$ using polyphase decomposition as shown in Section 7.4 and Figure 7.20, we can write the filter transfer function $H(z)$ as shown in Eq. (7.46).

$$H(z) = \sum_{l=1}^{L-1} z^{-l} E_l(z^L) = E_0(z^L) + z^{-1} E_1(z^L) + \cdots + z^{-(L-1)} E_{L-1}(z^L) \qquad (7.46)$$

In Eq. (7.46) if $E_k(z^L)$ is a constant, then we can rewrite Eq. (7.46) as shown in Eq. (7.47) where we have removed the kth term as it is a constant. The output from this filter also reflects the constant kth term as shown in Eq. (7.48); in both Eqs. (7.47) and (7.48) the constant is "c."

$$H(z) = cz^{-k} + \sum_{\substack{l=1 \\ l \neq k}}^{L-1} z^{-l} E_l(z^L) \qquad (7.47)$$

$$Y(z) = cz^{-k} X(z^L) + \sum_{\substack{l=1 \\ k \neq l}}^{L-1} E_l(z^L) X(z^L) z^{-l} \qquad (7.48)$$

Consider the effect of making the constant $c = 1$ and $k = 0$. Now, even though $H(z)$ is an image elimination filter that fills in all the in-between samples after $x[n]$ is upsampled by a factor of L, every Lth value from the image elimination filter is the same as the original input signal multiplied by a constant. Thus, the output sample for every Lth output is $y[nL + k] = c \cdot x[n]$. Therefore, all the nonzero input samples of $x[n]$ are first multiplied by the constant c and then passed through the filter without distortion and are present at the output. A filter with the above property is known as an Lth band filter or a Nyquist filter. The impulse response of such a filter for $L = 3$ is shown in Figure 7.29 and for $k = 0$ this filter satisfies the relation given in Eq. (7.49) for every Lth filter coefficient. Usually in practice, $c = 1$ so the exact value of the input is preserved in the output.

$$h[nL] = \begin{cases} c & n = 0 \\ 0 & \text{otherwise} \end{cases} \qquad (7.49)$$

7.7.1
Design of Linear Phase *L*-Band Filters

These filters are generally designed as FIR linear phase filters. Thus, the design of an image elimination linear phase filter will proceed in the same manner as the design of FIR low-pass filters except this time the pass band will be determined for the upsampled signal that is π/L after the signal is upsampled by L, so the design of the filter can be accomplished as shown in Eq. (7.50). In Eq. (7.50), we have used the window function that can represent any window we wish to use.

$$h[n] = \underbrace{\frac{\sin(n\pi/L)}{n\pi}}_{\text{Low-pass filter}} \cdot \underbrace{w[n]}_{\text{Window}} \tag{7.50}$$

In Eq. (7.50), the pass band frequency $\omega_p = \pi/L$. From the first term in Eq. (7.50), we see that the impulse response of the filter will have zero values every Lth sample except for the sample when $n = 0$. This is exactly what we wanted. Other expressions for the low-pass filter given in Eq. (7.50) are also possible. The MATLAB program below shows us the upsampling process and then the design of the filter to eliminate the image spectra.

MATLAB script

```
%% First up sample the signal by order L
L = 3;                        %Up sampling parameter
N = 25;                       %Number of samples in the signal
f1 = 3.25;                    %Frequencies that make up the signal.
f2 = 7.25;
n = 0:N;
x = 2* sin (2* pi* n* f1) +2* sin (2* pi* n* f2);  %Build the signal.
xu (1:L:L* length (x) ) = x;   %Up sample the signal
subplot (311) ; stem (xl)      %Plot the up sampled signal.
%% Design a seven point; L band interpolation filter
M = 15;                       %Build a 15 point filter
for n = 1:M
h (n) = sin (n* pi/L) / (n* pi) ;
end
subplot (312) ; stem (1:M, h) ;  %Plot the filter
%% Filter the up-sampled signal with the built filter
xint=filter (h, 1, xu) ;        %Filter the signal to interpolate.
subplot (313) ; stem (xint)
```

7.8
Chapter Summary

There is an enormous need for means of altering the sampling rate of a signal so that the same signal can be heard, or seen, or in general processed by different devices. There are basically two different sampling rate alteration devices, one is the upsampler used to increase the sampling rate of the signal and the other is the downsampler used to decrease the sampling rate of the signal.

The upsampler increases the sampling rate of the signal by inserting $L - 1$ zero-valued samples between two consecutive samples of the signal. Thus, each sample of the original signal is replaced by first itself and then it is followed by $L - 1$ zero-valued samples before the next sample. When we increase the sampling rate in this way, we introduce image spectra of the original signal. These spectra are in general unwanted and have to be eliminated, hence we usually follow the upsampler with an image elimination filter.

The downsampler, as the name implies, reduces the sampling rate of the signal by removing $M - 1$ consecutive samples from the original signal and keeping every Mth sample. When you decrease the sampling rate in this way, there is a good possibility that the resulting sampling rate would be less than the Nyquist rate for the signal and hence there would be aliasing present. Aliasing of the signal must be avoided when we are going to downsample the signal and hence the signal is first band limited to π/M radians and only then is the signal downsampled.

Using the upsampler and the downsampler by themselves (individually) will give us a change in the sampling rate by an integer amount. Very often, we need a fractional change in the sampling rate. In these cases, we resort to using both the upsampler and the downsampler in cascade. To use the two devices in cascade, we saw that the two integers L and M must be relative prime to each other. Generally, in such an arrangement, we place the upsampler first, then the filter, and finally the downsampler. The filter is placed between the up- and the downsampler serves the function of both the image elimination and the band limiting to avoid aliasing.

When a filter is placed after the upsampler or the filter is placed before the downsampler, we saw that the filter is operating at a higher sampling rate and on samples that are either zero-valued samples or are going to be thrown away. This leads to inefficiencies that can be eliminated by the use of two Noble identities. These identities permit us to exchange the location of the filter and the rate alteration device without altering the relation between the signal before and the signal after the exchange. The study of the Noble identities led us to examine the polyphase representation of the filters.

Using the polyphase representation for the filter, we saw that the filter can be broken down as either the analysis filter or as the synthesis filter. The two filters are closely related to each other and one can be derived from the other. With the saving in the computation effort that we were able to obtain by either moving the upsampler after the filter or moving the downsampler before the filter, we wanted to see if we could do this for both when we wanted to alter the sampling rate by a fraction. To do this, we had to make use of a special result from mathemetics that says that it is always possible to determine two integers such that $\mu M - \lambda L = 1$. With this result, we were able to interchange the location of the upsampler and the downsampler and arrange the filter as a polyphase filter with only time delays, thus achieving the most gain in reducing the computation effort required to alter the sampling rate by a fraction.

Finally to close the chapter, we examined a special type of filter that is used as an image elimination filter. This filter will estimate only the zero-valued samples and leave unchanged the original nonzero-valued samples. For the design of these filters, we saw that the standard window method of design works just fine.

8
Introduction to Stochastic Processes

8.1
Introduction

In signal processing, we often want to make inferences on data obtained by either experimental means or on signals that are corrupted by noise. These data or signals can be obtained from many varied backgrounds such as data received from an antenna corrupted by noise or a signal that is a radar echo used to estimate or predict some physical quantity. Irrespective of where the signals or the data are obtained, they fall into three broad groups of processes such as filtering, detection, or prediction. Irrespective of the group the data or the signal belongs to, making informed decisions based on the statistical properties of the signal is a very important branch of signal processing since most measured or received data is random in nature. This chapter tries to throw light on some fundamental principles of estimation theory and random processes.

In this chapter, we review the discrete time stochastic process. Among all the discrete time stochastic processes, we concentrate mostly on one subset of stochastic processes; the processes that we concentrate on have a special property that characterizes them as wide sense stationary (WSS) stochastic processes. To do this we first introduce the basic parameters of a random experiment and then quickly move on to a type of random process known as stochastic process. A stochastic process can be thought of as a two-dimensional experiment or as a group of experiments. Taken together, we can think of all the related experiments as an entire family of random experiments. With this point of view, we can say that a stochastic process is a process that is made of many different experiments running over time. Thus, the group of all the different experiments represents the entire ensemble of experiments and each of these experiments within the ensemble is developed over time.

When we are working with stochastic processes, we will find that often there are only a few experiments that can be observed. We do not have many experiments that we can call as a family of experiments. That is, an ensemble consists of only a few experiments, so the ensemble statistics are often hard or even impossible to come by, and we can measure and determine only the time statistics. In cases like these, if the process that we are examining is an ergodic process, then we can use the time statistics in place of the ensemble statistics. With this in mind, we present a review of stochastic processes that are discrete in nature and processes that are considered "stationary." This treatment is by no means complete, but it is

Digital Filters: Theory, Application and Design of Modern Filters, First Edition. Rajiv J. Kapadia.
© 2012 Wiley-VCH Verlag GmbH & Co. KGaA. Published 2012 by Wiley-VCH Verlag GmbH & Co. KGaA.

sufficient to understand and use the subject matter in the remaining chapters of this book. If you are more interested in this subject, please refer to one of the several excellent textbooks available on stochastic processes. One such book is "Introduction to Stochastic Processes" by Erhan Cinlar, published by Prentice Hall.

8.2
Types of Random Variables, Expected Value, and Moments

In the description of a random experiment and the nature of the random experiment, we find that the behavior can be represented by a continuous random variable, a discrete random variable, or mixed random variable. The type of random experiment relates to how the probability of an event is distributed. Each of these random variables can be either real valued or complex valued. In communication theory, the real part and the imaginary part of the variable are often referred to as the in-phase and the quadrature parts of the signal.

8.2.1
Continuous Random Variables

A random variable X is said to be a continuous random variable if the probability that the random variable takes on any specific value is zero. This can be expressed as shown in Eq. (8.1). This implies that a random variable is a continuous random variable if there is no discontinuity in its probability density function.

$$P[X = x] = 0 \tag{8.1}$$

The density function of a continuous random variable itself is a continuous function and it would look like the function shown in Figure 8.1a. The probability distribution function of a continuous random variable is also a continuous function of the random variable x. The probability distribution function is a monotonically increasing function. This function is also shown in Figure 8.1a.

8.2.2
Discrete Random Variables

A random variable X is said to be a discrete random variable if the random variable can take on only a finite number of different values and the probability that the random variable will take on any one particular value from the possible values is finite. This can be expressed as shown in Eq. (8.2). A discrete random variable will have discontinuities in its probability density function.

$$P[X = x_k] = p_k \tag{8.2}$$

The density function of a discrete random variable would look like the function shown in Figure 8.1b, notice how the density function is represented by impulses. The probability distribution function of a discrete random variable is a staircase type function of the random variable x. This is also shown in Figure 8.1b.

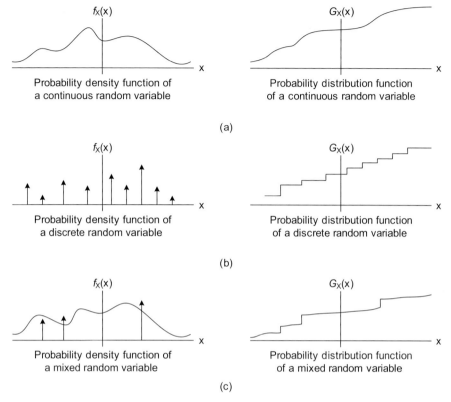

Figure 8.1   Different possible classes of random variables.

8.2.3
Mixed Random Variables

A random variable X is said to be a mixed random variable if the probability that the random variable takes on any specific value is zero except for a few countable finite values for which the probability is finite. The density function of a mixed random variable would look like the function shown in Figure 8.1c; notice how the density function is represented by a continuous function except for a few locations that are impulses. The probability distribution function of a discrete random variable is a continuous function except for a few locations where it has a discontinuity. Notice that the probability distribution function is a monotonically increasing function. The discontinuity in the probability distribution function occurs where there is a finite probability of a specific value of the function.

8.2.4
Expected Value Operator

In most of the experiments we come across, the probability density function is either not known or is very difficult to determine. In these cases also, we wish to characterize the experiments in some meaningful way. We do this by some of the more accessible statistics. One of these statistics is the mean or the expected value. The mean or the expected value is defined as shown in Eq. (8.3).

$$\mu_x = \mathcal{E}\{X\} = \int_{-\infty}^{\infty} xf(x)dx$$

$$\mu_x = \mathcal{E}\{X\} = \sum_x xf(x)dx$$

(8.3)

Equation (8.3) shows the expected value for both the continuous and the discrete random variable. The expected value can also be interpreted as a center of gravity of the distribution or as the most probable value you would expect to get in the next run of the experiment, or it represents the value that the random variable assumes on the average. Some obvious but useful facts about the expectation are as follows:

1) If a is a constant, then $\mathcal{E}[a] = a$.
2) If $v(x)$ is a function, then $\mathcal{E}[av(x)] = a\mathcal{E}[v(x)]$.
3) If $v1(x)$ and $v2(x)$ are functions, then $\mathcal{E}[v_1(x) + v_2(x)] = \mathcal{E}[v_1(x)] + \mathcal{E}[v_2(x)]$.

These properties of the expectation operator make this operator a linear operator.

8.2.5
Variance

In the definition of the expectation of Eq. (8.3), if we define the variable X as $(x - \mu_x)^2$ and then determine the expectation, we get what we call as the variance of the random experiment as shown in Eq. (8.4).

$$\sigma^2 = \mathcal{E}\left[(x-\mu_x)^2\right]$$

(8.4)

The variance of the random variable has a special interpretation. Say, you want to estimate the next value of the random variable from the experiment. You know that any value that you choose to estimate it with will be in error from the actual value. So, what would be the best estimate that you can make for the random variable? As it turns out, if you use the mean value for the estimate, the error of estimation will be minimum; this can be shown by differentiating Eq. (8.4) with respect to the mean value. It is customary to denote the positive root of the variance as σ, which is known as the standard deviation of x. The standard deviation represents the dispersion of the sample points that represent the variable x about the mean value. The mean value is often referred to as the first moment and the variance is referred to as the second moment of the variable x. The variance is also defined as an expectation and this expectation can be simplified as shown in Eq. (8.5).

$$\mathcal{E}\left[(x-\mu_x)^2\right] = \mathcal{E}[x^2 - 2\mu_x x + \mu^2]$$

$$\mathcal{E}\left[(x-\mu_x)^2\right] = \mathcal{E}[x^2] - 2\mu_x \mathcal{E}[x] + \mu^2$$

(8.5)

$$\mathcal{E}\left[(x-\mu_x)^2\right] = \mathcal{E}[x^2] - 2\mu^2 + \mu^2 = \mathcal{E}[x^2] - \mu^2$$

8.3
Correlation and Covariance

Consider now a pair of experiments that are somehow related. The relation could represent the signal received by two antennas that are located in the same vicinity. The signals from the two antennas are similar, but they are not exactly the same. We are often interested in knowing how the signals from the two antennas relate to each other; this is done by looking at the joint distribution function of the two signals.

8.3.1
Correlation

In the previous section we examined the expected value and the variance of a single random variable. Often, we will have a system that has signals that are dependent on two random variables. In these cases, we will be interested in determining how the two variables are related to each other. The correlation is one of the measures of the joint relation between the two variables. The correlation is the expected value of the product of the two random variables; it is defined as shown in Eq. (8.6).

$$\gamma_{xy} = \mathcal{E}(XY) = \int_{-\infty}^{\infty} \int_{-\infty}^{\infty} xy f_{XY}(xy) dx \, dy \tag{8.6}$$

The correlation provides a measure of the association between the two variables. An estimate of the correlation can be determined without the knowledge of the joint density function as shown in Eq. (8.7).

$$\hat{\gamma}_{xy} = \frac{1}{N} \sum_{n=0}^{N-1} x_n y_n \tag{8.7}$$

Equation (8.7) is an estimate of the correlation of the joint random variables. It can be shown that the estimate of Eq. (8.7) approaches the actual value of Eq. (8.6) as the number of samples N tends to infinite. Equation (8.7) suggests that there are three major types of correlations as shown by the scatter diagram in Figure 8.2. If the correlation is greater than zero, both the variables are said to have positive correlation and both the variables change in the same manner. When one variable increases, the other will also increase. If the correlation is less than zero, then both the variables are said to have negative

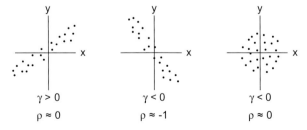

Figure 8.2 Scatter diagrams representing different regions of the correlation function.

correlation; in this experiment, both the variables change in the opposite manner. When one variable increases, the other will decrease. If the correlation is very close to zero, then both the variables are said to be uncorrelated and both the variables change in some unpredictable manner. When one variable increases, the other may increase, decrease, or stay the same.

The value of the correlation will depend on the magnitude of the values of the two sequences. This makes it very difficult to compare the correlation between two pairs of sequences. So, if we are going to use the correlation as a measure to make a meaningful inference concerning the two sequences, then we have to remove the dependence on the magnitudes of the two functions. We do this by normalizing the correlation.

To normalize the correlation coefficients, consider two sequences $x[n]$ and $y[n]$. We wish to compare the two sequences, and to do this we will define a cost function as shown in Eq. (8.8).

$$J = \sum_{n=0}^{N-1} \left[y[n] - a\,x[n] \right]^2 \tag{8.8}$$

Next, we minimize the cost function J by computing its derivative and setting it equal to zero. By doing this we get Eq. (8.9).

$$a = \frac{\sum_{n=0}^{N-1} x[n]y[n]}{\sum_{n=0}^{N-1} x^2[n]} \tag{8.9}$$

Using this value of the parameter a, the minimum value of the cost function is determined in Eq. (8.10).

$$J_{\min} = \left[1 - \varrho^2 \right] \sum_{n=0}^{N-1} y^2[n] \tag{8.10}$$

The coefficient ϱ is the correlation coefficient, it measures how closely the two sequences $y[n]$ and $x[n]$ track each other. The parameter ϱ is defined as shown in Eq. (8.11).

$$\varrho_{xy} = \frac{\sum_{n=0}^{N-1} x[n]y[n]}{\sqrt{\sum_{n=0}^{N-i} x^2[n]}\sqrt{\sum_{n=0}^{N-i} y^2[n]}} = \frac{\gamma_{xy}}{\sigma_x \sigma_y} \tag{8.11}$$

When the correlation coefficient is defined as shown in Eq. (8.11), the correlation coefficient is range bound between -1 and $+1$, is normalized, and is independent of the energy present in either of the two signals. When the two sequences are positively related, the correlation coefficient approaches 1 as shown in the first scatter diagram in Figure 8.2; when they are negatively related, the correlation coefficient approaches -1 as shown in the second scatter diagram in Figure 8.2; and when they are not related to each other, the coefficient approaches zero as shown in the third scatter diagram in Figure 8.2.

8.3.2
Covariance

The covariance of the two experiments can be defined as the covariance of the two variables as shown in Eq. (8.12). In Eq. (8.12), μ_x and μ_y represent the mean values of the random variables X and Y.

$$\varsigma_{xy} = \mathcal{E}\left\{ [Y-\mu_x]\left[Y-\mu_y\right]\right\}$$

$$\varsigma_{xy} = \int\limits_{-\infty}^{\infty}\int\limits_{-\infty}^{\infty} (X-\mu_x)\left(Y-\mu_y\right)f_{XY}(x,y)dx\,dy \tag{8.12}$$

Since we know that the expectation operator is a linear operator, we can use the linearity to rewrite the covariance as shown in Eq. (8.13). Equation (8.13) shows that the covariance is equal to the correlation minus the product of the two mean values. Notice that when the two means are zero, the covariance is the same as the correlation.

$$\varsigma_{xy} = \mathcal{E}\left\{ [X-\mu_x]\left[Y-\mu_y\right]\right\} = \mathcal{E}\left[XY-X\mu_y-Y\mu_x+\mu_x\mu_y\right]$$

$$\varsigma_{xy} = \mathcal{E}(XY)-\mu_x\mu_y = \gamma_{xy}-\mu_x\mu_y \tag{8.13}$$

8.3.3
Correlation Matrix

Consider the example of many different experiments (instead of two) that have the same basic phenomenon as the base. Now, we can define the process X that is made of many experiments as shown in Eq. (8.14).

$$X = [X_1, X_2, \ldots, X_n]' \tag{8.14}$$

Thus, we can say that the phenomenon X is represented by many experiments X_1–X_n. Each of these random experiments will have their own mean values, these mean values of the system can be written as a vector of mean values or $\mu_x = \mathcal{E}\{X\} = [\mu_1, \mu_2, \ldots, \mu_n]'$. Just as we defined the vector of mean values, we can define a matrix of correlations. This matrix of correlations will represent the correlation of any random experiment in Eq. (8.14) with every other random experiment. This correlation matrix is defined as shown in Eq. (8.15).

$$\Gamma_{xx} = \mathcal{E}\{XX'\} \tag{8.15}$$

The matrix inside the expectation operator in Eq. (8.15) computes the covariance of the mth random variable with the kth random variable. Using the definition of correlation given in Eq. (8.6), we can rewrite the correlation matrix as shown in Eq. (8.16)

$$\Gamma_{xx} = \begin{bmatrix} \gamma_{xx}(X_1X_1) & \gamma_{xx}(X_1X_2) & \cdots\cdots & \gamma_{xx}(X_1X_n) \\ \gamma_{xx}(X_2X_1) & \gamma_{xx}(X_2X_2) & \cdots\cdots & \gamma_{xx}(X_2X_n) \\ \vdots & \vdots & \ddots & \vdots \\ \gamma_{xx}(X_nX_1) & \gamma_{xx}(X_nX_2) & \cdots\cdots & \gamma_{xx}(X_nX_n) \end{bmatrix} \tag{8.16}$$

From the expression in Eq. (8.16), we see that the elements of the correlation matrix satisfy the relation $\gamma_{xx}(X_mX_k) = \gamma_{xx}(X_kX_m)$.

8.3.4
Complex Valued Random Experiments

In our study of signals and systems, we will come across signals that are complex so we should extend our random signals to be complex also. A complex valued random variable is made of both its real component and its imaginary component as shown in Eq. (8.17), where $\alpha_1[n]$ is the real part of the random $\mathbf{x}_1[n]$ and $\beta_1[n]$ is the imaginary part of the random variable $\mathbf{x}_1[n]$. In communication theory, we often refer to the real and the imaginary parts of a signal as the in-phase and the quadrature components of the signal.

$$\mathbf{x}_1[\mathbf{n}] = \alpha_1[n] + j\beta_1[n] \tag{8.17}$$

The mean value of the imaginary random sequence will be the mean value of the real sequence and the mean value of the imaginary sequence as shown in Eq. (8.18).

$$\mu_{x_1} - \mathcal{E}\{\mathbf{x}_1[\mathbf{n}]\} = \mathcal{E}\{\alpha_1[n] + j\beta_1[n]\}$$

$$\mu_{x_1} = \mathcal{E}\{\alpha_1[n]\} + j\mathcal{E}\{\beta_1[n]\} \tag{8.18}$$

$$\mu_{x_1} = \mu_{\alpha_1} + j\mu_{\beta_1}$$

8.4
The Notion of the Stochastic Process

Consider an experiment consisting of many similar experiments that are run over time. Now, we have a group of experiments or processes that are similar to each other and each of the experiments is run over time. This is shown in Figure 8.3. Taken together, the entire group of all the individual processes that are similar to each other are known as stochastic process. A stochastic process can be either continuous or discrete. In the remainder of this

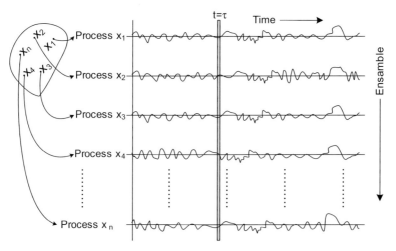

Figure 8.3 Different experiments of the same stochastic process.

book, we will examine signals that are functions of time on some observation interval and the signals are random; that is, before conducting the experiment, we do not know what the signal will be. Stochastic processes can be either continuous or discrete. In this book, we will concentrate only on discrete processes as we will be dealing with discrete signals. So all the processes that we will come across in this book are experiments that are "discrete and the samples occur at uniformly spaced instants in time." These samples can occur from speech signal that is sampled and stored or transmitted, from radar signal that is received by a radar gun, a digital TV stream broadcast by the TV station and received by your TV, or various other different and varied examples of signals.

How Flat Are My Pancakes? Suppose you own a chain of restaurants that serves breakfast all day and this chain of restaurants prides in serving pancakes that are the most uniform in thickness and flat pancakes. We all know that pancakes rise when cooked and their thickness varies from one pancake to another and the thickness also varies within the pancake from one coordinate on the pancake to another. In this example, the number of pancakes you make serves as different experiments that make up the ensemble, any one pancake serves as a single experiment. Here, the experiment does not run over time but it runs over space; this space is represented by the different coordinates where we will measure the thickness of the pancake.

So the time statistic will be obtained when you choose one pancake; this is like choosing one experiment. You measure the thickness of the pancake at many different predetermined coordinates, measuring the thickness at a whole lot of predetermined coordinates on the same pancake represents one experiment. Thus, you get information about one experiment over time (in our example time is replaced by space). You get one experiment that is represented as a process in Figure 8.3.

If instead of choosing one pancake, you had chosen one coordinate, say the center of the pancake, and determined the thickness at this coordinate across all the pancakes that are cooked. In doing this we have chosen to freeze the coordinate and are determining the statistics from all the experiments at the same coordinate. Thus, you get information about one instant in time (space in our example) but across the entire ensemble. This time we are determining the ensemble statistics. If we need more statistics, ensemble or time, then we can choose either another pancake to get another sample of the time statistics or another coordinate to get another sample of the ensemble statistics. The difference between a random experiment and a stochastic process then is that a stochastic process is a collection of related random experiments.

To visualize a stochastic process and the signals that make up a stochastic process, look at Figure 8.3. Figure 8.3 shows the entire space of all the possible processes. From the entire space of all the stochastic processes, the figure shows several different processes that are being executed. Each process develops over time; this represents one variable of the stochastic process. When we gather statistics, such as the mean, the variance, and so on, from any one process, we get the statistics over time. Now, suppose we take a snapshot of all the processes at some time say $t = \tau$, this is shown in Figure 8.3. This snapshot gives us the outcome of all the experiments at a specific instant in time but across all the experiments. From these observations also, we can determine the different statistics such as the mean and the variance. These statistics are known as the ensemble statistics. In general, the ensemble and the time statistic are not equal.

To characterize a process completely, we need to know the joint probability density functions (PDFs) of all the different processes. Most of the time, it will be either impossible or extremely difficult to determine the required joint PDF of all the processes; in these cases, we will rely on the first-order and the second-order statistics of the process. The first-order and the second-order statistics are the mean, the variance, and the correlation. So for a stochastic process, we can have the time mean and the ensemble mean, and in the same way, we can have the other statistics that are either time or ensemble.

The two different statistics – time and ensemble – can be different from each other or they may be equal to each other. Most, if not all, of the processes that we will encounter in this book are characterized by having the ensemble statistics equal to the time statistics. We first define what these statistics are and then we examine the implications when the time and the ensemble statistics are equal to each other.

8.4.1
The Mean Value

A discrete time process defined by a time series is represented by $x[n], x[n-1], \ldots, x[n-M]$, in which the individual samples can be complex. The mean value of this process is shown in Eq. (8.19).

$$\mu[n] = \mathcal{E}\{x[n]\} \tag{8.19}$$

In Eq. (8.19), the operator $\mathcal{E}\{\cdot\}$ is the statistical expectation operator. This mean value is the ensemble average of the process. We can also define the time mean value of the process as shown in Eq. (8.20).

$$\mu_t[N] = \frac{1}{N} \sum_{n=0}^{N-1} x[n] \tag{8.20}$$

In Eq. (8.20), N is the total number of samples used for this estimation.

8.4.2
The Autocorrelation

The autocorrelation function measures what is the relation of the present sample with another sample k sample units away. The parameter k is known as the lag. From this definition, we can write the autocorrelation function as shown in Eq. (8.21).

$$\gamma_x[n, n-k] = \mathcal{E}\{x[n]x^*[n-k]\} \quad k = 0, \pm 1, \ldots \tag{8.21}$$

Equation (8.21) represents the ensemble autocorrelation of the process. The definition of the time autocorrelation of the process is given in Eq. (8.22).

$$\gamma_{x_t}[k, N] = \frac{1}{N} \sum_{n=0}^{N-1} x(n)x^*(n-k) \quad 0 \le k \le N-1 \tag{8.22}$$

Note that when the autocorrelation is measured for zero lag ($k = 0$), the autocorrelation is the same as the expected value square of the sequence.

8.4.3
Properties of Autocorrelation

The autocorrelation function of the process plays a very important part in the study of a WSS process. The autocorrelation provides us a means to measure how rapidly the process is fluctuating. When a process fluctuates widely, the autocorrelation function will vanish to zero quickly. The autocorrelation has the following properties:

1) The autocorrelation function is an even function or $\gamma_x(l) = \gamma_x(-l)$. This property follows immediately from Eq. (8.22) when we substitute first $n - k = l$ and then $n = k + l$. This property implies that we can shift one sequence forward by l units or we can shift the second sequence backward by l units. In both the cases, we are correlating the same sample points.

2) The autocorrelation $\gamma_x(l)$ of the process is bounded by its mean square value of the process, so we can write $\gamma_x(l) \leq \gamma_x(0)$. Here, remember that $\gamma_x(0) = \mathcal{E}[x^2[n]]$. To prove this property, we proceed as shown in Eq. (8.23), where expanding and rearranging the terms and setting $l = 0$ will give you the desired result.

$$\mathcal{E}\left[(x[n] - x[n-l])^2\right] \geq 0 \tag{8.23}$$

When the input sequence is arranged backward, the autocorrelation sequence is also arranged backward.

Example 8.1

You are given an observation sequence $x[n] = [1\ 0.8\ 0.6\ 0.4\ 0.2]$. For this observation vector, determine the autocorrelation function for lag of $0 \leq k \leq 3$.

Solution 8.1: The autocorrelation vector for zero lag is given in Eq. (8.24).

$$\gamma(0) = \frac{1}{5}\left[x[n]x^T[n]\right] = \frac{1}{5}[1\ 0.8\ 0.6\ 0.4\ 0.2]\begin{bmatrix} 1 \\ 0.8 \\ 0.6 \\ 0.4 \\ 0.2 \end{bmatrix} = \frac{2.2}{5} = 0.44 \tag{8.24}$$

With similar calculations, we get $\gamma(1) = 0.32$, $\gamma(2) = 0.208$, $\gamma(3) = 0.112$, $\gamma(4) = 0.04$.

8.4.4
The Autocovariance

The autocovariance function, which is also known as the covariance function, measures the variance present from sample to sample. From this definition, we can write the covariance function as shown in Eq. (8.25).

$$\varsigma_x[n, n-k] = \mathcal{E}\{(x[n] - \mu[n])(x[n-k] - \mu[n-k])^*\} \quad k = 0, \pm 1, \ldots \tag{8.25}$$

Making use of Eqs. (8.19) and (8.21) in Eq. (8.25), we can rewrite the autocovariance as shown in Eq. (8.26).

$$\varsigma[n, n-k] = \gamma[n, n-k] - \mu[n]\mu^*[n-k] \qquad (8.26)$$

It is interesting to note that when the process has zero mean, the covariance of the function is the same as the autocorrelation of the function.

8.4.5
Stationary Process

Above we have defined the various parameters so that they are measured at some value of lag. There is a very important subclass of stochastic processes that have very nice properties. These sets of stochastic processes are known as stationary processes. A process is considered to be stationary when the joint PDF of the variables is invariant under shifts of the time origin. When a process is stationary, the parameters of the process such as the mean, the variance, the autocovariance, and so on are independent of the lag. Examine Figure 8.4 to get a picture of a stationary process. In Figure 8.4a, we represent the samples of various processes when the processes are operated on some timescale. All the samples from all the processes at the corresponding time will fall in the specified window at that instant in time. Consider now the same set of windows at a delayed time, all the windows are delayed by time t_0. This is shown in Figure 8.4b. Now, if all the processes are delayed and the samples from all the processes at the corresponding time still pass through the delayed windows, the process is considered to be a stationary process.

Most of the time, it is not necessary to have a process stationary in the strict sense. There are many processes that are stationary in the wide sense. A wide sense stationary process or a process that is stationary in the second order will be a process that has the mean, the variance, and the autocorrelation that are independent of the lag; so, for a WSS process the relations in Eq. (8.27) will be satisfied.

$$\mu[n] = \mu \qquad \text{for all } n$$
$$\gamma(n, n-k) = \gamma(k) \qquad (8.27)$$
$$\varsigma(n, n-k) = \varsigma(k)$$

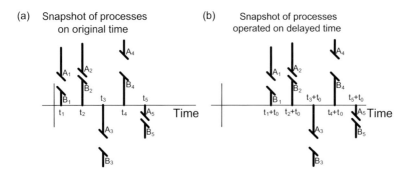

(a) Snapshot of processes on original time (b) Snapshot of processes operated on delayed time

Figure 8.4 Demonstrating the concept of a stationary process.

8.4.6
Ergodic Process

In stochastic processes, we have seen that the statistics of the process can be measured either along time or across the ensemble. In a stochastic process, the time statistic may or may not be equal to the ensemble statistic. A stochastic process is said to be ergodic when the error between the time statistics and the ensemble statistics tends to zero in the mean square sense, as the order of the process N tends to infinity. This is very useful as time statistics are relatively easy to measure and obtain, while ensemble statistics will be relatively difficult to obtain. For an ergodic process, the time mean is equal to the ensemble mean, so the mean value in Eq. (8.19) is equal to the mean value in Eq. (8.20). Similarly, the time autocorrelation is equal to the ensemble autocorrelation and the time variance is equal to the ensemble variance.

8.4.7
Power Spectrum Density

In all our analysis of signals and systems, we have related the time domain to the frequency domain. We would like to do the same with the stochastic signal. The autocorrelation that we saw in Section 8.1.2 gives us the second-order statistic of a stochastic process. The autocorrelation for a WSS process is dependent only on the lag. The *power spectrum density (PSD)* of the process represents the second-order statistic in the frequency domain. We define the PSD as the discrete time Fourier transform of the autocorrelation sequence as shown in Eq. (8.28).

$$S_x(\omega) = \sum_{n=-\infty}^{\infty} \gamma_x(n) e^{-j\omega n} \tag{8.28}$$

Similarly given the PSD, we can determine the autocorrelation sequence by determining the inverse Fourier transform of the PSD; this is shown in Eq. (8.29).

$$\gamma_x(n) = \frac{1}{2\pi} \int_{-\pi}^{\pi} S_x(\omega) e^{j\omega n} d\omega \tag{8.29}$$

The two relations given in Eqs. (8.28) and (8.29) together are known as the *Einstein–Weiner–Khintchine* relations.

8.4.8
Properties of the Power Spectrum Density

The PSD of a stochastic process has some very interesting properties that will enable us to deduce many different things about the stochastic process.

1) **The PSD of a real-valued WSS discrete time process is a real-valued even function of** ω:
   To prove this we split $e^{-j\omega n}$ into its even and its odd parts. Now since the autocorrelation $\gamma_x(n)$ is an even function, the odd part, which is the imaginary part must be zero. Hence, the PSD is a real-valued even function.

2) **The PSD is a periodic function in ω with a period of 2π**: We have defined the PSD as the DTFT of the autocorrelation function in Eq. (8.28). Since the DTFT is periodic in ω with the period being 2π, the PSD is a periodic function with period 2π. $S(\omega + 2k\pi) = S(\omega)$ for any integer k.

3) **The mean square value of the process is proportional to the total area under the curve of the PSD**: To prove this property, we set $n = 0$ in Eq. (8.29) to obtain the mean square value of the process $\gamma_x(0)$.

4) **The PSD of a WSS discrete time stochastic process is nonnegative**: This property is proved in Eq. (8.51).

8.5
The Correlation Matrix

The autocorrelation function as shown in Eq. (8.21) is measured with a lag between the two sequences. Computing the autocorrelation function for all the possible lags gives us an autocorrelation matrix. To get the autocorrelation matrix, we begin with the time sequence $x[n], x[n-1], \ldots, x[n-M+1]$, which can be written in vector notation as shown in Eq. (8.30).

$$\hat{x}[n] = [x(n), x(n-1), \ldots, x(n-M+1)]^{\mathsf{T}} \tag{8.30}$$

Next, we measure the autocorrelation for various values of lag beginning with $(-M+1)$ all the way to $M-1$. So, the correlation matrix of a stationary discrete time stochastic process is the expectation of the vector product for various values of lag and can be written as shown in Eq. (8.31).

$$\Gamma = \mathcal{E}\left[\hat{x}[n]\hat{x}^{\mathsf{H}}[n]\right] \tag{8.31}$$

In Eq. (8.31), the superscript H represents the Hermitian transpose of the vector. The Hermitian transpose is the combination of the transpose and the conjugation of the vector. Using Eq. (8.30) in Eq. (8.31), we get, for a wide sense stationary process, the correlation matrix as shown in Eq. (8.32).

$$\Gamma = \begin{bmatrix} \gamma(0) & \gamma(1) & \cdots & \gamma(M-1) \\ \gamma(-1) & \gamma(0) & \cdots & \gamma(M-2) \\ \vdots & \vdots & \ddots & \vdots \\ \gamma(-M+1) & \gamma(-M+2) & \cdots & \gamma(0) \end{bmatrix} \tag{8.32}$$

In the correlation matrix, the elements on the main diagonal are all $\gamma(0)$, this is always a real value, while the other elements may be complex if the elements of the original vector $\hat{x}[n]$ are complex. The correlation matrix has many properties that will be of interest to us and they are listed below.

Property 8.1
The correlation matrix of a stationary discrete time process is Hermitian. A complex valued matrix is a Hermitian matrix when the conjugate transpose of the matrix is equal to the matrix itself. This property is obvious from Eq. (8.31). This property is a consequence of the autocorrelation sequence that satisfies the relation $\gamma(-k) = \gamma^*(k)$. This also means that we

need to know the autocorrelation value for only M values of lag in order to completely define the correlation matrix. When the sequence $x[n]$ is a real-valued sequence, the autocorrelation is a real-valued function and the correlation matrix is a symmetric matrix. When $x[n]$ is complex, the correlation matrix can also be written as shown in Eq. (8.33).

$$\Gamma = \begin{bmatrix} \gamma(0) & \gamma(1) & \cdots & \gamma(M-1) \\ \gamma^*(1) & \gamma(0) & \cdots & \gamma(M-2) \\ \vdots & \vdots & \ddots & \vdots \\ \gamma^*(M-1) & \gamma^*(M-2) & \cdots & \gamma(0) \end{bmatrix} \tag{8.33}$$

Property 8.2

The correlation matrix of a stationary discrete time process is Toeplitz. A matrix is Toeplitz when all the elements along its main diagonal are equal and all the elements along any other diagonal parallel to the main diagonal are also equal. We can clearly see from Eqs. (8.32) and (8.33) that the correlation matrix does indeed have all the elements along any diagonal equal, so the matrix is clearly Toeplitz.

Property 8.3

The correlation matrix of a stationary discrete time process is always nonnegative definite and almost always positive definite. To prove this property, we define an arbitrary nonzero complex valued vector α. Next, define a random variable that is the inner product of the nonzero vector and the observation vector $x[n]$ as shown in Eq. (8.34).

$$y[n] = \alpha^H x[n] \quad \text{and} \quad y^*[n] = x^H[n]\alpha \tag{8.34}$$

The mean square value of the random variable y is given by Eq. (8.35).

$$\mathcal{E}\left[|y|^2\right] = \mathcal{E}[yy^*] = \mathcal{E}[\alpha^H x[n]x^H[n]\alpha]$$
$$\mathcal{E}\left[|y|^2\right] = \alpha^H \mathcal{E}[x[n]x^H[n]]\alpha = \alpha^H \Gamma \alpha \tag{8.35}$$

Since the expected value on the left-hand side of Eq. (8.35) is always equal to or greater than zero, it implies that the correlation matrix Γ is also always equal to or greater than zero, hence Γ is nonnegative definite.

Property 8.4

When the elements of the stationary discrete time process are arranged backward, the effect is equivalent to the transpose of the correlation matrix. When we rearrange the observation vector $x[n]$ backward, we can write it as shown in Eq. (8.36).

$$x^{BT}[n] = [x[n-M+1]x[n-M+2] \cdots x[n-1], x[n]] \tag{8.36}$$

In Eq. (8.36), the superscript B represents the fact that the vector is written backward, then the correlation matrix can be written as shown in Eq. (8.37) as a direct consequence of writing the observation vector arranged backward.

$$\mathcal{E}\left[x^B[n]x^{BT}[n]\right] = \begin{bmatrix} \gamma(0) & \gamma^*(1) & \cdots & \gamma^*(M-1) \\ \gamma(1) & \gamma(0) & \cdots & \gamma^*(M-2) \\ \vdots & \vdots & \ddots & \vdots \\ \gamma(M-1) & \gamma(M-2) & \cdots & \gamma(0) \end{bmatrix} = \Gamma^T \tag{8.37}$$

Property 8.5

The matrices Γ_M and Γ_{M+1} of a stationary discrete time process pertaining to M and $M + 1$ observations of the process, respectively, are related by Eq. (8.38).

$$\Gamma_{M+1} = \begin{bmatrix} \Gamma_M & \hat{\gamma}^{B*} \\ \hat{\gamma}^{BT} & \gamma(0) \end{bmatrix} = \begin{bmatrix} \gamma(0) & \hat{\gamma}^{H} \\ \hat{\gamma} & \Gamma_M \end{bmatrix} \tag{8.38}$$

Examine the matrix Γ_{M+1} and the matrix Γ_M as shown in Eq. (8.39). It shows the first relation in Eq. (8.38).

$$\Gamma_{M+1} = \begin{bmatrix} \gamma(0) & \gamma(1) & \cdots & \gamma(M-1) & \gamma(M) \\ \gamma^*(1) & \gamma(0) & \cdots & \gamma(M-2) & \gamma(M-1) \\ \vdots & \vdots & \ddots & \vdots & \vdots \\ \gamma^*(M-1) & \gamma^*(M-2) & \cdots & \gamma(0) & \gamma(1) \\ \gamma^*(M) & \gamma^*(M-1) & \cdots & \gamma^*(1) & \gamma(0) \end{bmatrix} = \begin{bmatrix} \Gamma_M & \gamma^{B*} \\ \gamma^{BT} & \gamma(0) \end{bmatrix} \tag{8.39}$$

Similarly, when we rewrite the expanded form of the matrix Γ_{M+1} as shown in Eq. (8.40), we find that the second relation in Eq. (8.38) is satisfied.

$$\Gamma_{M+1} = \begin{bmatrix} \gamma(0) & \gamma(1) & \gamma(2) & \cdots & \gamma(M) \\ \gamma^*(1) & \gamma(0) & \gamma(1) & \cdots & \gamma(M-1) \\ \gamma^*(2) & \gamma^*(1) & \gamma(0) & \cdots & \gamma(M-2) \\ \vdots & \vdots & \vdots & \ddots & \vdots \\ \gamma^*(M) & \gamma^*(M-1) & \gamma^*(M-2) & \cdots & \gamma(0) \end{bmatrix} = \begin{bmatrix} \gamma(0) & \gamma^{H} \\ \gamma & \Gamma_M \end{bmatrix} \tag{8.40}$$

Example 8.2

For the following observation sequence $x[n] = [1\ 0.8\ 0.6\ 0.4\ 0.2]$, we determined the autocorrelation vector in Example 8.1. Determine here the correlation matrix Γ_3.

Solution 8.2: By using the autocorrelation vector from Solution 8.1, we can write the correlation matrix Γ_3 as shown in Eq. (8.41).

$$\Gamma_3 = \begin{bmatrix} 0.44 & 0.32 & 0.208 & 0.112 \\ 0.32 & 0.44 & 0.32 & 0.208 \\ 0.208 & 0.32 & 0.44 & 0.32 \\ 0.112 & 0.208 & 0.32 & 0.44 \end{bmatrix} \tag{8.41}$$

Example 8.3

Using the autocorrelation vector from Example 8.1 and the relation for Γ_{M+1} from Eq. (8.38), determine the correlation matrix Γ_4.

Solution 8.3: Using the equality for Γ_{M+1} from Eq. (8.39), we first need to determine γ^{B*} and γ^{BT} from the autocorrelation sequence given in Solution 8.1 as shown in Eq. (8.42).

$$\gamma^{B*} = [0.04\ 0.112\ 0.208\ 0.32]$$
$$\gamma^{BT} = [0.04\ 0.112\ 0.208\ 0.32]^T \tag{8.42}$$

Using these results along with the value for $\gamma(0)$ and the correlation matrix Γ_3 computed in Solution 8.2, we get Eq. (8.43).

$$\Gamma_{3+1} = \begin{bmatrix} \Gamma_3 & \gamma^{B*} \\ \gamma^{BT} & \gamma(0) \end{bmatrix} = \begin{bmatrix} 0.44 & 0.32 & 0.208 & 0.112 & 0.04 \\ 0.32 & 0.44 & 0.32 & 0.208 & 0.112 \\ 0.208 & 0.32 & 0.44 & 0.32 & 0.208 \\ 0.112 & 0.208 & 0.32 & 0.44 & 0.32 \\ 0.04 & 0.112 & 0.208 & 0.32 & 0.44 \end{bmatrix} \tag{8.43}$$

8.6
White Noise Process

Among all the stochastic processes that we will come across, there is one process that is very special, this is, the *white noise process*. We call this process "white" in analogy to light where white light has a spectrum made of equal amounts of all the different colors; in the same way, the white noise process is made of equal energy at all frequencies. Since the white noise process has equal energy at all frequency values, the PSD of the white noise process is constant at all frequencies. If we let σ_ν^2 be the variance of the process of zero mean, then by definition of such a process the PSD is given in Eq. (8.44).

$$\Gamma_\nu(\omega) = \sigma_\nu^2 \quad \text{for all } \omega \tag{8.44}$$

Now that we know the PSD of the white noise process, we can determine the autocorrelation of the white noise process; this will be the inverse discrete time Fourier transform of the PSD as shown in Eq. (8.45).

$$\gamma_\nu(n) = \frac{\sigma_\nu^2}{2\pi} \int_{-\pi}^{\pi} e^{j\omega n}\, d\omega = \begin{cases} \sigma_\nu^2 & n = 0 \\ 0 & n = \pm 1, \pm 2, \ldots \end{cases} \tag{8.45}$$

Eq. (8.45) shows us another very special property of the white noise process. The white noise process has nonzero autocorrelation only for a lag of zero, implying that all the samples of the process are uncorrelated with each other. The PSD and the autocorrelation of the process are shown in Figure 8.5.

8.7
Stochastic Process through a Linear Shift-Invariant Filter

Consider a linear shift-invariant filter with impulse response $h[n]$. Also, assume that the input to this filter $x[n]$ is a WSS stochastic process. The output from the filter will also be a stochastic process $y[n]$. We wish to relate the input process and the output process in the arrangement shown in Figure 8.6.

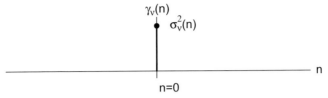

Autocorrelation Function of White Noise Process

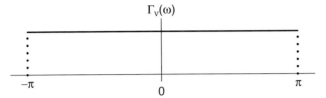

Power Spectrum Density of White Noise Process

Figure 8.5 Autocorrelation and the PSD of white noise process.

We begin with the understanding that the output $y[n]$ is the result of convolving the input sequence $x[n]$ with the impulse response $h[n]$ of the filter as shown in Eq. (8.46).

$$y[n] = \sum_{k=-\infty}^{\infty} h[k]x[n-k] \tag{8.46}$$

Similarly, the output from the filter after a delay of l sample times can be written as shown in Eq. (8.47).

$$y[n-l] = \sum_{i=-\infty}^{\infty} h[i]x[n-l-i] \tag{8.47}$$

Next, if we determine the autocorrelation of the output from the filter, we get the relation shown in Eq. (8.48) when we recognize that the filter coefficients $h[k]$ and $h[i]$ are constants.

$$\gamma_y(n, n-l) = \mathcal{E}\{y[n]y^*[n-l]\}$$

$$\gamma_y(n, n-l) = \mathcal{E}\left\{ \sum_{k=-\infty}^{\infty}\sum_{i=-\infty}^{\infty} h[k]h^*[i]x[n-k]x^*[n-l-i] \right\}$$

$$\gamma_y(n, n-l) = \sum_{k=-\infty}^{\infty}\sum_{i=-\infty}^{\infty} h[k]h^*[i]\mathcal{E}\{x[n-k]x^*[n-l-i]\} \tag{8.48}$$

When the input process is WSS, the autocorrelation can be used as

$$\gamma_y(n, n-l) = r_y(l) = \sum_{k=-\infty}^{\infty}\sum_{i=-\infty}^{\infty} h[k]h^*[i]\gamma_x(l+i-k)$$

Figure 8.6 Transmission of a stochastic process through a linear filter.

Equation (8.48) shows that the output process has properties that are similar to the properties of the input process. If the input process is stationary as assumed in Eq. (8.48), then the output of the filter will also be a stationary process as shown in the last equation. Now that we know the relation between the input and the output in the time domain, it is a simple process to determine the relation in the frequency domain. For this determine the DTFT of the last relation in Eq. (8.48). This is done in Eq. (8.49).

$$S_y(\omega) = \sum_{l=-\infty}^{\infty} \gamma_y[l] e^{-j\omega l}$$

$$S_y(\omega) = \sum_{l=-\infty}^{\infty} \sum_{k=-\infty}^{\infty} \sum_{i=-\infty}^{\infty} h[k] h^*[i] \gamma_x[l+i-k] e^{-j\omega l}$$

(8.49)

In Eq. (8.49) if we substitute $m = l + i - k$ and rearrange the various terms, we get Eq. (8.50).

$$S_y(\omega) = \sum_{k=-\infty}^{\infty} h[k] e^{-j\omega k} \sum_{i=-\infty}^{\infty} h^*[k] e^{j\omega i} \sum_{m=-\infty}^{\infty} \gamma_x[m] e^{-j\omega m}$$

$$S_y(\omega) = H(e^{j\omega}) H^*(e^{j\omega}) S_x(\omega)$$

$$S_y(\omega) = \left| H(e^{j\omega}) \right|^2 S_x(\omega)$$

(8.50)

Equation (8.50) relates the PSD of the output process to the PSD of the input process. The PSD of the output from a linear shift-invariant filter is the PSD of the input multiplied by the magnitude squared of the filter transfer function. This method also provides us a procedure of determining the PSD of an unknown process when it is a WSS stochastic process. The process that we will follow to determine the PSD of the unknown process is as follows: first design a very narrow band filter with uniform (ideally unity) pass band with an adjustable center frequency. The filter is tuned to the frequency at which you want to determine the PSD. Measure the mean square value of the process at the output. The mean square value of the output is proportional to the PSD of the input process. The constant of proportionality is determined in Eq. (8.51) where we determine the amplitude square value of the output from the filter. Figure 8.7 shows us how can do this. In Figure 8.7, we have a narrow band filter with unity pass band. When this filter is tuned to frequency ω_c, we will determine the power in the spectrum in the input process at frequency ω_c. Changing the tuned frequency to a different value will give us the power in the spectrum at this alternative frequency.

$$\mathcal{E}\{y^2[n]\} = \frac{1}{2\pi} \int_{-\pi}^{\pi} S_y(\omega) d\omega$$

$$\mathcal{E}\{y^2[n]\} = \frac{1}{\pi} \int_{-\omega_c-(\Delta\omega/2)}^{\omega_c+(\Delta\omega/2)} S_x(\omega_c) d\omega$$

(8.51)

$$\mathcal{E}\{y^2[n]\} = \frac{\Delta\omega}{\pi} S_x(\omega_c)$$

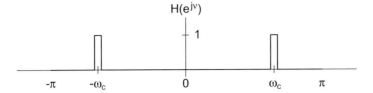

Figure 8.7 Amplitude response of a tunable narrow band filter used to determine the PSD of the input.

Notice that Eq. (8.51) is also an indirect proof that the PSD is always nonnegative. On the right-hand side, we have the PSD while on the left-hand side, we have a nonnegative expression, so the PSD is always nonnegative.

8.8
Stochastic Models

The representation of the stochastic process by a model dates back to about 1927 when Yule first proposed such a representation. The idea used was that the time series $x[n]$, which represents the process, can be obtained by applying a series of statistically independent shocks to a linear filter as shown in Figure 8.8. The series of shocks are applied from a white noise process with zero mean and σ^2_v variance, as shown in Eqs. (8.52) and (8.53).

$$\mathcal{E}\{v[n]\} = 0 \quad \text{for all } n \tag{8.52}$$

Equation (8.53) comes from the assumption of "white noise."

$$\mathcal{E}\{v[n]v^*[k]\} = \begin{cases} \sigma^2_v & n = k \\ 0 & \text{otherwise} \end{cases} \tag{8.53}$$

In general, the time domain representation of the output process can be thought of as the sum of the present input value or shock and a linear combination of the past values of the model as shown in Eq. (8.54).

$$x[n] + \sum_{k=1}^{p} a_k x[n-k] = \sum_{k=1}^{q} b_k w[n-k] \quad p > q \tag{8.54}$$

There are three different popular types of models depending on the three quantities:

1) Present value of the input.
2) Linear combination of all the past values of the output.
3) Linear combination of the present input and the past values of the output.

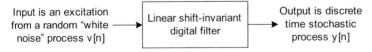

Figure 8.8 A stochastic model.

- **The autoregressive model (AR)**: No past values of the model input are used.
- **The moving average model (MA)**: No past values of the model output are used.
- **Combination of the AR and the MA models**: The model uses the linear combination of all the past inputs and the linear combination of all the past outputs from the model.

8.8.1
The Autoregressive Model

We say that the time series $x(n)$, $x(n-1)$, ..., $x(N-M)$ represents the realization of an autoregressive process of order M if the time series satisfies the differential equation.

$$x[n] + a_1 x[n-1] + \cdots + a_M x[n-M] = v[n] \tag{8.55}$$

In Eq. (8.55), the constants a_1, a_2, ..., a_M are all constants and are known as the autoregressive parameters and the term $v[n]$ is white noise process. Thus, the present output from the model $x[n]$ represents the linear combination of the present input and all the past $M-1$ outputs. This is shown in Figure 8.9b. We can rewrite Eq. (8.55) as shown in Eq. (8.56); this representation of the process is shown in Figure 8.9a.

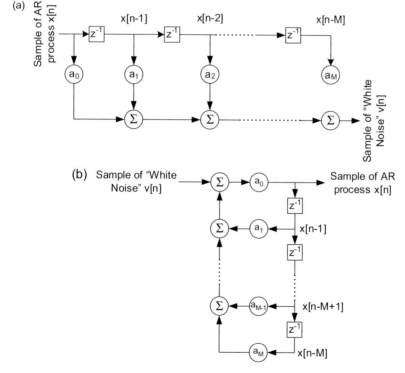

Figure 8.9 An autoregressive model analyzer. (a) Analyzer and (b) generator.

$$x[n] = -a_1 x[n-1] + \cdots -a_M x[n-M] + v[n]$$

$$x[n] = \sum_{k=1}^{M} -a_k x[n-k] + v[n] \qquad (8.56)$$

The left-hand side of Eq. (8.55) represents the convolution of the input sequence $x[n]$ with the filter coefficients a_n to give us the response $v[n]$ as shown in Eq. (8.57).

$$\sum_{k=0}^{M} a_k x[n-k] = v[n] \qquad (8.57)$$

In Eq. (8.57), the coefficient $a_0 = 1$. We can get an alternative interpretation for the autoregressive process by looking at the Z-transform of Eq. (8.57). The Z-transform of Eq. (8.57) can be written as $A(z)^* X(z) = V(z)$. The interpretations that we get are as follows:

1) If we rewrite the Z-transform as $A(z)^* X(z) = V(z)$, then the filter shown in Figure 8.9a can be used to produce white noise process. Such a filter is known as the process analyzer. The impulse response of the AR process analyzer is of a finite duration.
2) If we rewrite the Z-transform as $V(z)/A(z) = X(z)$, then we get the filter shown in Figure 8.9b. This filter produces the autoregressive sequence as the output with the white noise as the input. Such a filter is known as the process generator. The impulse response of the process generator has infinite duration.

8.8.2
The Moving Average Model

When the present and the past values of the white noise process are linearly combined, we get Eq. (8.58).

$$v[n] + b_1[n]v[n-1] + \cdots + b_K v[n-k] = x[n] \qquad (8.58)$$

The filter shown in Figure 8.10 is an all-zero filter. In this filter, the past $K-1$ and the present input from a white noise process are linearly combined to get the sequence that represents the moving average process. The filter in Figure 8.10 represents the process generator of a moving average process. The term moving average is used because each of the

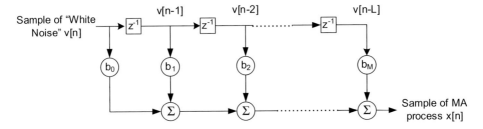

Figure 8.10 A moving average model generator.

past K inputs is multiplied by a weight and the combination is used to represent the output. Then, during the next sample instant, the inputs are moved forward, so we now have a different set of K input values that are used in computing the average. Since the set of K inputs used is moving forward, we get the name moving average. This filter is an all-zero filter and hence the impulse response of this filter is of finite length.

8.8.3
The Autoregressive Moving Average Process

When the process is a combination of the AR and the MA processes, we get the ARMA process. The filter with this transfer function contains both poles and zeros. So, the defining equation for an ARMA filter can be written as shown in Eq. (8.59).

$$x[n] + a_1 x[n-1] + \cdots + a_M x[n-M] = v[n] + b_1 v[n-1] + \cdots + b_K v[n-K] \quad (8.59)$$

Using Eq. (8.59), we can determine the block diagram for an ARMA process as shown in Figure 8.11. From Figure 8.11 and Eq. (8.59) we see that the ARMA model consists of both poles and zeros with the order of the model being M, K where $M > K$.

We now have three different models to represent a process, and among these three models the AR model is used most often as it leads to linear equations when it comes to computing the AR coefficients. The other two models require the solution of nonlinear equations. The wide use of the AR model is further justified by the discussion in the next section.

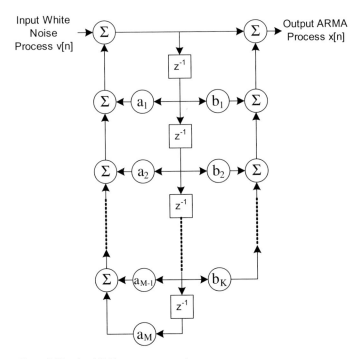

Figure 8.11 An ARMA process generator.

A stationary discrete time stochastic process can be thought of as a combination of two processes, one of the processes is a predictable process $p[n]$, while the other is a general linear process $x[n]$. The two processes are assumed to be uncorrelated with each other, so a stationary discrete time stochastic process can be expressed as shown in Eq. (8.60).

$$u[n] = x[n] + p[n] \tag{8.60}$$

The process $p[n]$ is the predictable process, so we know everything about this process and there is no error in knowing this process. The process $x[n]$ is a general linear process. The two processes $x[n]$ and $p[n]$ are uncorrelated with each other. The general linear process $x[n]$ can be represented as shown in Eq. (8.61).

$$x[n] = \sum_{k-0}^{K} b_k v[n-k] \tag{8.61}$$

The process $v[n]$ in Eq. (8.61) is a white noise process and it is uncorrelated with the predictable process $p[n]$, so the expectation of the product $\mathcal{E}[v[n]p^*[k]] = 0$ for all values of k. Furthermore, the predictable process $p[n]$ is a process that is completely known, so the next value of the process can be exactly predicted from the past values with zero error. This result was discovered by Wold and it is known as Wold decomposition.

The general linear process $x[n]$ can then be generated by using $v[n]$ the white noise process as input to an all-zero filter as shown in Figure 8.10. This filter has its zeros defined by the roots of Eq. (8.62).

$$B(z) = \sum_{k=0}^{K} b_k z^{-k} = 0 \tag{8.62}$$

The all-zero filter of Eq. (8.62) has many possible different filters. Of all the possible filters, an all-zero filter that has all its zeros inside the unit circle is of particular interest to us. Such a filter is a minimum phase filter. The minimum phase filter can be replaced by an all-pole filter that has the same impulse response as the all-zero filter. This ability to replace the all-zero filter with an all-pole filter with the same impulse response implies that the process may be represented as an AR process of an appropriate order.

The basic difference between an MA model and an AR model is shown in Figures 8.9 and 8.10. In Figure 8.9, the AR process analyzer uses the process $x[n]$ as the input to give us $v[n]$ the white noise as the output. In Figure 8.10, the MA process analyzer uses white noise as the input to give us the process as the output.

8.8.4
Correlation Function of a Stationary AR Process

When we have a stationary process, we can determine a very important relation for the autocorrelation function of the corresponding AR process. We begin by first multiplying the difference equation of the AR process given in Eq. (8.55) by $x^*[n-l]$ and then determine the expectation of both sides as shown in Eq. (8.63).

$$\mathcal{E}\left[\sum_{k=0}^{M} a_k x[n-k] x^*[n-l]\right] = \mathcal{E}[v[n] x^*[n-l]] \tag{8.63}$$

On the left-hand side, we can interchange the expectation and the summation operators. With the interchange of the two operators, we recognize that the expectation of the product x $[n-k]x^*[n-l]$ equals the autocorrelation function of the AR process for a lag of $(l-k)$. Next, when we look at the expectation on the right-hand side, we have the expectation of the product of a white noise process with the AR process. This product is uncorrelated and hence the expectation of the product is zero. Thus, with the stated values of the two expectations, we transform Eq. (8.63) into Eq. (8.64). In Eq. (8.64) the coefficient $a_0 = 1$.

$$\sum_{k=0}^{K} a_k \gamma[l-k] = 0 \quad l > 0 \tag{8.64}$$

From Eq. (8.64), we see that the autocorrelation function of an AR process satisfies the difference equation.

$$\gamma(l) = -a_1 \gamma(l-1) - a_2 \gamma(l-2) \cdots -a_M \gamma(l-M) \tag{8.65}$$

Note that the equation of the autocorrelation of the AR process is similar to the difference equation satisfied by the AR process itself. So by knowing the output sequence of a stationary process, we can determine the coefficients of the

8.8.5
Yule–Walker Equations

Examining the AR process generator in Figure 8.9(b), we see that we need to specify two sets of parameters to define the AR process. These two parameters are the variance of the white noise process σ_v^2 and the autoregressive coefficients $a_1, a_2, \ldots, a_M$. If we rewrite Eq. (8.65) for various values of the lag parameter "l," we get several simultaneous equations as shown in Eq. (8.66). If the process is a complex process, then we can replace $\gamma(-n)$ with $\gamma^*(n)$. The set of equations in Eq. (8.66) are known as the Yule–Walker equations.

$$\begin{bmatrix} \gamma(0) & \gamma(-1) & \cdots & \gamma(-M+1) \\ \gamma(1) & \gamma(0) & \cdots & \gamma(-M+2) \\ \vdots & \vdots & \ddots & \vdots \\ \gamma(M-1) & \gamma(M-2) & \cdots & \gamma(0) \end{bmatrix} \begin{bmatrix} -a_1 \\ -a_2 \\ \vdots \\ -a_M \end{bmatrix} = \begin{bmatrix} \gamma(1) \\ \gamma(2) \\ \vdots \\ \gamma(M) \end{bmatrix} \tag{8.66}$$

The Yule–Walker equations can be written in a compact matrix form as $\Gamma a = \gamma$. Then, by solving the Yule–Walker equations, we can determine the coefficients of the AR process as shown in Eq. (8.67).

$$a = \Gamma^{-1} \gamma \tag{8.67}$$

The solution to the Yule–Walker equations gives us the required coefficients. To determine the variance of the white noise process, examine Eq. (8.63) for $l = 0$. With $l = 0$, the right-hand side of Eq. (8.63) is transformed to Eq. (8.68).

$$\mathcal{E}[v[n]x^*[n-0]] = \mathcal{E}[v[n]x^*[n]] = \mathcal{E}[v[n]v^*[n]] = \sigma_v^2 \tag{8.68}$$

Using Eq. (8.68) in Eq. (8.63) with $l = 0$, we get Eq. (8.69) where $a_0 = 1$.

$$\sigma_v^2 = \sum_{k=0}^{M} a_k \gamma(k) \tag{8.69}$$

Hence, the AR process can be uniquely determined from Eq. (8.67) that gives us the process coefficients and Eq. (8.69) that gives us the variance of the input white noise process.

8.8.6
Relation between the Filter Parameters and the Autocorrelation Sequence

When we have a WSS process that is generated by one of the three processes we have just seen, there is a basic relation between the autocorrelation and the coefficients of the filter that generated the process. This relation can be obtained by multiplying the difference equation in Eq. (8.54) by $x^*[n-m]$ and then taking the expectation of both sides as shown in Eq. (8.70).

$$\mathcal{E}[x[n]x^*[n-m]] = -\mathcal{E}\left[\sum_{k=1}^{p} a_k x[n-k]x^*[n-m]\right]$$
$$+ \mathcal{E}\left[\sum_{k=1}^{q} b_k w[n-k]x^*[n-m]\right] \quad p > q \tag{8.70}$$

By using the autocorrelation sequence, we can rewrite Eq. (8.70) as shown in Eq. (8.71).

$$\gamma_{xx}[m] = -\left[\sum_{k=1}^{p} a_k \gamma_{xx}[m-k]\right] + \left[\sum_{k=1}^{q} b_k \gamma_{wx}[m-k]\right] \quad p > q \tag{8.71}$$

In Eq. (8.71), $\gamma_{wx}[m]$ is related to the impulse response of the filter. Using this fact, we can write the cross correlation as shown in Eq. (8.72) where we have made use of the fact that since the model is excited by white noise $w[n]$, its autocorrelation is zero everywhere except for a lag of zero. When the lag is zero, the autocorrelation is equal to the variance of the process.

$$\gamma_{wx}[m] = \mathcal{E}[x^*[n]w[n+m]] = \mathcal{E}\left[\sum_{k=0}^{\infty} h[k]w^*[n-k]w[n+m]\right] = \sigma_w^2 h[-m] \tag{8.72}$$

By substituting the result from Eq. (8.72) in Eq. (8.71), we get the relation that we are looking for as shown in Eq. (8.73). This relation applies in general to an ARMA process.

$$\gamma_{xx}[m] = \begin{cases} -\sum_{k=1}^{p} a_k \gamma_{xx}[m-k] & m > q \\ -\sum_{k=1}^{p} a_k \gamma_{xx}[m-k] + \sigma_w^2 \sum_{k=0}^{q-m} b_{k+m} h[k] & 0 \le m \le q \\ \gamma_{xx}^*[-m] & m < 0 \end{cases} \tag{8.73}$$

When we have an AR process only then can we modify Eq. (8.73) by setting $q=0$ and only the b_{k-m} coefficient for $q=0$, which is 1, will be present. Similarly, when we have an MA process, we can modify Eq. (8.73) by setting $p=0$ and all the coefficients a_k are zero.

Examine the following MATLAB script that first generates a WSS autoregressive process and then uses the AR process to compute the coefficients of the AR process.

MATLAB 8.1. Compute the coefficients of the underlying AR process.

```
%% First generate a WSS process.
N = 400;
x = randn(1,N);
H = [1 0.68 -0.45 -0.125 0.145 0.045] ;
y = filter(1,H,x);
%% Compute the 'biased' Autocorrelation for lags of 0 to 8.
lg = 8;
xs = zeros(lg,N + 1-lg);
for m = 1 : lg
   for n = 1 : (N + 1-m)
      xs(m,n) = y(n-1 + m);
   end
end
r1 = xs*y';
r1 = r1./N;
[rx lgs] = xcorr(y,y,15,'biased');
%% Compute the Autocorrelation Matrix, its inverse.
R1 = toeplitz(r1(1:6),r1(1:6)');
R = inv(R1);
%% Compute the parameters of the underling AR process.
a = sqrt(var(x))*R(2:6,1),
```

In the MATLAB script the vector x represents the white noise process. The vector y represents the AR process generated by the filter H. The vector rx represents the autocorrelation and the vector a represents all the coefficients of the AR model that are the same as the coefficients of the filter. The first filter coefficient, which is always 1, is not part of the vector a as we always know what it is.

8.9
Chapter Summary

There are three different types of random variables. In this book, we are concerned with only the discrete random variable. The random variable is characterized by its expected value, its variance, and the probability density function. The expected value operator is a linear operator.

The correlation and the covariance are two other second-order statistics that we examined for a pair of random variables. The correlation measured how the two variables behaved together. When the correlation is positive, the two random variables move together. When one variable increases, the other one will also increase. When the correlation is negative, the two random variables move opposite to each other. When one variable increases, the other one will decrease. The use of correlation depends on the magnitude of the two random variables. To remove the bias of the magnitude, we defined the correlation coefficient. The correlation coefficient lies in the range from -1 to $+1$.

When we have a family of random experiments, we have what we call a stochastic process. A stochastic process consists of an ensemble of experiments that run over time. So in a stochastic process, we can measure the statistics of the experiment either over time or across the ensemble. In a stochastic process when the time statistic is equal to the ensemble statistic, the process is known as an ergodic process.

Another important property of stochastic process that we examined is if the process was stationary or not. The property of a process being stationary is akin to the property of time invariance in a deterministic signal. With this idea and the concept that a stochastic process is a random process, we defined windows at each sampling instant. Then, if the process is a stationary process, then shifting the starting time of the process and at the same time shifting the windows will allow the process to go through the same windows. In all the work that we will be doing, we will assume that all the processes are wide sense stationary. When a process is stationary, the statistics that we compute for the process do not depend on the starting time of the computation.

The power spectrum density and the autocorrelation function of a process represent a DFT pair. This relation is known as the *Einstein–Weiner–Khintchine* relation. The PSD of a process is a real-valued even function of ω. The PSD is periodic with period 2π. The mean square value of the process is proportional to the total area under the curve of the PSD.

The correlation matrix of the autocorrelation function of a WSS process is rather special. It is a Hermitian matrix and it is a Toeplitz matrix. The matrix of the autocorrelation is a nonnegative and almost always a positive definite matrix. If the process is run backward, then the correlation matrix is the transpose of the correlation matrix. There is a definite relation between the correlation matrix of order n and order $(n + 1)$ as given in Eq. (8.38).

One special process that we examined is the white noise process. This process is known as white noise process since the PSD of this process is essentially constant across all frequencies and the autocorrelation of the process is zero for every value of lag except for a lag of zero, where it is equal to the variance of the process.

When a stochastic process is the input to a linear shift-invariant filter, the output process is also a stochastic process. The ratio of the PSD of the output process to the input process is equal to the magnitude squared of the filter frequency response. This relation is shown in

Eq. (8.50). This also gave us a means of determining the PSD of an unknown input process as shown in Eq. (8.51).

Next, we examined different modes that can be used to represent a process. The three models are the AR, the MA, and the ARMA models. Of the three models, we most frequently use the AR process as it leads to linear equations when we need to determine the model coefficients.

9
Weiner Filters

9.1
Introduction

A classic problem of estimating a signal is that of estimating the received signal when it has been corrupted by noise. This input signal that is observed gets corrupted by noise as it is transmitted through the channel from the transmitter to the receiver. As we try to estimate the input signal, there will be some error present in the estimate; we would like our estimate of the transmitted signal to be as close as possible to the original signal that was transmitted. We will do this by minimizing the error that represents the difference between the transmitted signal and the noisy signal that was received. To minimize this error, we will define a performance index that is formulated based on the statistics of both the signal and the noise that corrupts the signal. The most common performance index used is the mean square value of the error signal. We choose this error performance index for several reasons such as the following: the performance index has a unique minimum; it depends on only the first- and the second-order statistics of the signal.

9.2
The Principle of Orthogonality

The filtering operation that we will examine in this chapter is shown in Figure 9.1. In Figure 9.1, we have a received signal $x[n]$ that is corrupted by noise as it traveled through the channel; this received signal is passed through a transverse filter with impulse response $h^*[n]$ to get the estimate $y[n]$, which is the estimate of the desired signal $d[n]$. We use $h^*[n]$ just as a convenience as later it will make the other equations simpler to use. The output $y[n]$ is obtained by convolving the input signal with the impulse response of the filter as shown in Eq. (9.1):

$$y[n] = \sum_{k=0}^{\infty} x[n-k]h^*[k] \tag{9.1}$$

The output from the filter is the sequence $y[n]$; this output sequence is the estimate of the desired signal $d[n]$. Since we are interested in knowing how good our estimate of the desired

Digital Filters: Theory, Application and Design of Modern Filters, First Edition. Rajiv J. Kapadia.
© 2012 Wiley-VCH Verlag GmbH & Co. KGaA. Published 2012 by Wiley-VCH Verlag GmbH & Co. KGaA.

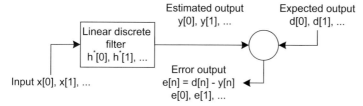

Figure 9.1 Optimum filtering problem.

output is, we next compare the estimate with the desired signal $y[n]$, which is the output from the filter, with the desired output signal $d[n]$. This comparison between the desired output and the estimate of the signal gives us the error of the estimate. Thus, error represents the difference between the desired signal and the estimate of the desired signal. Our goal is to minimize this error. This error is defined as the difference between the estimate and the expected output as shown in Eq. (9.2):

$$e[n] = d[n] - y[n] \tag{9.2}$$

Using Eq. (9.1) in Eq. (9.2), we can rewrite Eq. (9.2) as shown in Eq. (9.3):

$$e[n] = d[n] - \sum_{k=0}^{\infty} x[n-k]h^*[k] \tag{9.3}$$

It is the purpose of the filter to minimize the error defined in Eq. (9.2). In order to minimize the error, we must first define the criteria over which we will minimize the error. There are several possible criteria that we can possibly use:

- Minimum absolute value of error.
- Minimum mean square value of the error.
- Minimum average value of error.

From all the possible minimization criteria that are possible, the criterion most often used is the minimum mean square value. The reasons for choosing this criterion are that this criterion has a unique minimum, this minimum can be computed with little effort, and the other criteria do not lend them to such a definite result. In order to determine this minimum, we first need to define a cost function. The cost function that we define will represent the mean square error criterion. In order to keep the development as general as possible, we will assume that the input signal $x[n]$ and the filter coefficients are complex quantities and hence the error function will also be a complex quantity. With this understanding, we can choose the cost function as the first equation in Eq. (9.4). Replacing the error function $e[n]$ using Eq. (9.3), we get the second equation:

$$J_n = \mathcal{E}\{e[n]e^*[n]\}$$

$$J_n = \mathcal{E}\left\{\left(d[n] - \sum_{k=0}^{\infty} x[n-k]h^*[k]\right)\left(d^*[n] - \sum_{k=0}^{\infty} x^*[n-k]h[k]\right)\right\} \tag{9.4}$$

To get to the minimum value of the cost function, we will determine the gradient of the cost function and then set the gradient to zero. The gradient that we will compute is with respect to the filter coefficients as the filter coefficients are the only adjustable parameters. Since we have assumed that the filter has complex coefficients, then in general the gradient will have to be computed with respect to the real as well as the imaginary coefficient. Once we have the gradient, we will set this gradient equal to zero to give us the filter coefficients that will minimize the error function. Since the filter coefficients are complex, we can separate them into their real and imaginary parts and write $h_k[n] = a_k[n] + j\beta_k[n]$. The required gradient will be evaluated as shown in Eq. (9.5):

$$\nabla_k J = \nabla_k[e[n]e^*[n]] = \mathcal{E}\left[\left(\frac{\partial e[n]}{\partial a_k} + j\frac{\partial e[n]}{\partial \beta_k}\right)e^*[n] + \left(\frac{\partial e^*[n]}{\partial a_k} + j\frac{\partial e^*[n]}{\partial \beta_k}\right)e[n]\right] \quad (9.5)$$

Each of the required derivatives of Eq. (9.5) is computed in Eq. (9.6). To get to the results given in Eq. (9.6), we replace $h[k] = a[k] + j\beta[k]$ in Eq. (9.4) and then determine the required gradients. Your results should match the equations given in Eq. (9.6):

$$\nabla_k J = \mathcal{E}\left[\left(\frac{\partial e[n]}{\partial a_k} + j\frac{\partial e[n]}{\partial \beta_k}\right)e^*[n] + \left(\frac{\partial^* e[n]}{\partial a_k} + j\frac{\partial^* e[n]}{\partial \beta_k}\right)e[n]\right]$$

$$\frac{\partial e[n]}{\partial a_k} = -x[n-k], \qquad \frac{\partial e^*[n]}{\partial a_k} = -x^*[n-k] \quad (9.6)$$

$$\frac{\partial e[n]}{\partial \beta_k} = jx[n-k], \qquad \frac{\partial e^*[n]}{\partial \beta_k} = -jx^*[n-k]$$

Substituting the result of the various derivatives from Eq. (9.6) into Eq. (9.5), we get the gradient of the cost function as shown in Eq. (9.7):

$$\nabla_k J = \mathcal{E}((-x[n-k] + j(jx[n-k]))e^*[n] + e[n](-x^*[n-k] + j(-jx[n-k])))$$
$$\nabla_k J = -2\mathcal{E}(x[n-k]e^*[n]) \quad (9.7)$$

Setting the gradient in Eq. (9.7) equal to zero gives us the required condition to get the optimum filter; if we let $e_0[n]$ denote the minimum error when Eq. (9.7) is satisfied, then the filter with the coefficients $h^*[n]$ is operating in the optimum condition as defined by the minimum mean square criterion, which is equivalent to writing $\mathcal{E}[x[n-k]e^*[n]] = 0$; in other words, we can say that there is no correlation between the input vector and the error sequence, which we can state as follows. This is also the statement of the principle of orthogonality.

> The necessary and sufficient condition for the cost function to attain a minimum value is for the estimation error value $e[n]$ to be orthogonal to each of the input samples that enter into the estimation of the desired response at time n.

The principle of orthogonality has a very nice geometric interpretation; the projection of the error signal is zero in the direction of the input vector. The principle of orthogonality can be extended to the output sequence from the filter. The output from the filter is also orthogonal to the error as shown in Eq. (9.8); this can be viewed as a corollary to the principle of orthogonality.

$$
\begin{aligned}
\mathcal{E}[y[n]e^*[n]] &= \mathcal{E}\left[\sum_{k=0}^{\infty} x[n-k]h^*[k]e^*[n]\right] \\
&= \sum_{k=0}^{\infty} h^*[k]\ \underbrace{\mathcal{E}[x[n-k]e^*[k]]}_{\text{principle of orthogonality}}
\end{aligned}
\tag{9.8}
$$

On the right-hand side of Eq. (9.8) we have the expectation, which is the definition of the principle of orthogonality, which we have seen is zero when the filter is operating in its optimum condition. That is, due to the principle of orthogonality, the expectation $\mathcal{E}[x[n-k]e^*[n]]$ is zero. Thus, the left-hand side is also operating in the optimum condition and is also zero. This says that the output from the filter is also orthogonal to the error signal, which we can say is a corollary to the principle of orthogonality.

9.3
Weiner–Hopf Equations

Now that we know the criterion for optimum operation of the filter, we can apply this condition and compute the required filter coefficients to operate the filter in its optimum condition as shown in Eq. (9.9):

$$
\mathcal{E}[x[n-k]e^*[n]] = \mathcal{E}\left[x[n-k]\left(d[n]-\sum_{m=0}^{\infty} x[n-m]h^*[m]\right)^*\right] = 0, \quad m = 0, 1, 2, \ldots
\tag{9.9}
$$

In Eq. (9.9), $h^*[n]$ represents the filter that we want to design; we will determine the value of the filter coefficients by rewriting Eq. (9.9). To rewrite Eq. (9.9), we first expand the expectation and with a little rearrangement we get Eq. (9.10), after setting the left-hand side in Eq. (9.9) to zero.

$$
\begin{aligned}
\mathcal{E}[x[n-k]d^*[n]] &= \mathcal{E}\left[x[n-k]\left(\sum_{m=0}^{\infty} x^*[n-m]h[m]\right)\right], \quad k = 0, 1, 2, \ldots \\
\mathcal{E}[x[n-k]d^*[n]] &= \left(\sum_{m=0}^{\infty} h_0[m]\mathcal{E}[x^*[n-m]x[n-k]]\right), \quad k = 0, 1, 2, \ldots
\end{aligned}
\tag{9.10}
$$

The two expectations in Eq. (9.10) can be looked upon as follows: The expectation on the left-hand side, $\mathcal{E}[x[n-k]d^*[n]]$, represents the cross-correlation between the expected output and the input to the filter with a lag of k units; we will denote this expectation as $\gamma_{dx}(-k)$. The expectation on the right-hand side, $\mathcal{E}[x^*[n-m]x[n-k]]$, represents an autocorrelation of the input sequence with a lag of $(k - m)$ units; we will denote this expectation as $\gamma_{xx}(k - m)$. Replacing the expectations in Eq. (9.10) by these two new quantities, we get Eq. (9.11):

$$\sum_{m=0}^{\infty} h_0[m]\gamma_{xx}(k-m) = \gamma_{dx}(-k), \quad k = 0, 1, 2, \ldots \tag{9.11}$$

Equation (9.11) represents the most general form of solution of the optimum filter. It consists of the autocorrelation of the input and the cross-correlation between the input and the expected output; this equation is known as the *Weiner–Hopf equation*. What we now need is a solution of Eq. (9.11). Equation (9.11) can be viewed as a convolution of the autocorrelation function with the filter coefficients to get the cross-correlation between the input and the desired response. In order to determine the coefficients of the optimum filter, we need to solve the system of difference equations that are present in Eq. (9.11).

We have available two different methods to solve the difference equation given in Eq. (9.11). We can determine the solution in the time domain or we can transform the equation in the Z-domain and obtain the solution in the frequency domain. We first begin with the solution in the time domain and then obtain the solution in the frequency domain.

9.4
Solution of the Weiner–Hopf Equations in the Time Domain

We begin with the optimum filter designed as a finite impulse response filter. With the filter being a finite impulse response filter, the impulse response of the optimum filter $h_0[n]$ is finite, so the difference equations in Eq. (9.11) reduce to a finite set of equations instead of an infinite set of equations; hence, the Weiner–Hopf equations can be rewritten as shown in Eq. (9.12), where the coefficients $h_0[n]$ are the optimum filter coefficients.

$$\sum_{m=0}^{M-1} h_0[m]\gamma_{xx}(m-k) = \gamma_{dx}(-k), \quad k = 0, 1, 2, \ldots \tag{9.12}$$

We can rewrite Eq. (9.12) in matrix form once we recognize that the autocorrelation can be written as the autocorrelation matrix as shown in Eq. (8.34). Using the matrix notation, Eq. (9.12) can be written as shown in Eq. (9.13):

$$\Gamma h_0 = \gamma \tag{9.13}$$

With the knowledge that the autocorrelation matrix Γ is nonsingular, we can invert it and the solution of Eq. (9.13) can be written as shown in Eq. (9.14):

$$h_0 = \Gamma_{xx}^{-1}\gamma_{dx} \tag{9.14}$$

Equation (9.14) is a system of M linear equations that permit us to determine the M coefficients of the finite impulse response filter. In Eq. (9.14), using the inverse of the autocorrelation matrix of the input sequence and the cross-correlation vector between the input and the desired response, we can determine the filter coefficients that represent the optimum filter. If you will remember, in Eq. (9.1), we began with the filter written as a conjugate, so the coefficients that we obtain here will need to be conjugated before we use them. This is the convention that is used in the literature, so in this book we will continue to follow the convention of all the other authors and we will write the estimate of the input signal as shown in Eq. (9.15) and the solution to the Weiner filter as shown in Eq. (9.16).

$$y[n] = \sum_{k=0}^{\infty} x[n-k]h_0^*[k] \tag{9.15}$$

$$h_0 = \Gamma_{xx}^{-1}\gamma_{dx} \tag{9.16}$$

9.4.1
Error Performance Surface

The cost function that we defined in Eq. (9.4) can be rewritten as shown in Eq. (9.17). Using the expression for the error signal from Eq. (9.3), we get Eq. (9.18):

$$J_N = \mathcal{E}[e[n]e^*[n]] \tag{9.17}$$

$$J_N = \mathcal{E}\left[\left(d[n] - \sum_{k=0}^{M-1} h_0^*[k]x[n-k]\right)\left(d^*[n] - \sum_{k=0}^{M-1} h_0[k]x^*[n-k]\right)\right] \tag{9.18}$$

Expanding the expectation, we get Eq. (9.19):

$$J_N = \mathcal{E}\left[|d[n]|2\right] - \mathcal{E}\left[d[n]\sum_{k=0}^{M-1} h_0^*[k]x[n-k]\right] - \mathcal{E}\left[d^*[n]\sum_{k=0}^{M-1} h_0[k]x^*[n-k]\right]$$
$$+ \mathcal{E}\left[\sum_{m=0}^{M-1}\sum_{k=0}^{M-1} h_0[k]h_0^*[m]x^*[n-k]x[n-m]\right] \tag{9.19}$$

We next look at the four terms in the expression for the error performance index in Eq. (9.19). The first expectation is simply the variance of the desired response from filter. The second and third expectations are simply the cross-correlation between the input and the desired response and the fourth expectation is the autocorrelation matrix of the inputs for a lag of $(m - k)$. With this recognition, we can rewrite Eq. (9.19) as shown in Eq. (9.20):

$$J_N = \sigma_d^2 - \sum_{k=0}^{M-1} h_0^*[k]\gamma_{dx}[-k] - \sum_{k=0}^{M-1} h_0[k]\gamma_{dx}^*[-k]$$
$$+ \sum_{m=0}^{M-1}\sum_{k=0}^{M-1} h_0[k]h_0^*[m]\gamma_{xx}[m-k] \tag{9.20}$$

Equation (9.20) says that when the input sequence $x[n]$ and the desired response $d[n]$ are jointly stationary, then the cost function J is a quadratic function of the filter coefficients. Due to the quadratic nature of the cost function, we can visualize the error performance surface as a bowl-shaped figure in the M-dimensional space with M degrees of freedom.

(These M degrees of freedom are the M coefficients of the filter that we are going to adjust.) A bowl-shaped surface is characterized with a unique minimum. At the minimum that occurs at the bottom of the bowl, the cost function attains its minimum value. At this point, we know that the gradient of the vector will have to be zero. If we differentiate the minimum cost function from Eq. (9.20), we get the solution of the Weiner–Hopf equation as shown in Eq. (9.21). This solution is the same as we saw in Eq. (9.12), which is obtained when we set $\nabla J = 0$. This shows an alternative way to get to the solution of the Weiner–Hopf equations.

$$\nabla J = \frac{\partial J}{\partial \alpha_k} + j\frac{\partial J}{\partial \beta_k} = -2\gamma_{dx}(-k) + 2\sum_{m=0}^{M-1} h[m]\gamma_{xx}(m-k) \qquad (9.21)$$

Example 9.1

To demonstrate the optimum filter, consider a model with a filter of size 3. For demonstration purpose, we will limit this example to real-valued sequences. Assume that the input sequence $x[n]$ is $[1, 0.4, 0.1, -0.05]$; the cross-correlation between the input and the desired response is $\gamma = [0.5, 0.12, -0.1, -0.05]^T$, the variance of the noise that corrupts the signal is $\sigma_v^2 = 0.125$, and the variance of the desired signal is $\sigma_d^2 = 0.95$.

Solution 9.1: We first build the autocorrelation matrix from the input vector. The autocorrelation matrix for different amounts of lag is

$$\Gamma = \begin{bmatrix} 1.1 & 0.5 & 0.1 & -0.05 \\ 0.5 & 1.1 & 0.5 & 0.1 \\ 0.1 & 0.5 & 1.1 & 0.5 \\ -0.05 & 0.1 & 0.5 & 1.1 \end{bmatrix}$$

We have partitioned this matrix to show the effect if we use different size filters. Using the information given and using Eq. (9.16), we can determine the required filter. The various solutions are shown in Table 9.1. With the filter length of 1, we use a 1-by-1 autocorrelation matrix and cross-correlation vector of size 1. Similarly, with the filter length of 3, we use a 3-by-3 autocorrelation matrix and cross-correlation vector of size 3. Using various size autocorrelation matrix and the cross-correlation vector, we get the optimum filter of sizes 1,

Table 9.1 Weiner filters for various filter lengths.

Filter length	1	2	3	4
Optimum tap weights	[0.4736]	$\begin{bmatrix} 0.8360 \\ -0.7853 \end{bmatrix}$	$\begin{bmatrix} 0.8719 \\ -0.9127 \\ 0.2444 \end{bmatrix}$	$\begin{bmatrix} 0.8719 \\ -0.9129 \\ 0.2444 \\ 0 \end{bmatrix}$

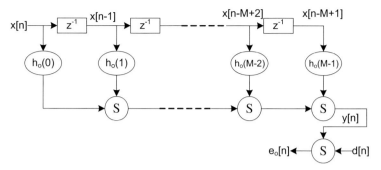

Figure 9.2 A finite impulse response optimum filter.

2, 3, and 4. In this example, since we are dealing with only real quantities, the filter is a real filter.

9.4.2
Minimum Mean Square Error

The linear transverse filter of Figure 9.2 computes the minimum error. To determine the value of this minimum error, we begin with Eq. (9.22):

$$e_0[n] = d[n] - y[n] \tag{9.22}$$

We will rewrite Eq. (9.22) as shown in Eq. (9.23), and then define the minimum error as shown in Eq. (9.24):

$$d[n] = y[n] + e_0[n] \tag{9.23}$$

$$J_{min} = \mathcal{E}\left[|e[n]|^2\right] = \mathcal{E}\left[|(d[n] - y[n])|^2\right] \tag{9.24}$$

Evaluating the mean square value of Eq. (9.23), we get Eq. (9.25):

$$\sigma_d^2 = \sigma_y^2 + J_{min} \tag{9.25}$$

While we are doing this, we should remember that the input process $x[n]$ is a WSS process with zero mean. This implies that both $d[n]$ and $y_0[n]$ are WSS processes with zero mean. Rewriting Eq. (9.25), we get Eq. (9.26):

$$J_{min} = \sigma_d^2 - \sigma_y^2 \tag{9.26}$$

Dividing both sides by the variance of the desired signal, we get the normalized minimum error as shown in Eq. (9.27). The normalized value of error is bounded and is $0 \le \varepsilon \le 1$.

$$\varepsilon = \frac{J_{min}}{\sigma_d^2} = 1 - \frac{\sigma_y^2}{\sigma_d^2} \tag{9.27}$$

Table 9.2 Minimum error for the filters designed in Example 9.1.

Filter length	1	2	3	4
Minimum mean square error	[0.6959]	[0.1576]	[0.1066]	[0.1066]

Next we determine the variance of the output from the transverse filter shown in Figure 9.2 with the output from the filter given in Eq. (9.15). This is done in Eq. (9.28):

$$\sigma_y^2 = \mathcal{E}\big[|y_0[n]|^2\big] = \mathcal{E}\big[(\mathbf{h}_0^H x[n])(x^H[n]\mathbf{h}_0)\big]$$

$$\mathbf{h}_0^H \mathcal{E}\big[x[n]x^H[n]\big]\mathbf{h}_0 = \mathbf{h}_0^H \mathcal{E}\big[x[n]x^H[n]\big]\mathbf{h}_0 \tag{9.28}$$

Substituting the result from Eq. (9.28) in Eq. (9.26), we get Eq. (9.29a):

$$J_{min} = \sigma_d^2 - \mathbf{h}_0^H \Gamma_{xx}\mathbf{h}_0 \tag{9.29a}$$

$$J_{min} = \sigma_d^2 - \gamma^H \mathbf{h}_0 \tag{9.29b}$$

$$J_{min} = \sigma_d^2 - \gamma^H \Gamma_{xx}^{-1}\gamma \tag{9.29c}$$

In Eq. (9.29a), we have first used the Hermitian transpose of the solution of the Weiner filter as shown here, $[\Gamma \mathbf{h}_0 = \gamma]^H = [\mathbf{h}_0^H \Gamma = \gamma^H]$, to get Eq. (9.29b). Next we have used the solution to the Weiner filter $\mathbf{h}_0 = \Gamma^{-1}\gamma$ in Eq. (9.29b) to get Eq. (9.29c). This gives us the desired result as all the quantities on the right-hand side of Eq. (9.29a) are known and they are independent of the filter coefficients. To demonstrate this, let us determine the minimum error for the system in Example .

Example 9.2

Determine the minimum error for the Weiner filters designed in Example .

Solution 9.2: The filters designed in Example are of order 1, 2, 3, and 4. Using the results that we obtained in Example , we can now determine the minimum error for each of the designed filters. For each of the filters, we use Eq. (9.29a) to get the following results given in Table 9.2.

Table 9.2 shows that the error reaches a stable or a minimum value rather quickly. The error for the filter of order 1 is rather large but when we have a filter of order 3 we have reached the minimum value that is an almost stable value.

9.5
Solution of the Weiner–Hopf Equations in the Frequency Domain

Previously when we have come across equations such as the one shown in Eq. (9.11), we have relied on the Z-transform to solve them; unfortunately in this case, we cannot use the Z-transform because the convolution summation in Eq. (9.11) is defined only for sample instants $m \geq 0$. When we use the Z-transform, the Z-transform summation is defined over the range $-\infty \leq m \leq \infty$; since the summation in Eq. (9.11) is not defined for the range $-\infty \leq m \leq \infty$, we cannot use the Z-transform method directly.

Since we cannot use the Z-transform, we will use a different method of solution; the method that we will adopt is known as spectral factorization. The method of spectral factorization uses the following basic idea. If the input signal $x[n]$ were a white noise process, then the autocorrelation function $\gamma_{xx}[k-n]$ would be the autocorrelation of a white noise process. The autocorrelation of a white noise process is shown in Eq. (9.30). Since the autocorrelation function of the white noise process is zero everywhere except for a lag of zero, we can use this fact in Eq. (9.11). Now since all the values for $k < 0$ are zero, we can extend the summation in Eq. (9.11) from $-\infty$ to $+\infty$. Extending this summation would now allow us to use the Z-transform and hence make the solution very easy to obtain; also there is only one nonzero term at every instant on the left-hand side as the autocorrelation of the input process is nonzero at only one sample time. With the input process being a white noise process, the solution to the Weiner–Hopf equations would be very simple as shown in Eq. (9.31):

$$\gamma_{\text{white noise}}[k-m] = \begin{cases} \sigma^2 & m = k \\ 0 & \text{otherwise} \end{cases} \tag{9.30}$$

$$h_0[k] = \begin{cases} \dfrac{\gamma_{dx}}{\sigma^2} & k = 0, 1, 2, \ldots \\ 0 & \text{otherwise} \end{cases} \tag{9.31}$$

In a typical filtering problem the input signal $x[n]$ is not a white noise process, so we cannot use the solution in Eq. (9.31) directly; however, if we can somehow transform the input process into a white noise process, then the solution to Eq. (9.11) would be very simple and trivial. The process of converting the input process into a white noise process is known as "prewhitening." Prewhitening implies that we convert the given process $\{x[n]\}$ into an innovation process $\{v[n]\}$ as discussed in the next section.

9.5.1
Innovation Representation of a Stochastic Process

We begin with a WSS process $\{x[n]\}$ with $\gamma_x[n]$ as its autocorrelation function and $\Gamma_x(\omega)$ as its power spectrum density. Here we will assume that $\Gamma_x(\omega)$ is a continuous function of ω. The assumption implies that the signal $x[n]$ does not have any pure sinusoidal signal. The two-sided Z-transform of the autocorrelation function can be written as shown in Eq. (9.32):

$$\Gamma_x(z) = \sum_{k=-\infty}^{\infty} \gamma_x[k] z^{-k} \tag{9.32}$$

With the substitution of $z = e^{j\omega}$, the function $\Gamma_x(z)$ can be looked upon as the power spectrum density of the process $x[n]$. We now assume that the natural logarithm of the PSD, $\ln\{\Gamma_x(\omega)\}$, is an analytic function in the annulus region $\varrho \leq |z| \leq 1/\varrho$, with $\varrho \leq 1$ and

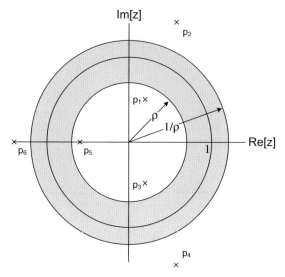

Figure 9.3 Annulus region of $\ln[S_x(z)]$ and centered at the origin.

centered at the origin as shown in Figure 9.3. The assumption that $\ln\{\Gamma_x(\omega)\}$ is an analytic function implies that all its derivatives are also continuous functions. Since all the derivatives are continuous functions, we can expand the function $\ln\{\Gamma_x(\omega)\}$ in the form of a Laurent's series as shown in Eq. (9.33). Note that Eq. (9.33) is defined from $k=-\infty$ to $k=\infty$.

$$\ln\{\Gamma_x(z)\} = \sum_{k=-\infty}^{\infty} c_k z^{-k} \tag{9.33}$$

Inverting the expression in Eq. (9.33), we get the PSD of the process $x[n]$ as shown in Eq. (9.34):

$$\Gamma_x(z) = \exp\left(\sum_{k=-\infty}^{\infty} c_k z^{-k}\right) = \exp\left(\sum_{k=-\infty}^{-1} c_k z^{-k}\right) e^{c_0} \exp\left(\sum_{k=1}^{\infty} c_k z^{-k}\right)$$

$$\Gamma_x(z) = \exp\left(\sum_{k=1}^{\infty} c_k z^{k}\right) e^{c_0} \exp\left(\sum_{k=1}^{\infty} c_k z^{-k}\right) \tag{9.34}$$

$$\Gamma_x(z) = G\left(\frac{1}{z}\right) \varrho^2 G(z) = \varrho^2 |G(z)|^2$$

We can interpret Eq. (9.34) as follows. The left-hand side represents the PSD of a WSS stochastic process, while the right-hand side represents a constant and the magnitude squared of a filter transfer function. The constant can be interpreted as the PSD of a white noise process as the PSD of a white noise process is indeed a constant. The magnitude squared of the filter transfer function represents the filter that we use to convert the WSS

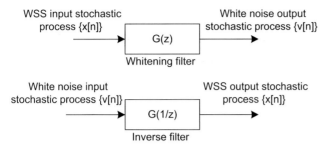

Figure 9.4 The whitening and inverse filters.

process to the white noise process. The last equation in Eq. (9.34) shows that the right-hand side can be interpreted as a product of the PSD of a white noise process and the magnitude squared of a filter. Equation (9.34) then can be looked upon as a recipe for converting a WSS stochastic process into a white noise process that is equivalent to the original stochastic process. Equivalent in this sense implies that the white noise process has the same statistical properties as the original process. This is shown in Figure 9.4.

The top of Figure 9.4 shows a WSS process as an input to the filter; this process when it goes through a whitening filter comes out as a white noise process. The filter $G(z)$ is a causal and a causally invertible filter. We know that this filter is a causal from its definition in Eq. (9.34), where we defined the filter as $G(z) = \exp\left(\sum_{k=1}^{\infty} c_k z^{-k}\right)$. The inverse filter is represented by another filter that is also defined in Eq. (9.34); this filter is represented by the reciprocal poles and zeros of the filter as seen also from Eq. (9.34), where we defined the filter as $G(1/z) = \exp\left(\sum_{k=1}^{\infty} c_k z^{k}\right)$. The two filters enable us to move back and forth between a WSS process and a white noise process that is equivalent to the original WSS process. Equivalent in this case implies that the statistical properties of the white noise process are the same as the statistical properties of the original WSS process. Thus, using the prewhitening approach we can solve Eq. (9.11) as shown in Eq. (9.31).

9.5.2
Solution to the Original Problem

The solution in Eq. (9.31) is for an input process that is a white noise process. The process that we have is a WSS process but it may not be a white noise process. In the previous section, we have seen how to get a white noise process that has the same statistical properties as the input WSS process. So our approach to get the Weiner filter will be to convert the WSS process to an equivalent white noise process with similar statistical properties. Even though the statistical properties of the white noise process are the same as the properties of the original WSS input process, the solution that we obtain using the white noise process is not for the original input process. So we will build a system that consists of two filters; the first filter will convert the input process into a white noise process and the second filter is now the Weiner filter. This is shown in Figure 9.5; now the filtering problem has been reduced to finding the filter $H_0'(z)$ in such a way that the output $y[n]$ is the optimum estimate of the expected output $d[n]$, for the input $x[n]$ in the mean square sense. We begin by looking at

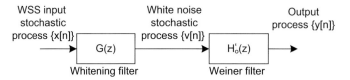

Figure 9.5 Solving the Weiner–Hopf equations.

the right-hand side of Figure 9.5, where we have the input process that is a white noise process and the filter is a Weiner filter, so the relation between the white noise process and the output can be written as shown in Eq. (9.35):

$$\sum_{m=0}^{\infty} h_0'[m]\gamma_{vv}[k-m] = \gamma_{dv}[-k], \quad k = 0, 1, 2, \ldots \tag{9.35}$$

Comparing the solution in Eq. (9.12) with Eq. (9.35), we see that the two solutions are the same if we think of the input process $v[n]$ as the input process $x[n]$. The input process $v[n]$, however, is a white noise process. We know that the autocorrelation of the white noise process is σ_v^2, so taking advantage of this fact we can rewrite the solution in Eq. (9.35) as shown in Eq. (9.36):

$$h_0'[m] = \frac{\gamma_{dv}[k]}{\sigma_v^2}, \quad k = 0, 1, 2, \ldots \tag{9.36}$$

Examine Eq. (9.36), which is the same as Eq. (9.31). Using the fact that now we can define $h_0'[k]$ from $-\infty$ to ∞ since $\gamma_{vv}[k] = \sigma_v^2$ is defined from $-\infty$ to ∞, we are now able to write the Z-transform of the optimum filter as shown in Eq. (9.37):

$$H_0'(z) = \sum_{k=0}^{\infty} h_0'[k]z^{-k} = \frac{1}{\sigma_v^2}\sum_{k=0}^{\infty} \gamma_{dv}[k]z^{-k} \tag{9.37}$$

Examine the summation on the right-hand side of Eq. (9.37). This summation represents the summation for the sequence of the cross-correlation between $\{d[n]\}$ and $\{v[n]\}$ from $0 \leq k \leq \infty$. The cross-correlation is of course defined for both the positive and the negative values of the index k, so we can write the PSD of the cross-correlation sequence, which is its Z-transform of the cross-correlation as shown in Eq. (9.38):

$$\Gamma_{dv}(z) = \sum_{k=-\infty}^{\infty} \gamma_{dv}[k]z^{-k} \tag{9.38}$$

We cannot use the PSD of Eq. (9.38) in the equation defining the optimum filter in Eq. (9.37) as it is, since the summation in Eq. (9.37) is defined only from 0 to ∞ while the summation in Eq. (9.38) is defined from $-\infty$ to ∞. To use the PSD, we will define another function that represents the PSD for only the portion $0 \leq k$. This new function is defined as

$$[\Gamma_{dv}(z)]_+ = \sum_{k=0}^{\infty} \gamma_{dv}[k]z^{-k} \tag{9.39}$$

With this definition of the PSD $[\Gamma_{dv}(z)]_+$, we can use this new function in Eq. (9.37) to obtain the optimum filter $H'_0(z)$ as shown in Eq. (9.40):

$$H'_0(z) = \frac{1}{\sigma_v^2}[\Gamma_{dv}(z)]_+ \tag{9.40}$$

This equation gives us the desired solution except for the knowledge of the PSD function $\Gamma_{dv}(z)$ between the sequences $\{d[n]\}$ and $\{v[n]\}$. We obtain the cross-correlation between the two sequences $\{d[n]\}$ and $\{v[n]\}$ as shown in Eq. (9.41):

$$v[n] = \sum_{k=0}^{\infty} \hat{g}[k]x[n-k] \tag{9.41}$$

In Eq. (9.41), $\hat{g}[k]$ is the impulse response of the whitening filter. This filter can be written as shown here:

$$\frac{1}{G(z)} = \sum_{m=0}^{\infty} \hat{g}[m]z^{-m} \tag{9.42}$$

Now the cross-correlation between $\{d[n]\}$ and $\{v[n]\}$ can be written as shown in Eq. (9.43); here we have determined the relation between the cross-correlation γ_{dv} in terms of the cross-correlation γ_{dx}, which we will use in Eq. (9.40) to eventually replace Γ_{dv}:

$$\gamma_{dv}[k] = \mathcal{E}[d[n]v[n-k]]$$
$$\gamma_{dv}[k] = \sum_{m=0}^{\infty} \hat{g}[m]\mathcal{E}[d[n]x[n-k-m]] \tag{9.43}$$
$$\gamma_{dv}[k] = \sum_{m=0}^{\infty} \hat{g}[m]\gamma_{dx}[k+m]$$

Equation (9.43) gives us a relation between the cross-correlations γ_{dv} and γ_{dx}. Substituting this cross-correlation γ_{dv} between $\{d[n]\}$ and $\{v[n]\}$ in Eq. (9.38), we get the PSD $\Gamma_{dv}(z)$ as shown here:

$$\Gamma_{dv}(z) = \sum_{k=-\infty}^{\infty}\sum_{m=0}^{\infty} \hat{g}[m]\gamma_{dx}[k+m]z^{-k}$$
$$\Gamma_{dv}(z) = \sum_{m=0}^{\infty} \hat{g}[m] \sum_{k=-\infty}^{\infty} \gamma_{dx}[k+m]z^{-k}$$
$$\Gamma_{dv}(z) = \sum_{m=0}^{\infty} \hat{g}[m] \sum_{k=-\infty}^{\infty} \gamma_{dv}[k]z^{-(k-m)} \tag{9.44}$$
$$\Gamma_{dv}(z) = \underbrace{\sum_{m=0}^{\infty} \hat{g}[m]z^{m}}_{\text{filter } G(z^{-1})} \underbrace{\sum_{k=-\infty}^{\infty} \gamma_{dx}[k]z^{-k}}_{\text{PSD } \Gamma_{dx}}$$

In Eq. (9.44), we see that the PSD $\Gamma_{dv}(z)$ is a product of two Z-transform functions: the first one is the Z-transform of the whitening filter and the second is the PSD $\Gamma_{dx}(z)$, where $\Gamma_{dx}(z)$

is defined as $\Gamma_{dx}(z) = \sum_{n=-\infty}^{\infty} \gamma_{dx}[k]z^{-k}$. So the PSD $\Gamma_{dv}(z)$ between $\{d[n]\}$ and $\{v[n]\}$ can now be written in terms of the PSD $\Gamma_{dx}(z)$ as shown in Eq. (9.45).

$$\Gamma_{dv}(z) = \frac{\Gamma_{dx}(z)}{G(z^{-1})} \tag{9.45}$$

Having obtained the relation between the PSD of the two cross-correlation sequences, we can substitute this in Eq. (9.40) to get the optimum filter as shown in Eq. (9.46):

$$H_0'(z) = \frac{1}{\sigma_v^2} \left[\frac{\Gamma_{dx}(z)}{G(z^{-1})} \right]_+ \tag{9.46}$$

The filter in Eq. (9.46) after it is multiplied by the transfer function of the whitening filter will give us the Weiner filter for the input sequence $x[n]$ as shown in Figure 9.5 and given in Eq. (9.47):

$$H_0(z) = \frac{H_0'(z)}{G(z)} = \frac{1}{\sigma_v^2 G(z)} \left[\frac{\Gamma_{dx}(z)}{G(z^{-1})} \right]_+ \tag{9.47}$$

This equation represents the Z-domain solution of the Weiner–Hopf equations that give us the causal Weiner filter for the WSS input sequence $x[n]$. The steps that we go through to determine the Weiner filter are as follows. The example that follows demonstrates the process.

1) Given the PSD of the WSS process $x[n]$, we factorize the PSD $\Gamma_x(z)$ into three functions as σ^2, $G(z)$, and $G(z^{-1})$. The function $G(z)$ contains all the poles of $\Gamma_x(z)$ that lie inside the unit circle and the function $G(z^{-1})$ consists of all the poles of $\Gamma_x(z)$ that lie outside the unit circle. The factor σ^2 is evaluated by making sure that the value $G(\infty) = 1$, so the factor σ^2 is indeed a scaling factor.
2) The function $\Gamma_{dx}(z)$ is formed by determining the two-sided Z-transform of the cross-correlation $\{\gamma_{dx}[k]\}$ between sequences $\{x[n]\}$ and $\{d[n]\}$.
3) Once the PSD $\Gamma_{dx}(z)$ is determined, it is divided by the function $G(z^{-1})$ obtained in step 1. After the division is completed, only the poles inside the unit circle are retained. The poles outside the unit circle are discarded. That is, we retain only the causal part of the function $[\Gamma_{dx}(z)/G(z^{-1})]$ to get $H_0'(z)$.
4) Finally to get the desired filter $H_0(z)$, we divide the filter $H_0'(z)$ by the scaling constant σ^2 and the whitening filter $G(z)$ both of which are obtained in step 1 of the process.

Example 9.3

Consider a signal $x[n]$ that is defined as $x[n] = d[n] + s[n]$, where $\{d[n]\}$ is the desired signal that is corrupted by noise $\{s[n]\}$. The PSD of the desired signal $\{d[n]\}$ is $\Gamma_d(\omega) = 0.34/[1.16 - 0.8\cos(\omega)]$, and the noise signal $\{s[n]\}$ has unit variance and consists of uncorrelated samples.

Solution 9.3: First determine the PSD of the input signal $\{x[n]\}$: $\Gamma_x(\omega) = \Gamma_d(\omega) + 1$; using the known expression for $\Gamma_d(\omega)$, we can determine $\Gamma_x(\omega)$ as shown here:

$$\Gamma_x(z) = \frac{0.34}{1.16-0.4(z+z^{-1})} + 1$$

$$\Gamma_x(z) = \frac{1.50-0.4(z+z^{-1})}{1.16-0.4(z+z^{-1})} = 1.3884\underbrace{\frac{1-0.2889z}{1-0.4z}}_{G(z^{-1})}\underbrace{\frac{1-0.2889z^{-1}}{1-0.4z^{-1}}}_{G(z)} \qquad (9.48)$$

$$(9.49)$$

The last part of Eq. (9.48) shows the PSD factored as required in step 1. Next we determine the cross PSD between the sequences $\{d[n]\}$ and $\{x[n]\}$ as $\gamma_{dx} = \gamma_d$ as $\{s[n]\}$ is an uncorrelated sequence, so the PSD can be written as

$$\Gamma_{dx}(\omega) = \Gamma_d(\omega) = \frac{0.34}{(1.0.4z)(1-0.4z^{-1})} \qquad (9.50)$$

This completes the work to be done in step 2. In step 3, determine the filter $H_0'(z)$ as follows:

$$\left[\frac{\Gamma_{dx}(z)}{G(z^{-1})}\right]_+ = \left[\frac{0.34}{(1-0.4z)(1-0.4z^{-1})}\frac{(1-0.4z)}{(1-0.2889z)}\right]_+$$

$$\left[\frac{\Gamma_{dx}(z)}{G(z^{-1})}\right]_+ = \left[\frac{0.34}{(1-0.2889z)(1-0.4z^{-1})}\right]_+ \qquad (9.51)$$

$$\left[\frac{\Gamma_{dx}(z)}{G(z^{-1})}\right]_+ = \left[\frac{0.3844}{(1-0.4z^{-1})} + \frac{0.088z}{(1-0.2889z)}\right]_+ = \left[\frac{0.3844}{(1-0.4z^{-1})}\right]$$

The required filter is the last fraction in Eq. (9.51); this completes step 3. We are now ready to determine the complete filter as shown here by dividing the filter $H_0'(z)$ by $\sigma_x^2 G(z)$ from Eq. (9.49) to get the required filter as shown here:

$$H_0(z) = \frac{H_0'}{\sigma_x^2 G(z)} = \frac{1}{1.3884}\left(\frac{1-0.4z^{-1}}{1-0.2889z^{-1}}\right)\left(\frac{0.3844}{1-0.4z^{-1}}\right) = \frac{0.276}{1-0.2889z^{-1}} \qquad (9.52)$$

9.5.3
The Minimum Mean Square Error

We have determined the optimum filter in the previous section, which is the Weiner filter. When this filter is used on a WSS input process, then the error between the input process and the desired response will be a minimum. In this section, we will determine what the minimum value of this output is. To determine the minimum error, we will set the filter impulse response coefficients to be equal to the Weiner filter from Eq. (9.11) in the expression of the error in Eq. (9.22) to get Eq. (9.53).

$$e_0 = \gamma_d(0) - \sum_{k=0}^{\infty} h_0[k]\gamma_{dx}(k), \quad k = 0, 1, 2, \ldots \qquad (9.53)$$

To compute the minimum error, this time we will use the frequency domain. So we begin first by knowing that since the ideal filter is a causal filter, $h_0[m]$ is zero for $m < 0$; so we can extend the summation in Eq. (9.53) so that it covers the range from $-\infty \leq m \leq \infty$ instead of from $0 \leq m \leq \infty$ as shown in Eq. (9.54).

$$e_0 = \gamma_d(0) - \sum_{k=0}^{\infty} h_0[k]\gamma_{dx}[k], \quad k = 0, 1, 2, \ldots \tag{9.54}$$

Next we look at each of the terms on the right-hand side of Eq. (9.54) one by one. First, the PSD of the desired response is $\Gamma_d(z) = \sum_{k=-\infty}^{\infty} \gamma_d(k)z^{-k}$. Since $\Gamma_d(z)$ is analytic in the annular region $\varrho \leq |z| \leq 1/\varrho$, we can evaluate $\gamma_d(k)$ on the unit circle by using the inversion integral on the PSD. When we evaluate the inversion integral for the PSD and determine the autocorrelation for zero lag ($k = 0$) as required in Eq. (9.54), we get Eq. (9.55):

$$\gamma_d(0) = \oint_c \Gamma_d(z)z^{-1} \, dz \tag{9.55}$$

The closed contour of integration is the unit circle in the z-domain as the function $\Gamma_d(z)$ converges on the unit circle. Next we look at the summation in Eq. (9.54). This summation is a product of two time domain functions. We know from the Parseval's relation that we can relate the product summation in the time domain to the product integral in the frequency domain. So using Parseval's relation we can write the summation as shown in Eq. (9.56):

$$\sum_{k=-\infty}^{\infty} h_0[k]\gamma_{dx}[k] = \frac{1}{2\pi j}\oint_c H_0(z)\Gamma_{dx}(z)z^{-1} \, dz \tag{9.56}$$

In Eq. (9.56), $H_0(z)$ represents the z-transform of the Weiner filter and $\Gamma_{dx}(z)$ is the PSD of the cross-correlation between the input signal and the desired response. The closed contour over which the integral takes place is the unit circle as both the functions $H_0(z)$ and $\Gamma_{dx}(z)$ converge on this contour. Both the functions converge on the unit circle as the unit circle lies within the region of convergence for the product function. Since all the terms on the right-hand side of Eq. (9.54) converge on the unit circle, we can combine them into one integral as shown in Eq. (9.57):

$$e_0 = \frac{1}{2\pi j}\oint_c [\Gamma_d(z) - H_0(z)\Gamma_{dx}(z^{-1})]z^{-1} \, dz \tag{9.57}$$

To evaluate the integral in Eq. (9.57) around the unit circle, we will use Cauchy's residue theorem. Cauchy's residue theorem says that the result of the integration equals the sum of the residues of the function $[\Gamma_d(z) - H_0(z)\Gamma_{dx}(z^{-1})]z^{-1}$ at each of the poles of the function that can be written as shown in Eq. (9.58):

$$\frac{1}{2\pi j}\oint_c [\Gamma_d(z) - H_0(z)\Gamma_{dx}(z^{-1})]z^{-1} \, dz$$

$$= \sum_{k=1}^{M} \text{Residues}\left[[\Gamma_d(z) - H_0(z)\Gamma_{dx}(z^{-1})]z^{-1}; z_k\right] \tag{9.58}$$

Substituting Eq. (9.58) into Eq. (9.57), we get a method of computing the minimum value of the error when the Weiner filter is used to estimate the desired response from the given input as shown in Eq. (9.59):

$$e_0 = \sum_{k=1}^{M} \text{Residues}\left[\left[\Gamma_d(z) - H_0(z)\Gamma_{dx}(z^{-1})\right]z^{-1}; z_k\right] \tag{9.59}$$

For the filtering problem considered in Example 9.3, determine the minimum error.

Solution 9.4: For the problem considered in Example 9.3, we have

$$\Gamma_d(z) = \Gamma_{dx}(z) = \left[\frac{0.34}{(1-0.4z)(1-0.4z^{-1})}\right]$$

and

$$H_0(z) = \frac{0.276}{(1-0.2889z^{-1})}$$

Next we form the function $[\Gamma_d(z) - H_0(z)\Gamma_{dx}(z^{-1})]z^{-1}$ as shown here:

$$[\Gamma_d(z) - H_0(z)\Gamma_{dx}(z^{-1})]z^{-1}$$

$$= \left[\frac{0.34}{(1-0.4z)(1-0.4z^{-1})} - \frac{0.276}{1-0.2889z^{-1}}\frac{0.34}{(1-0.4z)(1-0.4z^{-1})}\right]z^{-1}$$

$$= \left[\frac{0.6154}{(z-0.2889)(2.5-z)}\right]$$

Next we determine the residues at the poles of the function for all the poles inside the unit circle. This is easy to evaluate as there is only one pole inside the unit circle. Evaluating the residue at the pole at $z = 0.2889$, we get $\text{Res}[z = 0.2889] = 0.278$, so the minimum error evaluates to 0.278.

It is interesting to compare the Weiner filters and the minimum error using the frequency domain method and the time domain method. We present the comparison by another example. Consider a signal whose desired response is given by the PSD $\Gamma_d(z) = 0.46/[1.04-0.2(z+z^{-1})]$ and the received signal is corrupted by white noise with unit variance. So the PSD of the received signal can be written as $\Gamma_d(z) = [1.5-0.2(z+z^{-1})]/[1.04-0.2(z+z^{-1})]$.

9.5.3.1 Solution in the Frequency Domain

We first rewrite the PSD of the received signal in terms of the whitening filter and the inverse filter as shown in Eq. (9.60). This completes step 1 in the design of the Weiner filter.

$$\Gamma_x(z) = \sigma^2 G(z)G(z^{-1}) = 1.478\left[\frac{1-0.1358z^{-1}}{1-0.2z^{-1}}\right]\left[\frac{1-0.1358z}{1-0.2z}\right] \tag{9.60}$$

Next we determine the PSD of the cross-correlation. Since the noise that corrupts the desired signal is white noise, the PSD of the cross-correlation is the same as the PSD of the desired signal as shown in Eq. (9.61):

$$\Gamma_{dx}(z) = \Gamma_d(z) = \frac{0.46}{(1-0.2z^{-1})(1-0.2z)} \tag{9.61}$$

Next we determine the causal part of the function described in step 3 and shown in Eq. (9.62):

$$\left[\frac{\Gamma_{dx}(z)}{G(z^{-1})}\right]_+ = \left[\frac{0.46}{(1-0.2z^{-1})(1-0.2z)} \quad \frac{(1-0.2z)}{(1-0.1358z)}\right] \tag{9.62}$$

$$\left[\frac{\Gamma_{dx}(z)}{G(z^{-1})}\right]_+ = \frac{0.0.32}{(1-0.1358z^{-1})}$$

Finally, we can determine the Weiner filter according to step 4 and shown in Eq. (9.63):

$$H_0(z) = \frac{1}{\sigma^2 G(z)}\left[\frac{\Gamma_{dx}(z)}{G(z^{-1})}\right]$$

$$= \frac{1}{1.478}\frac{(1-0.2z^{-1})}{(1-0.1358z^{-1})} \quad \frac{0.473}{(1-0.2z^{-1})} = \frac{0.32}{(1-0.1358z^{-1})} \tag{9.63}$$

$$h_0[n] = 0.32(0.1358)^n$$

Equation (9.63) gives us the required Weiner filter. For this filter, we next determine the minimum error in the frequency domain. To compute the minimum error, we first determine the function $[\Gamma_d(z)-H_0(z)\Gamma_{dx}(z^{-1})]z^{-1}$ as shown in Eq. (9.64):

$$[\Gamma_d(z)-H_0(z)\Gamma_{dx}(z^{-1})]z^{-1} = \frac{1.564}{(5.0-z)(z-0.1358)} \tag{9.64}$$

Evaluating the sum of residues at all the poles inside the unit circle, we get the minimum error $e_0 = 0.3215$.

9.5.3.2 Solution in the Time Domain

Next we repeat the same problem in the time domain. For this we first determine the autocorrelation for the desired response by determining the inverse Z-transform of the PSD. Determining the inverse Z-transform, we get the result given in Eq. (9.65):

$$\gamma_d = 0.46(0.2)^{|k|} = [0.46; 0.092; 0.0184; 0.00368]^T \tag{9.65}$$

Next we need to determine the autocorrelation of the input signal. Since the input signal autocorrelation is equal to the autocorrelation of the desired signal and the additive noise, the autocorrelation can be written as shown in Eq. (9.66):

$$\gamma_x = \gamma_d + \gamma_v = [1.46; 0.092; 0.0184; 0.00368]^T \tag{9.66}$$

Knowing the autocorrelation sequence, we can write the autocorrelation matrix that is given in Eq. (9.67):

$$\Gamma_x = \begin{bmatrix} 1.46 & 0.092 \\ 0.092 & 1.46 \end{bmatrix}, \qquad \Gamma_x^{-1} \begin{bmatrix} 0.6877 & -0.04333 \\ -0.04333 & 0.6877 \end{bmatrix} \tag{9.67}$$

Next using Eq. (9.14) we can determine the Weiner filter of order 2 as shown in Eq. (9.68). This filter compares very nicely with the frequency domain filter given in Eq. (9.63):

$$h_0 = \Gamma_x^{-1}\gamma_{dx} = \begin{bmatrix} 0.6877 & -0.04333 \\ -0.04333 & 0.6877 \end{bmatrix} \begin{bmatrix} 0.46 \\ 0.092 \end{bmatrix} = \begin{bmatrix} 0.3123 \\ 0.04333 \end{bmatrix} \tag{9.68}$$

To determine the minimum error in the time domain, we use Eq. (9.29a). This is evaluated in Eq. (9.69):

$$e_0 = \sigma_d^2 - \gamma_{dx}^T \Gamma^{-1} \gamma_{dx}$$

$$= \begin{bmatrix} 0.46 & 0.092 \end{bmatrix} \begin{bmatrix} 06877 & -0.04333 \\ -0.04333 & 0.6877 \end{bmatrix} \begin{bmatrix} 0.46 \\ 0.092 \end{bmatrix} = 0.3123 \tag{9.69}$$

As you can see from the results, the two methods are equivalent to each other.

9.6
Canonical Form of the Error Surface

The equation for the mean square error in the time domain is written in Eq. (9.20). In this section, we examine this equation further to get a little more insight into the mean square error produced by the Weiner filter. To do this we first rewrite the equation using matrix notation as shown in Eq. (9.70):

$$J(\mathbf{h}) = \sigma_d^2 - \mathbf{h}^H \gamma - \gamma^H \mathbf{h} + \mathbf{h}^H \Gamma \mathbf{h} \tag{9.70}$$

In Eq. (9.70), if we write $\mathbf{h}^H \gamma = \mathbf{h}^H \Gamma \Gamma^{-1} \gamma$ and write $\gamma^H \mathbf{h} = \gamma^H \mathbf{h} \Gamma^{-1} \Gamma \mathbf{h}$ and we add $\gamma^H \mathbf{R}^{-1} \gamma$ and subtract the same term but we write it as $\gamma^H \Gamma^{-1} \Gamma \Gamma^{-1} \gamma$, then Eq. (9.70) will become a perfect square. Rewriting it as a perfect square, we get Eq. (9.71):

$$J(\mathbf{h}) = \sigma_d^2 - \gamma \Gamma^{-1} \gamma + (\mathbf{h} - \Gamma^{-1}\gamma)^H \Gamma (\mathbf{h} - \Gamma^{-1}\gamma) \tag{9.71}$$

Examining Eq. (9.71) carefully we see that the left-hand side represents the error function. Since this is represented as a perfect square, we can visualize which terms add to the error and which terms subtract from the error. The first term σ_d^2 is a constant and it is independent of the filter, so this term cannot be altered. It is the variance of the desired signal. The second term $\gamma \Gamma^{-1} \gamma$ is also dependent on the cross-correlation and the autocorrelation of the signal and hence we cannot alter these terms. The third term adds to the error. This term is dependent on the coefficients of the Weiner filter. If we are able to make this term equal to zero, then the error will be the smallest possible. To make the third term zero, we must have $\mathbf{h} = \Gamma^{-1}\gamma$ to make the two brackets equal to zero. Examine this condition carefully and you

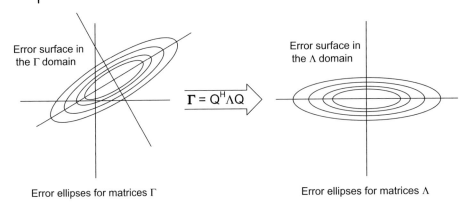

Error ellipses for matrices Γ Error ellipses for matrices Λ

Figure 9.6 Transforming the error surfaces using the eigenvalue decomposition of the Γ matrix.

will see that the condition for minimum error is the same as Eq. (9.14), which was the solution to the Weiner–Hopf equations and gave us the Weiner filter. So in effect we have derived the Weiner filter in a rather straightforward manner. This is an alternative way of deriving the Weiner filter. When we make the third term zero, then we get the minimum error that can be written as shown in Eq. (9.72):

$$J_{\min}(\mathbf{h}) = \sigma_d^2 - \gamma\Gamma^{-1}\gamma \tag{9.72}$$

The quadratic form of Eq. (9.71) is very interesting and gives us the minimum error and the Weiner filter, but if we change the basis on which the various matrices are defined we get further insight into the Weiner filter. To do this we will use the eigendecomposition of the square matrix Γ (Figure 9.6). Since we know that Γ matrix is a nonsingular matrix, we can rewrite it as shown in Eq. (9.73):

$$\Gamma = \mathbf{Q}\Delta\mathbf{Q}^H \tag{9.73}$$

In Eq. (9.73), the matrix $\mathbf{Q}$ is a square matrix with its columns made up of all the eigenvectors. The matrix $\mathbf{D}$ is a diagonal matrix with all the eigenvalues of the matrix Γ on the main diagonal. Using this transformation in Eq. (9.72), we get Eq. (9.74):

$$J(\mathbf{h}) = \sigma_d^2 - \gamma\Gamma^{-1}\gamma + (\mathbf{h} - \mathbf{h}_0)^H \mathbf{Q}\Delta\mathbf{Q}^H(\mathbf{h} - \mathbf{h}_0)$$
$$J(\mathbf{h}) = J_{\min} + (\mathbf{h} - \mathbf{h}_0)^H \mathbf{Q}\Delta\mathbf{Q}^H(\mathbf{h} - \mathbf{h}_0) \tag{9.74}$$

Next we define a new vector that transforms the difference between the parameters of the filter and the optimum parameters that make the filter into a Weiner filter as shown in Eq. (9.75):

$$\mathbf{v} = \mathbf{Q}^H(\mathbf{h} - \mathbf{h}_0) \tag{9.75}$$

Using this new vector in Eq. (9.74), we get the error cost function as shown in Eq. (9.76):

$$J(\mathbf{h}) = J_{\min} + \mathbf{v}\Delta\mathbf{v} \tag{9.76}$$

Equation (9.76) defines the error at any time. The first term represents the minimum error when the filter coefficients are at their optimum values while the second term represents the error that is increased when the filter coefficients are not optimum. Another fact about Eq. (9.76) that is interesting is that there are no cross-product terms in Eq. (9.76), so we can rewrite Eq. (9.76) as shown in Eq. (9.77).

$$J(h) = J_{\min} + \sum_{k=1}^{M} \lambda_k v_k v_k^*$$

$$J(h) = J_{\min} + \sum_{k=1}^{M} \lambda_k |v_k|^2 \tag{9.77}$$

This representation tells us that we can determine the amount of error along the various principal axes of the error performance surface.

9.7
Weiner Filters with Additional Constraints

The Weiner filter that we have seen so far minimizes the error between the input and the expected output in the mean square sense. In getting to the optimum filter, we have not imposed any constraint on the filter coefficients, so the filter coefficients are free to take on any values. Often, however, we want a particular response in one frequency band and a different response in a different frequency band. For example, we will want a passband that has unit gain, while in the stop band we want a large attenuation. So in these situations it will be mandatory to design the filter that minimizes the mean square error subject to the constraint that we will specify. To study these types of requirements, consider the output from a filter shown in Figure 9.2. The output from this filter is given in Eq. (9.1).

For the filter given in Figure 9.2, we will consider the input to be a sinusoid $e^{j\omega n}$. With this input, the output from the filter can be written as shown in Eq. (9.78):

$$y[n] = e^{-j\omega n} \sum_{k=0}^{\infty} e^{-j\omega k} h^*[k] \tag{9.78}$$

So the constrained optimization problem can be stated as shown in Eq. (9.79):

$$\sum_{k=0}^{\infty} e^{-j\omega_0 k} h^*[k] = g \tag{9.79}$$

where ω_0 is the prescribed frequency at which we want the response to be equal to the gain g, which may be a complex value. To use the constrained minimization problem, we will use the method of *Lagrange multipliers*. To use this method, we first write the real-valued cost function J that combines both the parts of the constrained optimization problem as shown in Eq. (9.80):

$$J = \underbrace{\sum_{k=0}^{M-1} \sum_{i=0}^{M-1} h^*[k]h[i]\gamma(i-k)}_{\text{output power from the filter}} + \underbrace{\mathrm{Re}\left[\lambda^* \sum_{k=0}^{M-1} e^{-j\omega_0 k} h^*[k] - g\right]}_{\text{added linear constraint}} \tag{9.80}$$

In Eq. (9.80), the parameter λ is the complex *Lagrange multiplier*. Note that there is no desired output in Eq. (9.80); instead, we have the linear constraint that has to be satisfied at a required frequency band. This linear constraint that requires us to maintain the specified response at a required frequency preserves the desired response to the filter. So the minimization of the cost function will attenuate the noise and give us the desired response. To determine the optimum filter in the presence of noise, we need to determine the gradient of the cost function J in Eq. (9.80) and set it to zero. So we determine the gradient of the cost function as we did earlier in Section 9.1. For the present case, the gradient of the cost function can be written as shown in Eq. (9.81):

$$\nabla_k J = 2 \sum_{i=0}^{M-1} h[i] \gamma[i-k] + \lambda^* e^{-j\omega_0 k} \tag{9.81}$$

Setting this gradient to zero, we get Eq. (9.82):

$$\sum_{i=0}^{M-1} h[i] \gamma[i-k] = -\frac{\lambda^*}{2} e^{-j\omega_0 k}, \quad k = 0, 1, 2, \ldots, M-1 \tag{9.82}$$

Equation (9.82) is still very similar to the Weiner–Hopf equations that we saw earlier. This equation represents a system of M simultaneous equations that when solved will give us the optimum filter tap weights. Rewriting Eq. (9.82) using matrix notations, we get Eq. (9.83); the solution is given in the second equation of Eq. (9.83):

$$\begin{aligned} \Gamma \mathbf{h}_0 &= -\frac{\lambda^*}{2} \mathbf{x}(\omega_0) \\ \mathbf{h}_0 &= -\frac{\lambda^*}{2} \Gamma^{-1} \mathbf{x}(\omega_0) \end{aligned} \tag{9.83}$$

In Eq. (9.83), we still have the unknown parameter λ^* that has to be determined; also we have the inverse of the autocorrelation matrix that we know is not singular. To determine the parameter λ, we will use the definition of the linear constraint that when written in matrix notation can be stated as $\mathbf{h}_0^H \mathbf{x}(\omega_0) = g$. Taking the Hermitian transpose of Eq. (9.83) and then substituting the constraint equation in the Hermitian transpose, we can solve for the parameter λ as shown in Eq. (9.84), where we have used the constraint equation in the last step.

$$\mathbf{h}_0^H = \left[-\frac{\lambda^*}{2} \Gamma^{-1} \mathbf{x}(\omega_0) \right]^H = -\frac{\lambda}{2} \mathbf{x}^H(\omega_0) \Gamma^{-H}$$

$$\lambda = -\frac{2\mathbf{h}_0^H}{\mathbf{x}^H(\omega_0) \Gamma^{-H}} = -\frac{2\mathbf{h}_0^H \mathbf{x}(\omega_0)}{\mathbf{x}^H(\omega_0) \Gamma^{-H} \mathbf{x}(\omega_0)} = -\frac{2g}{\mathbf{x}^H(\omega_0) \Gamma^{-H} \mathbf{x}(\omega_0)} \tag{9.84}$$

Finally, substituting this value of λ in the optimum solution given in Eq. (9.83), we get the optimum solution for the Weiner filter with additional constraints as given in Eq. (9.85):

$$\mathbf{h}_0 = -\frac{\lambda^*}{2}\boldsymbol{\Gamma}^{-1}\mathbf{x}(\omega_0) = \frac{g\boldsymbol{\Gamma}^{-1}\mathbf{x}(\omega_0)}{\mathbf{x}^H(\omega_0)\boldsymbol{\Gamma}^{-H}\mathbf{x}(\omega_0)}$$

$$\lambda = -\frac{2\mathbf{h}_0^H}{\mathbf{x}^H(\omega_0)\boldsymbol{\Gamma}^{-H}} = -\frac{2\mathbf{h}_0^H\mathbf{x}(\omega_0)}{\mathbf{x}^H(\omega_0)\boldsymbol{\Gamma}^{-H}\mathbf{x}(\omega_0)} = -\frac{2g}{\mathbf{x}^H(\omega_0)\boldsymbol{\Gamma}^{-H}\mathbf{x}(\omega_0)} \tag{9.85}$$

We have minimized the output power subject to the linear constraint, so the signals that are in the frequency band ω_0 will have a gain of g while the signals that are outside the band are minimized. Since we have minimized the power and the input signal is a zero mean signal, this filter is known as *minimum variance linearly constrained filter.*

9.8
Chapter Summary

We began this chapter with the definition of the problem. In the problem under consideration, we have a signal that is transmitted through a noisy channel. At the receiver, we receive this noisy signal and from this signal we want to recreate the original transmitted signal. We assumed that the transmitted signal is a WSS signal with zero mean and the noise signal is white noise with zero mean. We solve this problem in the time domain first and then in the frequency domain.

We first defined the error signal in Eq. (9.2) as the difference between the desired signal and the estimate of the desired signal as shown here: $e[n] = d[n]-y[n]$. Under the conditions of the problem, we first found that the error signal and the received noisy signal were orthogonal to each other. As a corollary to this principle, we also found that the output signal from the filter and the error signal are also orthogonal to each other. Next we defined a cost function. The cost function that we chose is the mean square value of the error signal. Minimizing the error will give us the desired filter that will give us the estimate of the signal as close to the original signal as we can get.

The Weiner–Hopf equations consist of a system of equations that when solved will give us the coefficients of the optimum filter. The Weiner–Hopf equations are also known as the normal equations. The system of equations is given in Eq. (9.12). The system of equations that represent the Weiner–Hopf equations represents a bowl-shaped surface in the M-dimensional space of the Weiner filter.

Since this is a bowl-shaped surface, it has a unique minimum that can be readily found. Once we have the solution of the Weiner filter, the minimum value of the error between the desired signal and the estimate of the desired signal can be computed using Eq. (9.29a).

We next turned our attention to the solution of the Weiner–Hopf equations in the frequency domain. The general method of solution that we use in the frequency domain is to use the Z-transform. In this case, however, we are not able to use this method since all the terms in the Weiner–Hopf equations are not defined for $k < 0$ and the Z-transform requires the values for $k < 0$, so we are not able to use this method directly. To be able to use the Z-transform, we had to resort to a new process known as the *innovation* process. This is a white noise process that has the same statistical properties as the input WSS process. The white noise process is zero everywhere except for $k=0$, so this process is defined for all values of $k < 0$. So when we transform the input process into the innovation process, we are able to use the Z-transform to solve the Weiner–Hopf equations. The solution of the Weiner–Hopf equations using the Z-transform is given in Eq. (9.47) and it is described as a four-step process in Section 9.4.2. Next we determined the minimum mean square value of the error using the Z-transform. Interestingly, the minimum error is just the sum of residues of the function $[\Gamma_d(z) - H_0(z)\Gamma_{dx}(z^{-1})]z^{-1}$ for all the poles inside the unit circle. Now that we have two different methods to solve the same problem, we next demonstrated that these two methods are identical to each other by solving one example using both the methods.

The error or more specifically the surface is interlinked with all the filter coefficients. By using the eigenvalue transform of the autocorrelation matrix, we were able to show that the error can be decoupled so that the error contributed by each eigenvalue can be observed individually.

The Weiner filters studied so far all have optimized the filter coefficients with no additional constraints. In the previous section, we showed that the Weiner filter can be designed with additional external constraints. These constraints serve the same purpose as the desired signal and to solve the Weiner filter we had to resort to the method of Lagrange multipliers. In this method, we first defined the error function using the constraint and minimized the error function. This gave us the optimum solution with an unknown, which is the Lagrange multiplier. Then using the constraint equation we solved for the Lagrange multiplier. Substituting this solution for the Lagrange multiplier gives us the desired optimum filter subject to an additional constraint.

10
Adaptive Filters

10.1
Introduction

The Weiner filter that we studied in the previous chapter required us to have knowledge of the first- and the second-order statistics of the received signal and the cross-correlation of the received and the desired signal. This is often unknown or more importantly the statistics may be drifting slowly. When the statistics are drifting slowly, the Weiner filter designed is no longer an ideal filter. The bottom of the bowl-shaped region no longer coincides with the minimum error point. In these cases, we need to both redesign the Weiner filter and use the new filter; most of the time, however, the statistics of the signal drift slowly rather than undergoing a step change. Due to the gradual shift, it is very difficult to know when the statistics shift. In cases where the statistics drift slowly, it would be indeed nice if we are able to shift the bottom of the bowl-shaped surface as the input statistics shift. This way we are always operating under optimum conditions.

So the idea of the adaptive filter is to first design a filter that operates under optimum conditions when we do not know the statistics of the input and yet we want the filter to operate under optimum conditions. So this filter is very similar to the Weiner filter except the coefficients are not designed directly by the user. We begin with one set of filter coefficients and the filter coefficients are adapted in such a way that the error is minimized. So in a stationary environment, after the filter has adapted to the minimum error, the adaptive filter will converge to the Weiner solution.

In this chapter, we study the adaptive filters; these filters use a recursive algorithm to gradually shift the operating point of the filter toward the optimum location. There are many recursive methods; the most basic among these is the gradient method. Adaptive filters based on the gradient algorithm make a class of filters that are sought after by engineers because of their simplicity, flexibility, and robustness. They are easy to design and it is easy to determine the characteristics of these filters.

As we study the filters in detail, we will see that because of the loop characteristics of the filter it is extremely difficult to determine the exact analysis of the filter. Even though this is a difficulty, simple approximate results can be deduced and for a vast majority of cases these simple approximate results are sufficient. Our emphasis in this chapter is to present the results and the information necessary to design and successfully use an adaptive filter.

Digital Filters: Theory, Application and Design of Modern Filters, First Edition. Rajiv J. Kapadia.
© 2012 Wiley-VCH Verlag GmbH & Co. KGaA. Published 2012 by Wiley-VCH Verlag GmbH & Co. KGaA.

While doing this we will also point out various options available that make the design of the filter as flexible as possible.

10.1.1
Applications of Adaptive Filters

As we will see soon, the adaptive filter is able to operate in an unknown environment satisfactorily. This makes this filter a very useful device in many different applications from diverse fields such as radar, sonar, image processing, finance, and biomedical engineering. In all these applications, when the adaptive filter is used we have to somehow determine an error vector that in turn is used to adjust the filter coefficients that will then reduce the error itself. This arrangement leads to several different types and classes of filters and applications. There are many different filter structures that are possible. In this book, we will concentrate only on FIR filters that are organized as direct form. The adaptation that we will do is in adjusting the coefficients of the direct form FIR filter like the filter structure shown in Figure 10.5.

In the following section, we examine several different applications of adaptive filters. In this section, we only demonstrate how the adaptation can take place. The mathematics of adaptation will be covered in the next section. This development will only serve as a foundation for what is developed in the next section. The basic classes of applications are provided below.

10.1.2
System Identification

The first class of application for an adaptive filter is to determine the system model. In this application, the unknown plant and the adaptive filter are both driven with the same signal. The output from the adaptive filter is compared with the output from the unknown system. The difference between the two represents the error. This error is then used to adjust the coefficients of the filter. When the error reaches its minimum steady-state value, we have identified a linear model of the unknown system. This application is shown in Figure 10.1. System identification using adaptive filter will keep track of the system even when the system changes as long as these changes are occurring slowly. If we use the Weiner filter solution as we did in one of the examples in the previous chapter, instead of the adaptive filter solution, then when the system changes the solution will no longer be correct. Examples 10.1 and 10.3 both demonstrate the adaptive filter in this arrangement. The two filters are designed using different algorithms.

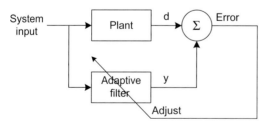

Figure 10.1 Adaptive filter used for system identification.

Some typical applications in this area come from seismology, where we are interested in modeling the various layers underground. An adaptive filter can reveal the changes in the density of the layers under the Earth. Another application comes from the field of control systems where the model is slowly changing. In this case, we are interested in designing a controller that will finally be used to control the plant.

10.1.3
Inverse Modeling

Sometimes it is necessary to cancel the adverse effects of a channel. To cancel the effects of the channel, we need to determine the inverse model of the channel, and then passing the received signal through the inverse model will reverse the adverse effects of the channel. This is especially true when the transmitted signal is digital pulses; the pulse at the transmitting end is perfect but the channel that can be modeled as a low-pass filter will distort the pulse in both the amplitude and the phase. The distortion will cause the pulse to smear and hence introduce intersymbol interference. Generally, the inverse modeling of the channel is done by using a known training sequence that is transmitted before the beginning of the information transmission. This training sequence is a short sequence lasting a fraction of a second. This training process can be repeated if it is found that the channel characteristics are changing due to channel drift.

For inverse modeling, an arrangement like the one shown in Figure 10.2 can be used. The signal is passed through the plant that is typically the transmission channel and then through the adaptive filter at the receiving end. This signal is then compared with the desired signal (this is done during the training sequence) that is delayed by an appropriate amount of time. The comparison gives us an error that is then used to adjust the coefficients of the adaptive filter. In this case, the transfer function of the adaptive filter will be inverse of the plant transfer function. In applications such as these, the plant can be the transmission channel that degrades the signal, so the adaptive filter will invert the adverse effects of the plant. Examples 10.2 and 10.4 both demonstrate the adaptive filter in this arrangement. In the examples, we assume that noise gets added to a signal as it travels through the channel and the job of the adaptive filter is to remove the noise and recreate the transmitted signal. The two filters are designed using different algorithms.

A typical application in this area is equalization. In wireless communication or radar imaging, the impulse response of the channel is unknown. The purpose of the adaptive filter is to operate on the received signal such that the cascade connection of the channel and

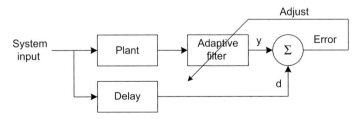

Figure 10.2 Adaptive filter used for inverse modeling.

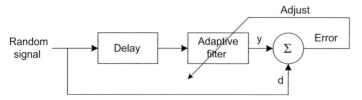

Figure 10.3   Adaptive filter used for predicting the input.

the adaptive filter will cancel the effect of the channel and hence provide an ideal transmission medium.

10.1.4
Prediction

In this application, the adaptive filter is used to represent the best estimate of the present *value* of the input random signal (Figure 10.3). In this case, the present value is estimated from the past N values of the random input and hence the required delay before the comparison. Here the present value represents the desired input and the output from the filter represents the estimate of the present value. Again the error between the desired value and the estimate of the value represents the error that is used to adjust the filter coefficients. In this application, the output can be the output from the adaptive filter when the arrangement is that of the predictor or the output may be the error signal in which case the arrangement is operating as the prediction error filter.

Predictive coding is often used for transmission of the speech signals. For transmitting of speech, we want to use as few bits as possible. This has led to many different low bit rate (10 000 bits/s) modeling, encoding, and transmission methods. Adaptive filtering finds application in these model-based speech coding systems.

One of the methods of speech coding is differential pulse code modulation; in this method, instead of transmitting the speech signal we transmit the error signal. Since the error signal has a much smaller variance than the speech signal itself, we are able to transmit the signal using fewer bits and hence achieve compression of the signal.

Since we can predict what the next event would be, would it not be nice if we can take a string of stock closing prices of a particular company, find the difference between today's and yesterday's closing price, and predict tomorrow's closing! Unfortunately this we cannot do with any regularity or any accuracy; can you think of a reason why?

10.1.5
Interference Canceling

In this application, the signal that we need is corrupted in some way. The filter is used to cancel out the interference. In this application, the primary signal *serves* the function of the desired signal and the output from the adaptive filter serves the purpose of the estimate of the desired signal (Figure 10.4). The difference between the two is the error signal that is used to adjust the filter coefficients.

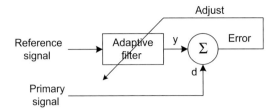

Figure 10.4 Adaptive filter used to cancel interference.

Noise canceling using an adaptive filter subtracts the additive noise from the signal. This way we can improve the signal to noise ratio of the signal. A special application of this application is echo canceling in telephone circuits. In transmission of data over the telephone system, a modem is used. The modem is used to provide an interface between the digital information and the analog information on the telephone channel. Examples 10.2 and 10.4 both demonstrate the adaptive filter in this arrangement. In the examples, we assume that noise gets added to a signal as it travels through the channel and the job of the adaptive filter is to remove the noise and recreate the transmitted signal. The two filters are designed using different algorithms.

10.2
Adaptive Direct Form FIR Filters

In all the typical applications of the adaptive filters above, the filter is organized as a typical Weiner filter with the output from the filter being compared to the desired output. The difference between the output from the filter and the desired signal represents the error that we will minimize. This was the exact same setup that we saw when we were looking at the Weiner filter. So as a first guess we can expect that the solution that we will expect when we minimize the error will be a solution that is similar to the Weiner solution. This time, however, we have added an adjustment to the filter. This adjustment tries to adjust or adapt the coefficients of the filter so that the error between the desired signal and the estimate of the desired signal is minimized. Since we are solving the same problem that we solved when we talked about the Weiner filter, we should expect that the equations will be same as those for the Weiner filter. The equations are repeated for you in Eq. (10.1). In this equation, $\gamma_{xx}(l)$ represents the autocorrelation and $\gamma_{dx}(l)$ represents the cross-correlation.

$$\sum_{k=0}^{N-1} h(k)\gamma_{xx}(l-k) = \gamma_{dx}(l), \qquad l = 0, 1, 2, \ldots \tag{10.1}$$

In Eq. (10.1), we need to compute $\gamma_{xx}(l)$ and $\gamma_{dx}(l)$. The computation of $\gamma_{xx}(l)$ and $\gamma_{dx}(l)$ depends on the length of the received data. When we were designing Weiner filters, the quantities $\gamma_{xx}(l)$ and $\gamma_{dx}(l)$ were computed from the data using expectations. This was not a problem when we were designing the Weiner filter since we had all data that we were going to use and these data were assumed to be from a WSS process. Computation has to be considered very carefully when designing an adaptive filter as we will have to recompute the autocorrelation and the cross-correlation matrix with every new sample. We will be computing the quantities $\gamma_{xx}(l)$ and $\gamma_{dx}(l)$ again and again with every new sample that arrives. Recomputing $\gamma_{xx}(l)$ and $\gamma_{dx}(l)$ with every new sample implies that we are computing the instantaneous values of $\gamma_{xx}(l)$ and $\gamma_{dx}(l)$. So when we are designing the adaptive filters we will be computing these quantities instantaneously from the actual received data. Once we have these quantities computed with every new sample, we will solve the Weiner–Hopf equation to get an instantaneous estimate of what the Weiner filter will be. This will allow us to obtain the filter coefficients $h(k)$ that are the estimates of the true coefficients. How close the filter coefficients come to the true coefficients depends on how good our computation of the quantities $\gamma_{xx}(l)$ and $\gamma_{dx}(l)$ is since we are using instantaneous computations.

A second problem that we need to consider is the signal itself. The underlying statistical properties of the signal may be drifting; that is, the signal is not from a WSS process. As a consequence of the signal being nonstationary, the autocorrelation and the cross-correlation of the sequences are changing with time. This implies that the coefficients of the filter must vary with the drifting of the two correlations. Due to the drifting we cannot just take a very large number of samples to compute the two correlations and then use the filter designed from then on. So we have to design a different way to determine the filter coefficients when the underlying process is not stationary.

There are two different approaches used to adjust the filter coefficients. First, we can vary the filter coefficients with every new sample that arrives. In this method, we compute the two correlations $\gamma_{xx}(l)$ and $\gamma_{dx}(l)$ with every new sample that arrives and use the result in Eq. (10.1) to determine the filter coefficients. The second method uses a small block of inputs to compute the two correlations $\gamma_{xx}(l)$ and $\gamma_{dx}(l)$ and use them to compute the filter coefficients. Then we start again with a new block and compute the required quantities recursively. When the block is received, we compute the two correlations $\gamma_{xx}(l)$ and $\gamma_{dx}(l)$ again and solve Eq. (10.1) to get a new set of filter coefficients.

In what follows, we will look at two classes of adaptive filter algorithms. The first one is generally known as the gradient algorithm that includes the more popular LMS algorithm. The other one is known as recursive least squares algorithm that is more popularly known as the RLS algorithm. The recursive least squares algorithm is slightly more complex than the gradient algorithm but the recursive least mean square algorithm provides faster convergence to changes in statistical properties in the signal.

10.3
The Gradient Algorithm

In this group of algorithms, we try to optimize the filter coefficients by estimating the minimum mean square error. To do this estimation, let us begin with the assumption that we have the sequence $x[n]$ that consists of samples from a WSS process. This sequence will then have its autocorrelation sequence as $\gamma_{xx}(l) = \mathcal{E}\left[x[n]x^*[n-l]\right]$. We can also pass the sequence

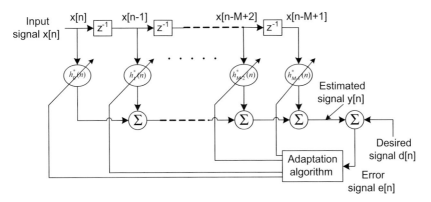

Figure 10.5 Structure of an adaptive filter.

$x[n]$ through an FIR filter like the one shown in Figure 10.5 and obtain the output that is the estimate of the desired signal as shown in Eq. (10.2):

$$y[n] = \sum_{k=0}^{N-1} h[k]x[n-k] \tag{10.2}$$

Now that we have the estimate of the desired signal, we can compute the estimation error between the desired signal $d[n]$ and the estimate of the desired signal $y[n]$ as

$$e[n] = d[n]-y[n] \tag{10.3}$$

From this error we can define an error function as the mean square value of the error as shown in Eq. (10.4):

$$J_N = \mathcal{E}\left[|e[n]|^2\right] = \mathcal{E}\left[\left|d[n] - \sum_{k=0}^{N-1} h[k]x[n-k]\right|^2\right] \tag{10.4}$$

We have come across Eq. (10.4) earlier in Chapter 9 as Eq. (9.17), where we got the expanded version of the equation as shown in Eq. (10.5):

$$J_N = \sigma_d^2 - 2\mathrm{Re}\left[\sum_{k=0}^{N-1} h^*[k]\gamma_{dx}[k]\right] + \sum_{l=0}^{N-1}\sum_{k=0}^{N-1} h^*[l]h[k]\gamma_{xx}[l-k] \tag{10.5}$$

Equation (10.5) is a quadratic equation with a unique minimum that we can determine by differentiating and setting the result equal to zero. Doing this gives us a system of equations as shown in Eq. (10.6):

$$\sum_{k=0}^{N-1} h[k]\gamma_{xx}[l-k] = \gamma_{dx}[l], \quad l = 0, 1, 2, \ldots, N-1 \tag{10.6}$$

The filter with the coefficients determined using Eq. (10.6) is the Weiner filter that we studied in Chapter 9. So what is the difference between the Weiner filter and the filter that we are after here? The difference comes by in how we will arrive at the solution of Eq. (10.6). Remember we began by claiming that we will estimate the mean square error. In what follows, we present some of the more popular numerical methods of solving Eq. (10.6) to obtain the filter coefficients.

10.3.1
The Method of Steepest Descent

Using the method of steepest descent to arrive at an optimum value is very straightforward. It is a recursive method in which we begin with a guess of the filtercoefficients. So as a first guess we will call the filter coefficients to be $h^{(0)}[0]$, $h^{(0)}[1]$, $h^{(0)}[2]$, . . .; here the superscript indicates the guess number. A zero indicates a beginning guess. Using this guess we compute the output from the filter. The output from the filter represents the current estimate of the desired output. From the difference of the two, we will determine the current value of error. It is this error that we want to minimize. This error is dependent on the filter coefficients, so adjusting the filter coefficients as we are about to do will change the error and we hope to make the error smaller. To minimize the error, we will define a cost function of the error that is dependent on the filter coefficients as $J(h^{(0)})$. From this error cost function, we will compute the gradient of the error in all directions. Using the computed gradients, we will move the filter coefficients for the next estimate in the direction opposite to the steepest gradient, hence the name of the method. Adjusting the filter coefficients gives us another estimate of the filter. We hope that every new estimate of the filter moves us toward the optimum filter that has the smallest error that we can attain. Using this new estimate for the filter coefficients, we hope and expect that the error between the desired output and the estimate of the desired output will have reduced some, so we will have a relation like the one shown in Eq. (10.7), which shows that the error estimate at time $(k + 1)$ is less than the estimate at time k. We continue this estimation process till there is no further change in the estimation error where we claim that the process has converged and no more reduction of the error is possible.

$$J\left(h^{(k+1)}\right) < J\left(h^{(k)}\right)$$

(10.7)

There is a distinct possibility that if proper care is not taken in developing the algorithm then Eq. (10.7) will not be satisfied, the error during the next iteration increases instead of decreasing, and the entire adaptation process diverges instead of converging to minimum error. In this case, the estimate will not be close to the desired response and we have an unstable system. We begin the investigation of this process with the calculation of the gradient vector as determined in Eq. (10.8). In Eq. (10.8), we demonstrate the computation of the gradient at the kth iteration and hence the superscript (k).

$$g^{(k)} = \nabla J\left(h^{(k)}\right) = \frac{\partial J\left(h^{(k)}\right)}{\partial h^{(k)}}$$

$$g^{(k)} = 2\left[\Gamma^{(k)} h^{(k)}[n] - \gamma^{(k)}\right]$$

(10.8)

Note that Eq. (10.8) represents the gradient vector of the error function given in Eq. (10.5). This gradient vector of the error function can also be written as shown in Eq. (10.6), which we have used in the second equation in Eq. (10.8). In Eq. (10.8), we compute the gradient $g^{(k)}$ using the coefficients of the filter after the kth iteration when the filter coefficients are $h^{(k)}$. Now that we have the gradient we will use the gradient to adjust the filter coefficients as shown in Eq. (10.9). We use this gradient to get the coefficients for the next iteration as shown in Eq. (10.9):

$$h^{(k+1)} = h^{(k)} - \frac{1}{2}\mu g^{(k)}$$

$$h^{(k+1)} = h^{(k)} - \mu\left[\Gamma^{(k)}h^{(k)}[n] - \gamma^{(k)}\right]$$

$$(10.9)$$

In Eq. (10.9), the superscript (k) denotes the iteration number for both the gradient and the filter coefficients. The superscript $(k + 1)$ represents the next iteration values for the filter coefficients. The parameter μ represents the step size. It determines how big a step we are going to take when we adjust the filter coefficients. The step size parameter, as we will see soon, has two conflicting requirements. If we make it large, then the algorithm will move the filter coefficients in the direction indicated by the gradient faster. This means we will get to the optimum solution faster. That is what we want. However, making it too large could make the entire process unstable, or oscillate around the optimum value, which can happen with a large step size as we overshoot the optimum value and have to back up; this is not what we want. So in the gradient methods the design of the step size parameter is a major consideration. The constant $1/2$ is not necessary but it makes things nice later, so we will use it anyway. Using the adjustment equation, we compute the weight adjustment as shown in Eq. (10.10):

$$\delta\left(h^{(k)}\right) = h^{(k+1)} - h^{(k)} = -\frac{1}{2}\mu g^{(k)} = -\mu\left[\Gamma^{(k)}h^{(k)}[n] - \gamma^{(k)}\right] \qquad (10.10)$$

The next question that we need to address is whether the algorithm suggested so far satisfies the requirement in Eq. (10.7) that says that with each iteration of the algorithm the mean square error must reduce. To answer this question, we use the Taylor series expansion of the cost function around the tap weight vectors to obtain an expression as shown in Eq. (10.11):

$$J\left(h^{(k+1)}\right) \approx J\left(h^{(k)}\right) + g^{(k)}\delta\left(h^{(k)}\right) \qquad (10.11)$$

Substituting Eq. (10.10) into Eq. (10.11), we get Eq. (10.12). In Eq. (10.12), the new value of the cost function is approximately equal to the old value of the cost function from which we have subtracted $\frac{1}{2}\mu\|g^{(n)}\|^2$. If this term $\frac{1}{2}\mu\|g^{(n)}\|^2$ is always positive, then Eq. (10.7) can be satisfied as the updated value of the cost function is less than the previous value of the cost function. We can make sure that this term is always positive by choosing the step size as a positive value and noting that $\|g^{(k)}\|^2$ is always positive. Now the second term in Eq. (10.12) is always positive, so with every iteration we are subtracting a positive value from the previous error value to arrive at the next error value. This implies that the condition in Eq. (10.7) will

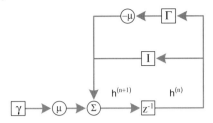

Figure 10.6 Signal flow graph to represent the steepest descent algorithm.

always be satisfied as the cost function after $(k + 1)$ iterations $\left(J\left(h^{(k+1)}\right)\right)$ will be less than the cost function after k iterations $\left(J\left(h^{(k)}\right)\right)$.

$$J\left(h^{(k+1)}\right) \approx J\left(h^{(k)}\right) - \frac{1}{2}\mu|g^{(k)}|^2 \tag{10.12}$$

Equation (10.9) is an interesting equation as it offers us a different perspective on the whole adaptation process. Note that we can write Eq. (10.9) as shown in Eq. (10.13). We can represent Eq. (10.13) as a signal flow graph as shown in Figure 10.6.

$$h^{(k+1)}[n] = h^{(k)}[n] - \mu\left[\Gamma^{(k)}h^{(k)}[n] - \gamma^{(k)}\right]$$
$$h^{(k+1)}[n] = \left[I - \mu\Gamma^{(k)}\right]h^{(k)}[n] + \mu\gamma^{(k)} \tag{10.13}$$

To get further insight into the adaptation process, let us look at the gradient vector in Eq. (10.8). If we use the orthogonality condition that we came across in Chapter 9, we can rewrite Eq. (10.8) as shown in Eq. (10.14):

$$g(n) = \frac{\partial J(n)}{\partial h(n)} = 2[\Gamma_{xx}h[n] - \gamma_{dx}] = -2\mathcal{E}\left[e[n]X^*[n]\right] \tag{10.14}$$

Using Eq. (10.14) we get another way to compute the gradient vector. Carefully compare the two ways of computing the gradient in Eq. (10.14). We can compute the gradient by computing the autocorrelation matrix and the cross-correlation vector or by computing the expected value of the product of the error vector and the input sequence. Computing the expected value of the product is just as difficult as the computation of the autocorrelation matrix and the cross-correlation vector. Instead of computing the expected value as required in Eq. (10.14), an estimate of the expected value of the product can be used in the place of the actual expected value. The instantaneous value of the product can be used as the estimate of the expected value of the product. The instantaneous value of the product is an unbiased estimator of the expected value of the product. So we can estimate the gradient vector as shown in Eq. (10.15). This estimate is an unbiased estimate of the gradient vector.

$$\hat{g}[n] = -2e[n]X^*[n] \tag{10.15}$$

Equation (10.15) tells us that the gradient can be estimated by the instantaneous value of the product of the error vector and the input vector. Now we do not have to compute the autocorrelation matrix and the cross-correlation vector. With every new input, we compute the error vector and the product of the error vector and the input will serve as an unbiased estimate of the gradient. Using this estimate of the gradient, we get the coefficient update equation as shown in Eq. (10.16):

$$h^{(k+1)}[n] = h^{(k)}[n] + \mu e[n]X^*[n] \tag{10.16}$$

Equation (10.16) is known as stochastic gradient algorithm. It uses the error vector $e[n] = d[n] - y[n]$ as shown in Figure 10.5, and $X^*[n]$ is the vector of n sample values in the filter at the present instant.

In all the work so far, we have not made any mention of the step size. It is possible to have either a fixed step size or a varying step size. If we do choose a varying step size, then there are two criteria that the step size has to satisfy. First, the sequence of step sizes should be absolutely summable, and second, as the number of iterations $n \to \infty$, the step size $\mu \to 0$. It can be shown that the filter will converge to its optimum value if the step size satisfies the above condition. The step size, however, is usually chosen as a fixed value. This is done for two very important reasons. First, it is much easier to implement the algorithm in either hardware or software if the step size is chosen as a fixed value. Second, as $n \to \infty$, the step size $\mu \to 0$; now if the underlying process drifts a little, then the filter will no longer be operating at its optimum value since with the step size being zero there is no adjustment to the filter coefficients in Eq. (10.16). If the step size is not zero, then the filter will be able to track the slowly changing process and will operate at its optimum value.

This algorithm is rather simple to implement and for this reason it is widely used in many different filtering applications. In the next section, we examine some of the variations to the stochastic gradient algorithm and the properties and the limitations of the algorithm. Specifically, we will examine the important properties of noise, stability, and the speed of convergence.

10.4
Other Related Stochastic Gradient Algorithms

There are many algorithms that are similar to the stochastic gradient algorithm that are also popular. Each algorithm tries to maintain the ease with which we can compute the gradient but tries to improve on some parameter such as reducing the gradient noise or increasing the speed of convergence.

10.4.1
The Average Stochastic Gradient Algorithm

In this variation, we compute the new coefficients for the filter not after each new input but after k inputs have been received. So here we are averaging the gradient over k iterations as shown in Eq. (10.17):

$$\hat{g}[nk] = -\frac{2}{k}\sum_{k=0}^{K-1} e[nK+k]X^*[nK+k] \tag{10.17}$$

$$h^{(k+1)}[(n+1)] = h^{(k)}[n] - \frac{1}{2}\mu\overleftrightarrow{g}[nk] \tag{10.18}$$

The averaging process reduces the gradient noise and hence it is hoped that there will be fewer or smaller fluctuations from the optimum filter coefficients. With the gradient computed as shown in Eq. (10.17), the tap weight update equation can be written as shown in Eq. (10.18). This method also tends to reduce noise in the gradient vector. The averaging process acts as a low-pass filter that reduces the noise in the gradient vector. It is thought that since we are not adjusting the filter coefficients for several different samples the convergence rate could be slower.

10.4.2
Low-Pass Filter of the Gradient

In this approach, we pass the gradient vector through a low-pass filter. With the low-pass filtering of the gradient vector, we use the output of the low-pass filter as an estimate of the gradient vector. A simple first-order filter that filters the gradient vector can be written as shown in Eq. (10.19):

$$\hat{S}[n] = \beta\hat{S}[n-1] - \hat{g}[n], \quad 0 < \beta < 1$$
$$\text{with} \quad \hat{S}[0] = \hat{g}[0] \tag{10.19}$$

$$h^{(k+1)}[n] = h^{(k)}[n] + \frac{1}{2}\mu\hat{S}[n] \tag{10.20}$$

The filtered stochastic gradient algorithm will then update the filter coefficients as shown in Eq. (10.20). The filtering process is very similar to the averaging process. When β is very small, the bandwidth of the filter is large. Large bandwidth implies high-frequency signals are allowed to go through. This is similar to averaging only a few gradient vectors. On the other hand, when β is close to 1 then the pass band is very small and only the low-frequency signal is allowed to go through. This is equivalent to averaging many different values of the gradient vector. This method also tends to reduce noise in the gradient vector. In this method, we adjust the filter coefficients with each new input, so the adaptation process is not particularly slow.

10.4.3
The Conjugate Gradient Algorithm

This algorithm provides faster convergence. In this variation, we compute the gradient vector as shown in Eq. (10.21), where $\beta(n)$ is a scalar function of the gradient vectors. This algorithm is also very similar to the low-pass filter gradient algorithm.

$$\hat{S}[n] = \beta[n-1]\hat{S}[n-1] - \hat{g}[n] \tag{10.21}$$

10.4.4
The Error Sign Algorithm

The sign of the error is used to either add or subtract the adjustment from the filter weights. The idea here is that if we use only the sign of the error then we reduce the number of multiplications we have to perform. The weight adjustment equation for this algorithm is given in Eq. (10.22), where the $\text{sign}(e[n])$ is $+1$ when $e[n] > 0$, it is 0 when $e[n] = 0$, and it is -1 when $e[n] < 0$.

$$h^{(k+1)}[n] = h^{(k)}[n] + 2\mu\{\text{sign}(e[n])\}X^*[n] \tag{10.22}$$

The sign algorithm can be considered to be a reduced complexity algorithm as multiplication can be avoided when implementing the algorithm by choosing the step size to be a power of 2. Then the adjustment to the filter weight can be determined by shifting the input sample by appropriate number of places and then adding. The convergence rate of this algorithm, however, is slower than the standard LMS algorithm.

10.4.5
The Normalized LMS Algorithm

This algorithm is frequently used in practice. In this algorithm, the step size is divided by the norm of the data vector. This way the step size is smaller when the energy in the signal is large while the step size is larger when the signal has only a small amount of energy. Thus, the step size is not constant and it changes with the energy in the signal. It is hoped that the energy in the signal suggests how fast the signal is changing, so if the signal is changing fast we will move slow so as to not miss any important information that is present in the signal. Whereas if the energy is small, then there is almost no change in the signal properties and we can take a larger step toward the optimum solution. So if the signal is not changing quickly, then a larger step will not make the algorithm unstable. The step size is adjusted as $\mu_n = \mu/\|X_M(n)\|^2$. The filter weight adjustment equation then will be as shown in Eq. (10.23):

$$h^{(k+1)}[n] = h^{(k)}[n] + 2\frac{\mu}{\|X_M[n]\|^2}(e[n])X^*[n] \tag{10.23}$$

In this variation, the step size is dependent on the energy of the signal. Hence, this variation is popular when the dynamic range of the energy in the signal is large. Such a signal occurs in wireless communications where the channel may fade away quickly and return quickly. It is thought advantageous to add a small positive constant to the norm of the signal before using it to adjust the step size. So the step size would be determined by $\mu_n = \mu/\left(\delta + \|X_M(n)\|^2\right)$. In this method of adjusting the step size, δ is always a positive constant and it determines the largest value the step size will be. This is done to avoid the condition that occurs when the energy gets very small; with the added constant, the step size is limited to some reasonable value. This way problems with instability caused by a large step size are avoided.

In all the different modifications, we have seen several different ways of computing the gradient vector. First, we have the exact gradient vector as computed in Eq. (10.8). To compute this gradient, we need to compute the autocorrelation as well as the cross-correlation. Next, we saw that we can use an estimate of the gradient vector as shown in Eq. (10.15). This reduced the computation effort required. Next, we saw several modifications of the gradient method where we computed the gradient vector in different ways; one way was to average several gradient vectors and then we used this average gradient vector to update the filter coefficients. This reduced the noise. Next, we saw the filter algorithm that is similar to the averaging algorithm but the control of how many different vectors are averaged is controlled by a parameter. Finally, we saw several different gradient algorithms that allow us to achieve convergence faster. In the next section, we examine the properties of the gradient algorithms.

10.5
Properties of the Gradient Algorithms

To study the convergence and the stability of the LMS algorithms, we begin with the weight update equation as written in Eq. (10.13). This equation represents a system of linear equations that are interlinked with each other. To see how the system behaves, we will first separate and isolate the various modes of the system so that we can study each mode individually. To do this we diagonalize the autocorrelation matrix Γ_{xx}. The diagonalization of the autocorrelation matrix is done as shown in Eq. (10.24). We did this in Chapter 9 in Eq. (9.23).

$$\Gamma_{xx} = Q\Lambda Q^H \tag{10.24}$$

In Eq. (10.24), the matrix Q is the matrix of eigenvectors of the matrix Γ_{xx}, the matrix Q^H is the Hermitian transpose of the Q matrix, and the matrix Λ is the diagonal matrix that consists of the eigenvalues of the matrix Γ_{xx} on the main diagonal and all the other terms in the matrix are zeros. With this representation, when we substitute Eq. (10.24) in the weight update equation (10.13) we have in effect decoupled all the equations in the weight update equation, so we can investigate how each of the modes of the system reaches its steady-state value. Using the transformation from Eq. (10.24) in Eq. (10.13), we get Eq. (10.25):

$$h^{(k+1)}[n] = [I - \mu Q\Lambda Q^H]h^{(k)}[n] + \mu\gamma^{(k)}$$
$$Q^H h^{(k+1)}[n] = [I - \mu\Lambda]Q^H h^{(k)}[n] + \mu Q^H \gamma^{(k)} \tag{10.25}$$
$$h_t^{(k+1)}[n] = [I - \mu\Lambda]h_t^{(k)}[n] + \mu\gamma_t^{(k)}$$

To get the second equation in Eq. (10.25), we have premultiplied the first equation by Q^H and we make use of the fact that the product of the two matrices QQ^H equals the identity matrix. In Eq. (10.25), $Q^H h^{(k)}$ modifies the filter coefficients and often many texts design the adaptive filter with these modified coefficients. We will also take this same approach and use modified notations to write $h_t = Q^H h$ and $\gamma_t = Q^H \gamma$, which represents the two quantities in the transform domain. This is shown in the third equation in Eq. (10.25). Now since the matrix Λ is a diagonal matrix, all the modes of the system in Eq. (10.25) are decoupled; now we can study how the individual weights of the adaptive filter are adjusted and what are the requirements for convergence. The convergence of each of all the modes now can be studied from the homogeneous equation for each individual mode as shown in Eq. (10.26), where $h_t^{(k)}[n, i]$ represents the ith filter coefficient from a total of n coefficients during the kth update of the transformed filter coefficient.

$$h_t^{(k+1)}[n, i] = (1-\mu\lambda_i)h_t^{(k)}[n, i] \tag{10.26}$$

The solution of this homogeneous equation can be written as shown in Eq. (10.27):

$$h_t^{(k)}[n, i] = C(1-\mu\lambda_i)^k u[i] \tag{10.27}$$

In Eq. (10.27), C is a constant to be determined. Looking at the solution of the homogeneous equation we see that the filter coefficient will converge if the absolute value of the geometric progression ratio is less than 1 or $|(1 - \mu\lambda)| < 1$. Using this condition, we can determine the bounds on the step size μ to guarantee that the tap filter coefficients will converge as shown in Eq. (10.28):

$$0 < \mu < \frac{2}{\lambda_i}, \quad i = 0, 1, 2, \ldots, N-1 \tag{10.28}$$

In Eq. (10.28), if we choose λ_i as the largest eigenvalue, then we have an upper bound on the step size. If the step size is less than two times the reciprocal of the largest eigenvalue, then all the filter coefficients will converge and the algorithm will be stable. Here we see that the step size also controls the convergence rate of the algorithm not just the stability of the algorithm. This is one condition that controls the convergence rate and the stability of the adaptive algorithm. We obviously want the algorithm to be stable and to converge to its optimum value as quickly as possible. These are two conflicting requirements as fast convergence implies that we make the step size large while the convergence requirement says that the step size is limited to be less than two times the reciprocal of the largest eigenvalue, which is a relatively small value. So to determine the upper bound on the step size we need to determine the autocorrelation matrix and then determine its eigenvalues.

Very often, however, this is difficult to accomplish as the autocorrelation matrix is either difficult to obtain or as suggested by Eq. (10.16) we have decided to use the estimate of the gradient, which is the product $e[n]X^*[n]$, instead of the gradient itself, which is $[\Gamma_{xx}h[n]-\gamma_{dx}]$. The reason we decided to use the estimate of the gradient is so that we did not have to compute the autocorrelation matrix. Thus, requiring that we use the eigenvalue as the upper bound for the step size defeats this purpose. To get around this difficulty, we first note that since Γ_{xx} is an autocorrelation matrix and all its eigenvalues will be positive, we can define an upper limit on the eigenvalues as shown in Eq. (10.29):

$$\lambda_{\max} \leq \sum_{i=1}^{M} \lambda_i = \text{trace}(\Gamma_{xx}) = M\gamma_{xx}(0) \tag{10.29}$$

Using Eq. (10.29) we can define the upper limit of the step size using the energy in the signal, so the step size can be limited to $\mu \leq 2/(M\gamma_{xx}(0))$, where M is the number of eigenvalues of the autocorrelation matrix Γ_{xx}. Using this as the upper limit for the step size, we do not need to compute the autocorrelation matrix.

10.5.1
Examples of Adaptive Filters Using the LMS Algorithm

Here we demonstrate the use of adaptive filters using the LMS algorithm. We use two different and distinct examples to demonstrate the LMS algorithm. In the first example, we use the adaptive filter to match the unknown filter. In this example, we do not know what the filter coefficients are, so we use the input to the filter as the input to the adaptive filter. This way when the error is minimized, the coefficients of the adaptive filter will match the coefficients of the unknown filter. In the second example, we have received a noisy signal, from which we want to extract the original signal. In this example, we use the noisy signal as the input to the adaptive filter and the output from the adaptive filter is an estimate of the desired signal.

Example 10.1

Using the LMS algorithm we want to match the system filter to the adaptive filter. The system diagram is shown in Figure 10.7.

Solution 10.1: In Figure 10.7, we see first that the input signal $x[n]$ is filtered using a system filter. After the noise signal is added to the output from the filter, we get the desired signal. This signal is labeled as $d[n]$. Noise is added to this signal, which corrupts the signal. The input signal to the filter to be matched is also the input to the adaptive filter. The output from the adaptive filter is compared with the desired signal to get the error signal. The error signal is used to adjust the filter coefficients. Figure 10.7 compares the coefficients of the system filter used and the adaptive filter coefficients. From the plot we can see that the two are indeed very similar to each other and hence the adaptive filter minimizes the error between the desired signal and the output from the adaptive filter. A MATLAB script is added that simulates the adaptive filter as given in MATLAB 10.1.

MATLAB 10.1 Matching a system filter using LMS algorithm.

```
N = 900;                  %Number of sample points.
w_cf = 0.25;              %Cutoff frequency used to design system filter to be
t = 0:2/N:(2-2/N);        %matched.
Nf = 21;                  %Number of filter coefficients.
x = randn(1,N);           %Random number sequence used as input.
n = 0.3* randn(1,N);      %Noise corrupting the signal.
w = fir1(Nf,w_cf);        %System filter to be matched.
d = filter(w,1,x) + n;    %Desired signal to be matched by the filter.
mu = 0.005;               %Step size used for weight update.
y = zeros(1,N);           %Initial vector of estimates of signal to be
                          matched.
```

```
e = zeros (1,N);          %Initial vector of errors.
h = zeros (1,Nf);         %Initial guess of the adaptive filter.
for i = Nf:1:N
   y(i) = h*x(i:-1:(i-Nf + 1))'; %Compute the next estimate from the
   e(i) = d(i) - y(i);            %Adaptive filter. Error signal.
   h= h + mu*e(i)*x(i:-1:(i-Nf + 1));%Update the adaptive filter
coefficients.
end
stem(1:(Nf + 1),w.','fill');hold;stem(1:(Nf),h.','color','r');hold
legend('Filter to be matched','Adaptive filter designed');
title('Comparing the coefficients of the adaptive filter with the filter to
be matched');
xlabel('Filter coefficient number');
ylabel('Filter coefficient value');
```

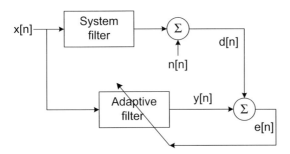

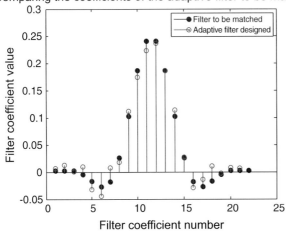

Figure 10.7 Comparing the filter coefficients of the system filter and a filter designed by adaptive filter to match the system filter.

Example 10.2

As a second example we will examine an application of an adaptive filter to remove noise from a noisy signal. The system used this time is shown in Figure 10.8.

Solution 10.2: In Figure 10.8, we have generated the signal $d[n]$, which is the desired signal. We assume that this signal is transmitted over a noisy channel and as it travels it picks up noise $v[n]$. It is desired to recreate the original signal that was transmitted, so we pass the received signal through the adaptive filter. From the plot we see that the signal $y[n]$, which is the output of the adaptive filter, is very close to the desired signal.

The MATLAB script given in MATLAB 10.2 shows how the adaptive filter is developed; first we build the desired signal and then corrupt it with noise that is Gaussian and zero mean. This time we have chosen the desired signal so that the energy in the signal decreases as time goes on. The plot of the desired signal and the noisy signals is shown in the top plot in Figure 10.8. Next we build three vectors that are all set to zero initially. In this particular

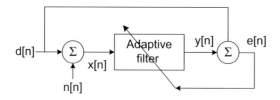

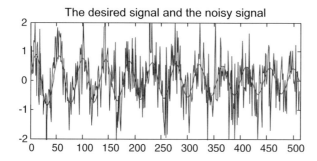

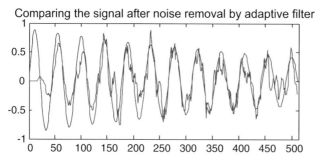

Figure 10.8 Comparing the signal after noise removal by adaptive filter to match the desired signal.

example, we have chosen the size of the filter to be 17 taps. You should try the algorithm with different size filters. For this demonstration, we have chosen the step size to be 0.005. You should try the algorithm with different step sizes.

As the algorithm runs, the estimate of the desired signal quickly catches up with the desired signal and from there on the error between the two is very small till the energy in the desired signal becomes very small so that the signal to noise ratio is very small. Experiment with different energies for the signal and the noise. Increasing the signal energy will give you better results while increasing the noise energy will give you an estimate that does not resemble the signal very closely.

MATLAB 10.2 Noise removal from a received signal.

```
N = 0:511;              %Sample number
Nl = length(N);         %length of the signal vector.
Nf = 17;                %Size of the filter to be used.
d = .985.^(0.1*N).*sin(0.045*pi*N)*0.9;      %Build the desired
signal.
mu = 0.005;             %The step size used.
v = 0.7*randn(1,length(d));    %Noise vector with zero mean.
x = d + v;              %Corrupt the desired signal with the noise.
subplot(211);plot(N,x,'b',N,d,'r'); %Plot the desired and noise
signal.
%% Next we design the adaptive filter that will remove the noise.
h = zeros(1,Nf);        %Build zero vector for filter coefficients.
e = zeros(1,Nl);        %Build zero vector for error signal.
y = zeros(1,Nl);        %Build zero vector for estimated signal.
for n = Nf:Nl %The Weight update algorithm starts here.
    y(n) = h*x(n:-1:(n-Nf + 1))';  %Calculate the next output.
    e(n) = d(n) - y(n);  %Determine the Error for this estimate.
    h = h + mu*e(n)*x(n:-1:(n-Nf + 1));  %Update the tap weights.
end
subplot(212);plot(N,d,'b',N,y,'r'); %Compare desired & filtered
signals
```

10.5.2
Computing Time to Convergence

Equation (10.27) shows us how the various transformed filter coefficients converge to the optimum value. The transformed filter coefficients individually exhibit an exponential curve as they approach the optimum value; with the exponential curve we can determine the time constant that can be fitted to the exponential curve. This gives us the time constant for the LMS algorithm or the time it takes for the adaptive filter to reach its steady-state value.

$$\exp\left(\frac{-kT}{\tau_{\text{lms}}}\right) = (1-\mu\lambda_i)^k$$

$$\tau_{\text{lms}} = \frac{-T}{\ln(1-\mu\lambda_i)} \approx \frac{T}{\mu\lambda_i}$$

(10.30)

To convert the time constant to number of samples, we assume that the sampling time T is 1; then the exponential decay in Eq. (10.30) will give us the number of samples it takes for convergence. From this equation we see that the convergence depends on the smallest eigenvalue of the autocorrelation matrix as that eigenvalue will give us the longest settling time.

10.5.3
Summary of the LMS Algorithm

The LMS algorithm is the real workhorse of adaptive filtering. As we have developed the algorithm we have seen that it is very simple to implement; it is independent of the model that is used to implement the algorithm. It is, however, slow to converge. The convergence rate of the LMS algorithm is governed by the step size parameter and the eigenvalue spread of the autocorrelation matrix $\Gamma_{xx}(n)$.

The step size parameter μ is limited by the largest eigenvalue of the autocorrelation matrix while the convergence time τ_{lms} is governed by the smallest eigenvalue and the step size. If the step size is small, then the convergence rate is slow. A small step size also implies that we will arrive at a filter as close to the optimum as possible. A slow convergence rate is equivalent to having a long memory of the process. When we have a long memory, we are able to use the error from many past inputs to arrive at the optimum solution. Having a large memory means that we are able to correct more of the imbalance between the desired value and the estimated value. So with a small step size the minimum error that is always present when the filter is operating in its optimum state will be smaller. In effect we can view the reciprocal of the step size as the memory of the LMS filter.

When the eigenvalues of the autocorrelation matrix are widely spread, then the step size is controlled by the largest eigenvalue while the time taken to converge is governed by the smallest eigenvalue. So when the eigenvalue spread is large, then the convergence rate is very slow since the step size is also very small. We examine the effect of wide eigenvalues at the end of this chapter.

The gradient descent algorithm provides a simple procedure for computing the coefficients of the filter, which is the Weiner filter. In computing the coefficients, we have solved the Weiner–Hopf equations in a method that does not require us to compute either the autocorrelation matrix or the cross-correlation vector. Instead, we use estimates of these quantities.

10.6
The Recursive Least Squares Algorithm

In the previous section, we used the least squares error criteria and the steepest descent algorithm to design a transverse adaptive filter of the type shown in Figure 10.5. In this section, we once again use the least squares algorithm to design an adaptive filter like the one shown in Figure 10.5. This time, however, we will solve the Weiner–Hopf equations using the recursive algorithm. A recursive algorithm uses the knowledge that you have from the previous computation to update the values for the next computation. In the recursive method, we begin with assumption that the tap weights of the adaptive filter are at their optimum value at iteration $N-1$. This would represent the optimum filter from all the inputs received up to time $N-1$.

At time instant N, a new sample is received; using this new sample and the previous $M-1$ inputs, we will compute a new optimum filter. We use only the previous $M-1$ inputs because these are the only inputs that are still present in the filter that has M taps and hence a memory of only M input samples. This new filter is the modified version of the optimum filter at time $N-1$ so that it uses the information from the Nth sample and also from the previous $M-1$ samples. Thus, we are using the previous optimum filter to compute the new optimum filter. We refer to this new algorithm as a *recursive least squares (RLS)* algorithm.

The biggest advantage of the gradient algorithm from the previous section is the computational simplicity of the algorithm. The algorithm is, however, slow to converge as shown in Eq. (10.30). The convergence rate is slow in this algorithm as it has only one parameter that we can adjust to control the stability of the algorithm. This is the step size parameter. We use the same step size no matter what the small eigenvalues are. The step size parameter is determined from the largest eigenvalue. As a result, when the spread in the eigenvalues is large then we are waiting for a long time for the large eigenvalues to converge while the small eigenvalues will have converged a long time ago. This is the result of having only one parameter – the step size – that controls the convergence of all the eigenvalues. If we had a separate parameter for each eigenvalue, then the convergence rate would be about the same for every eigenvalue. Hence, the algorithm will converge relatively faster.

In this algorithm, we use the least squares criteria to get the optimum filter. In using the least squares criteria, we are using the input data sequence directly to get the estimates of the statistical properties of the input data.

10.6.1
Definitions

To understand the RLS algorithm, we need to define a few quantities that are unique to the understanding of the algorithm. We define the filter coefficients at time instant n as

$$\boldsymbol{h}_M(\boldsymbol{n}) = [h(0, n) \quad h(1, n) \quad \cdots \quad h(M-1, n)]^T \tag{10.31}$$

Equation (10.31) says that there are M filter coefficients and the values of the filter coefficients are given in Eq. (10.31) are at time instant n. In this context, the parameter n represents the time that has elapsed before. Similarly, the input vector is defined as

$$X_M(n) = [x(n) \quad x(n-1) \quad \cdots \quad x(n-M+1)]^T \tag{10.32}$$

Equation (10.32) says that the most recent M inputs represent the input vector that the filter is going to see at sample time n. $X_M(n)$ then represents the snapshot of the most recent inputs from the entire input vector. This is also the vector that we are going to use for the current iteration. This process of looking at only the most recent M values from the entire input vector is known as *prewindowing*. Prewindowing is one of the modifications from the LMS algorithm. Another modification in the RLS algorithm is the use of a weighted error cost function. The weighted error cost function is shown in Eq. (10.33):

$$J_M(n) = \sum_{i=1}^{M} \lambda^{M-i} |e(i)|^2 \tag{10.33}$$

Equation (10.33) represents the weighted error function where exponential weighting is used. In this weighting function, the parameter λ lies between 0 and 1 and it is known as the *forgetting factor*. The weighting factor weighs or emphasizes the most recent error more while the earlier errors are gradually forgotten over time. An exponential weighting factor of this kind effectively gives us a memory. This length of memory represents when the effect of the earlier inputs is forgotten. When executing the RLS algorithm you want to make sure that the memory is at least as long as the filter, so if the filter has M taps then the weighting factor should be chosen so that the memory exists for at least M samples.

$$N = \frac{\sum_{n=0}^{\infty} n\lambda^n}{\sum_{n=0}^{\infty} \lambda^n} = \frac{\lambda}{1-\lambda} \tag{10.34}$$

The error in the RLS algorithm, just like the LMS algorithm, is the difference between the desired response $d(n)$ and the estimate of the desired response $y(n)$ at sample time n as shown in Figure 10.5. The estimate of the desired response $y(n)$ is produced by the filter with tap weights $h_M(n)$ and the input vector $X_M(n)$, so the error can be written as shown in Eq. (10.35):

$$\begin{aligned} e(n) &= d(n) - y(n) \\ &= d(n) - h_M^T(n) X_M(n) \end{aligned} \tag{10.35}$$

When we substitute the error from Eq. (10.35) in the error cost function in Eq. (10.33), the Weiner–Hopf equations are slightly altered. To account for these alterations, we need to modify some of the terms that we used in developing the LMS algorithm. First modification is the autocorrelation term in the Weiner–Hopf equations. The equivalent to the autocorrelation function with the forgetting factor is shown in Eq. (10.36):

$$\hat{\Gamma}_{xx}(n) = \sum_{i=1}^{M} \lambda^{M-i} X_M(i) X_M^T(i) \tag{10.36}$$

The modified autocorrelation type of matrix in Eq. (10.36) is not a true autocorrelation. First, it is not a *Toeplitz matrix*. Since it is not a true autocorrelation matrix, we are not guaranteed that it will be a nonsingular matrix. This means that we cannot assume that the inverse of the matrix will always exist. One solution to making sure that the inverse exists is to keep on increasing M (the size of the filter) till we find $\hat{\Gamma}_{xx}$ that is nonsingular. If we attempt to do this, then a lot of effort will be wasted in examining the various modified autocorrelation matrices till we find one that is not a singular matrix and that the matrix $\hat{\Gamma}_{xx}$ will be invertible. To get around this trial and error method, we generally modify Eq. (10.36) as shown in Eq. (10.37). In Eq. (10.37), we have added a small constant to the modified autocorrelation matrix. This small constant will become our initial of beginning modified autocorrelation matrix and now the modified autocorrelation matrices will always be invertible.

$$\hat{\Gamma}_{xx}(n) = \sum_{i=1}^{M} \lambda^{M-i} X_M(i) X_M^T(i) + \delta I$$

$$\hat{\Gamma}_{xx}(0) = \delta I$$

$$(10.37)$$

The small bias that is introduced in the modified autocorrelation matrix is forgotten over time due to the effect of the forgetting factor. Another term that we have in the Weiner–Hopf equations is the cross-correlation vector between the input and the desired output. The cross-correlation vector is modified as shown in Eq. (10.39). With these two modifications, the normal equations can be written as shown in Eq. (10.38):

$$\hat{\Gamma}_{xx}(n) h_M(n) = D_M(n)$$

$$(10.38)$$

In the modified normal equations, the modified cross-correlation vector is modified and it is written as shown in Eq. (10.39):

$$D_M(n) = \sum_{i=1}^{M} \lambda^{M-i} X_M^T(i) d(i)$$

$$(10.39)$$

In the modified normal equation since $\hat{\Gamma}_{xx}$ is known to be invertible, due to the introduction of the small bias at iteration number zero, we can invert it and get the solution for the optimum filter coefficients at sample time n as shown in Eq. (10.40):

$$h_M(n) = \hat{\Gamma}_{xx}^{-1}(n) D_M(n)$$

$$(10.40)$$

The solution for the optimum filter in Eq. (10.40) requires that we obtain the vector $D_M(n)$, the $\hat{\Gamma}_{xx}$ matrix, and its inverse with every new input sample that is received. This computation would be too much time consuming and prohibitive if these computations are done directly. To reduce the computation requirement, the modified cross-correlation vector and the inverse of the modified autocorrelation vector are computed recursively as shown in the next section.

10.6.2
Recursive Computation of $D_M(n)$ and $\hat{\Gamma}_{xx}$

The computation of the modified autocorrelation matrix is done recursively and hence the method gets its name. The computation is regularized by using an initial value of the inverse of the $\hat{\Gamma}_{xx}$ matrix and then computing the next value of the inverse of the $\hat{\Gamma}_{xx}$ matrix by using very well-known lemma for the inversion of the matrix. To use the matrix inversion lemma, we first rewrite Eq. (10.37) as shown in Eq. (10.41), where we have isolated the term $i = M$:

$$\hat{\Gamma}_{xx}(n) = \lambda \left(\sum_{i=1}^{M-1} \lambda^{M-i} X_M(i) X_M^T(i) + \delta I \right) + X_M(n) X_M^T(n)$$

$$\hat{\Gamma}_{xx}(n) = \lambda \hat{\Gamma}_{xx}(n-1) + X_M(n) X_M^T(n)$$

(10.41)

In Eq. (10.41), the term $\hat{\Gamma}_{xx}(n-1)$ represents the "old" value of the modified autocorrelation matrix at iteration $(n - 1)$. To get the present value of the modified autocorrelation matrix, we multiply the previous value of the matrix with the forgetting factor and then to this matrix we add the matrix product $X_M(n) X_M^T(n)$. This term plays the role of correction and will update the modified autocorrelation matrix from iteration $(N - 1)$ to iteration N. Note that with every new sample we multiply the old modified autocorrelation matrix with λ, the forgetting factor. So over time the bias δI that is introduced initially is forgotten. For the same reason, the early values of the correction factor $X_M(n) X_M^T(n)$ are also forgotten. Equation (10.41) shows us how to compute the present value of the modified autocorrelation matrix from the previous value of the autocorrelation matrix. We can similarly compute the modified cross-correlation vector as shown in Eq. (10.42). Note that here also we are computing the present value of the modified cross-correlation vector using the previous value and an update term or we are computing the cross-correlation vector recursively.

$$D_M(n) = \lambda \left(\sum_{i=1}^{M-1} \lambda^{M-i} X_M^T(i) d(i) \right) + X_M^T(n) d(n)$$

$$= \lambda D_M(n-1) + X_M^T(n) d(n)$$

(10.42)

Now we have to see how we can invert the modified autocorrelation matrix recursively; that is, we begin with the inverse at time $(N - 1)$ and use it to compute the inverse of the matrix at time N. To do this we turn to a very well-known *matrix inversion lemma*. The matrix inversion lemma can be stated as follows, given a matrix that is written as

$$A = B^{-1} + CD^{-1}C^H$$

The inverse of the matrix A can be obtained as

$$A^{-1} = B - BC(D + C^H BC)^{-1} C^H B$$

Using the matrix inversion lemma, we can invert the modified autocorrelation matrix when we set the following equivalencies using Eq. (10.41):

$$A^{-1} = \Gamma_{xx}^{-1}(n)$$
$$B = \lambda^{-1} \Gamma_{xx}^{-1}(n-1)$$
$$C = X_M(n)$$
$$C^H = X_M^H(n)$$
$$D = 1$$

So using the matrix inversion lemma we can write the inverse of the modified autocorrelation matrix at iteration time N as shown in Eq. (10.43). The advantage of using the matrix inversion lemma becomes apparent when we examine the denominator of the second term. Note that the denominator of the second term is a scalar. Being a scalar, the division is a simple arithmetic operation and does not require the inversion of a matrix. So to compute the new inverse of the modified autocorrelation matrix we do not need to invert any matrix. This leads to very simpler computations.

$$\hat{\Gamma}_{xx}^{-1}(n) = \lambda^{-1}\hat{\Gamma}_{xx}^{-1}(n-1) - \frac{\lambda^{-1}\hat{\Gamma}_{xx}^{-1}(n-1)X_M(n)\lambda^{-1}X_M^H(n)\hat{\Gamma}_{xx}^{-1}(n-1)}{1+\lambda^{-1}X_M^H(n)\hat{\Gamma}_{xx}^{-1}(n-1)X_M(n)} \tag{10.43}$$

Usually in using the algorithm we compute an in-between value of gain as shown in Eq. (10.44). This definition is of a gain vector known as the *Kalman gain vector*. The Kalman gain vector is defined as shown in Eq. (10.44). The Kalman gain is a vector with M elements.

$$K(n) = \frac{\lambda^{-1}\hat{\Gamma}_{xx}^{-1}(n-1)X_M(n)}{1+\lambda^{-1}X_M^H(n)\hat{\Gamma}_{xx}^{-1}(n-1)X_M(n)} \tag{10.44}$$

When we substitute this definition of the Kalman gain vector in Eq. (10.43), the recursive computation of the inverse of the modified autocorrelation can be written as shown in Eq. (10.45). Equation (10.45) is known as the *Riccati equation*.

$$\hat{\Gamma}_{xx}^{-1}(n) = \lambda^{-1}\hat{\Gamma}_{xx}^{-1}(n-1) - \lambda^{-1}K(n)X_M^H(n)\hat{\Gamma}_{xx}^{-1}(n-1) \tag{10.45}$$

10.6.3
The RLS Algorithm

Now we are ready to see how the RLS algorithm is executed. We begin with the modified normal equations given in Eq. (10.40). We first use the recursive relation from Eq. (10.42) for the cross-correlation vector as shown in Eq. (10.46):

$$\begin{aligned}
\boldsymbol{h}_m(n) &= \hat{\Gamma}_{xx}^{-1}(n)D_M(n) \\
&= \lambda\hat{\Gamma}_{xx}^{-1}(n)D_M(n-1) + \hat{\Gamma}_{xx}^{-1}(n)X_M^T(n)d(n) \\
&= \lambda\hat{\Gamma}_{xx}^{-1}(n)D_M(n-1) + K(n)d(n)
\end{aligned} \tag{10.46}$$

In Eq. (10.46), we have used the relation $\hat{\Gamma}_{xx}^{-1}(n)X_M^T(n) = K(n)$. It is left for you to show that this relation is indeed true. As a hint to do this, rewrite Eq. (10.44) as two terms where the second term has $K(n)$ as one of the factors. Next we substitute for the inverse of the modified autocorrelation matrix from Eq. (10.45) to get Eq. (10.47):

$$\boldsymbol{h}_m(n) = \lambda\left(\lambda^{-1}\hat{\Gamma}_{xx}^{-1}(n-1) - \lambda^{-1}K(n)X_M^H(n)\hat{\Gamma}_{xx}^{-1}(n-1)\right)D_M(n-1) + K(n)d(n)$$

$$= \lambda\lambda^{-1}\hat{\Gamma}_{xx}^{-1}(n-1)D_M(n-1) - \lambda\lambda^{-1}K(n)X_M^H(n)\hat{\Gamma}_{xx}^{-1}(n-1)D_M(n-1) + K(n)d(n)$$

$$= \hat{\Gamma}_{xx}^{-1}(n-1)D_M(n-1) - K(n)\left[X_M^H(n)\ \underbrace{\hat{\Gamma}_{xx}^{-1}(n-1)D_M(n-1)}_{\text{optimum filter at iteration } n-1} - d(n)\right]$$

$$= \boldsymbol{h}_m(n-1) - K(n)\left[X_M^H(n)\boldsymbol{h}_m(n-1) - d(n)\right]$$

$$= \boldsymbol{h}_m(n-1) + K(n)\left[\underbrace{d(n) - X_M^H(n)\boldsymbol{h}_m(n-1)}_{\text{error at iteration } n}\right]$$

$$= \boldsymbol{h}_m(n-1) + K(n)e(n)$$

$$(10.47)$$

In developing the evolutions in Eq. (10.47), we have made use of the fact that the product $\hat{\Gamma}_{xx}^{-1}(n-1)D_M(n-1)$ is equal to $\boldsymbol{h}_m(n-1)$, the optimum filter at iteration $(n-1)$ and $\left[d(n) - X_M^H(n)\boldsymbol{h}_m(n-1)\right]$ represents the error using the optimum filter at iteration $(n-1)$ and the desired output at instant n. So the last equation in Eq. (10.47) represents the update to the filter coefficients. In this equation, the Kalman gain vector is a vector of M terms; hence, there is a unique correction for each of the M filter coefficients.

10.6.4
Summary of the RLS Algorithm

Equations (10.33), (10.35), (10.44), (10.45), and (10.47) collectively represent the RLS algorithm, which is summarized in Table 10.1. As noted in Eq. (10.47), the error is computed using the present input and the previous optimum filter. This error is known as a priori error.

Table 10.1 Summary of RLS algorithm.

Initialize

$h_M(0) = 0$ $\qquad\qquad$ $\hat{\Gamma}_{xx}(0) = \delta I$ $\qquad\qquad$ $\delta = $ small for high SNR

$\qquad\qquad\qquad\qquad\qquad\qquad\qquad\qquad\qquad\qquad\qquad\qquad\qquad$ $\delta = $ small for high SNR

Recursively compute for $n = 1, 2, \ldots$

$$K(n) = \frac{\hat{\Gamma}_{xx}^{-1}(n-1)X_M(n)}{\lambda + X_M^T(n)\hat{\Gamma}_{xx}^{-1}(n-1)X_M(n)}$$

$$e(n) = d(n) - h_M^T(n-1)X_M(n)$$

$$h_M(n) = h_M(n-1) + K(n)e(n)$$

$$\hat{\Gamma}_{xx}^{-1}(n) = \lambda^{-1}\hat{\Gamma}_{xx}^{-1}(n-1) - \lambda^{-1}K(n)X_M^T\hat{\Gamma}_{xx}^{-1}(n-1)$$

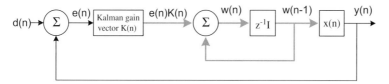

Figure 10.9 Signal flow graph of the RLS algorithm.

So in Eq. (10.47) we are updating the filter weights using a priori error and the Kalman gain vector. Equation (10.44) allows us to update the gain vector itself for the next go around. This systematic approach (using the previous value and the correction term to get the next value) describes an important feature of the RLS algorithm. The required inversion of the modified autocorrelation matrix is replaced by a scalar division as shown in Eq. (10.43). The RLS algorithm can be represented as shown in Figure 10.9. In the signal flow graph, the error is computed in the first summation, and this error is multiplied by the Kalman gain vector; this product represents the correction term that computes the filter to be used for the next iteration. The output from the filter is the estimate of the desired output that is a scalar value and is used as negative feedback to once again compute the error. Note the blocks that produce scalar and vector outputs. This time the correction term for the filter coefficients is a vector and not a scalar as it was in the LMS algorithm.

Here we will once again demonstrate the use of the RLS algorithm using the same two examples that we used in the gradient descent algorithm. The first example is the identification of the filter.

Example 10.3

Using the LMS algorithm we want to match the system filter to the adaptive filter. The system diagram is shown in Figure 10.10.

Solution 10.3: In Figure 10.10, we see first that the input signal $x[n]$ is filtered using a system filter. Noise is added to this filtered signal. After the noise signal is added to the output from the filter, we get the desired signal. This signal is labeled as $d[n]$. The input signal to the filter $x[n]$ is also the input to the adaptive filter. The input signal to the filter to be matched is also the input to the adaptive filter. The output from the adaptive filter is compared with the desired signal to get the error signal. The error signal is used to adjust the filter coefficients. Figure 10.10 compares the coefficients of the system filter used and the adaptive filter coefficients. From the plot we can see that the two are indeed very similar to each other and hence the adaptive filter minimizes the error between the desired signal and the output from the adaptive filter. A MATLAB script is added that simulates the adaptive filter as given in MATLAB 10.3.

Compare this figure with Figure 10.7 where we have designed the same filter using the LMS algorithm. In this figure, we have used the RLS algorithm. Note that the block diagrams of the system are identical no matter what algorithm we use to design the adaptive filter. The change occurs in the algorithm, so the change is inside the block labeled adaptive filter.

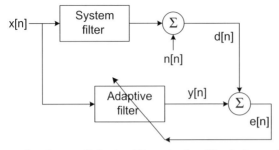

Comparing the coefficients of the adaptive filter to be matched.

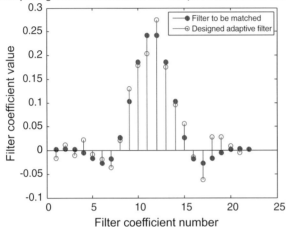

Figure 10.10 Comparing the filter coefficients of the system filter and a filter designed by adaptive filter using the RLS algorithm to match the system filter.

MATLAB 10.3 RLS algorithm to match a system filter.

```
%% Initialize Setup the constants and the signal
N = 440;                    % Size of the signal
w_cf = 0.25;                % Cutoff frequency of the filter to be matched.
t = 0:2/N:(2-2/N);          % Time vector
Nf = 21;                    % Size of the filter.
x = randn(1,N);             % Signal input to the filter.
v = 0.6*randn(1,N);         % Noise signal corrupting the output from the
                              filter
h = fir1(Nf,w_cf);          % Filter to be matched.
d = filter(h,1,x) + v;      % Desired signal.
%% Build the required zero vectors to store intermediate results
y = zeros(1,N);             % Vector to store output from the filter.
```

```
e = zeros(1,N);          % Vector to store the error terms
w = zeros(1,Nf+1);       % Vector to store adaptive filter coefficients
xd = zeros(1,Nf+1);      % Vector to store current input to the filter.
lamda = 0.9975;          % Error Weighting factor.
delta = 1;               % Constant
R_inv = (1/0.01*delta)*eye(Nf+1); % First autocorrelation matrix
stem(1:(Nf+1),h,'fill');hold; %Plot the filter coefficients
%% Start the RLS algorithm to compute the Adaptive filter.
for i = (Nf+1):N
    xd =[ x(i) xd(1:length(xd)-1)]; %Next input vector
    K = (R_inv*xd')/(lamda + xd*R_inv*xd'); % Compute Kalman Gain
    y(i) = xd*w';                   % Next estimate of desired signal.
    e(i) = d(i) - y(i);             % Next error signal
    w = w + e(i)*K';                % Next filter
    R_inv = (1/lamda)*(R_inv - K*xd*R_inv); %Next autocorrelation
matrix
end
stem(1:(Nf+1),w,'r');    %Plot the Adaptive filter.
legend('Filter to be Matched','Designed Adaptive Filter');
title('Comparing the coefficients of the Adaptive filter with the
filter to be matched')
xlabel('Filter coefficient number')
ylabel('Filter coefficient value');
```

Example 10.4

The second example that we demonstrated was to eliminate the noise from a signal. The system diagram is shown in Figure 10.11.

  Solution 10.4: In Figure 10.11, we begin with the signal $d[n]$, which is the desired signal. We assume that this signal is transmitted over a noisy channel and as it travels it picks up noise $v[n]$. It is desired to recreate the original signal that was transmitted, so we pass the received signal through an adaptive filter. From the plot we see that the signal $y[n]$, which is the output of the adaptive filter, is very close to the desired signal.

  The MATLAB script given in MATLAB 10.4 shows how the adaptive filter is developed; first we build the desired signal and then corrupt it with noise that is Gaussian and zero mean. This time we have chosen the desired signal so that the energy in the signal increases as time goes on. The plot of the desired signal and the noisy signals is shown in the top plot in Figure 10.11. Next we build three vectors that are all set to zero initially. In this particular example, we have chosen the size of the filter to be 7 taps. You should try the algorithm with different size filters. For this demonstration we have chosen the forgetting factor to be 0.9275. You should try the algorithm with the forgetting factor smaller than this value. When you do try the algorithm with smaller forgetting factors, you should find that the estimate of the desired signal follows the desired signal only when the energy in the signal is larger.

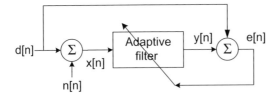

Using the RLS algorithm to remove noise from a signal

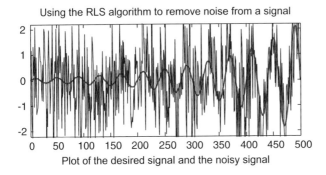

Plot of the desired signal and the noisy signal

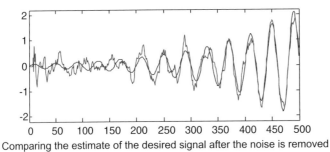

Comparing the estimate of the desired signal after the noise is removed

Figure 10.11 Comparing the signal after noise removal by adaptive filter, using the RLS algorithm to match the desired signal.

As the algorithm runs, the estimate of the desired signal initially is noisy but as the energy in the signal increases the estimate is able to follow it very closely. The initial modified autocorrelation matrix is set by the ratio of the energies in the desired signal and the noise signal. Experiment with different energies for the signal and the noise. Increasing the signal energy will give you better results while increasing the noise energy will give you an estimate that does not resemble the signal very closely initially but as the energy in the desired signal increases the estimate catches up with the desired signal.

MATLAB 10.4 Noise removal from a received signal using the RLS algorithm.

```
%% Initialize the required vectors
N = 0:499;              % Length of the signal vector.
```

```
L = 7;                    % Length of the Adaptive filter to be used.
lamda = 0.9725;      % The forgetting factor.
d = 0.08*1.1.^(0.07*N).*sin(0.05*pi*N);% Signal with increasing
energy.
v = 1.25*randn(1,length(d));     % Noise signal to be removed.
delta = sum(d*d')/sum(v*v');     % Initial value used for R_inv.
R_inv = (1/0.1/delta)*eye(L);    % Initial R_inv matrix.
w = zeros(L,1);                  % Build a zero vector for filter
                                    coefficients
x = d + v;                       % Corrupt desired signal with the noise
                                    signal
xd = zeros(1,L);             % Vector of current signals.
y = zeros(length(N),1);      % Build zero vector for estimated signal.
e = zeros(length(N),1);      % Build zero vector for error signal.
subplot(211);plot(N,d,'r','linewidth',2);hold;% Plot of
desired signal.
axis([ 0,500,-2.2,2.2] );
subplot(211);plot(N,x,'b');       % Plot of the Noisy signal.
title('Using the RLS algorithm to remove noise from a signal')
xlabel('Plot of the desired signal and the Noisy signal')
%% Execute the algorithm
for i = L:length(N)
    xd =[ d(i) xd(1:length(xd)-1)] ; % Build the next input vector.
    K = (R_inv*xd')/(lamda + xd*R_inv*xd') ; % Compute Kalman Gain.
    y(i) = xd*w; % Compute the next estimate of desired signal.
    e(i) = x(i) - y(i); % Compute the next error.
    w = w + e(i)*K;% Update the filter coefficients to be used next
    R_inv = (1/lamda)* (R_inv-K*xd*R_inv) ; %Inverse of Auto Matrix
end
subplot(212);plot(N,[ d' y] ); %Plot the desired signal and the
estimate
axis([ 0,500,-2.2,2.2] );
xlabel('Comparing the estimate of the desired signal after Noise
removal')
```

When you examined Example 10.4 and tried it with different values of the forgetting factor and the energy in the signal, you will have seen that there is more or less time to convergence. In general, we can make the following statements about the relation between the noise and the forgetting factor:

a) **When the signal to noise ratio is high**: When the noise level in the signal at the tap inputs is low (SNR is at least 30 dB), the RLS algorithm exhibits a very fast rate of convergence if the modified autocorrelation matrix is initialized with a small initial value. This can be chosen as the reciprocal of the signal to noise ratio.

b) **When the signal to noise ratio is medium**: When the noise level in the signal at the tap inputs is higher but not too high (SNR is between 10 and 30 dB), then the convergence rate is slower. The convergence rate has almost no effect due to the initial value of the modified autocorrelation matrix.

c) **When the signal to noise ratio is low**: When the noise level in the signal at the tap inputs is very low (SNR is less than -10 dB), then the convergence rate is much slower. To improve the convergence rate, we should initialize the initial value of the modified autocorrelation matrix to a very large value.

The above remarks hold for stationary input signal or a signal that is changing very slowly. If there is an abrupt or a step change in the energy of the noise in the signal, then the adaptive filter will not be able to follow. If such a change does occur, then generally we reset the algorithm and start with a new modified autocorrelation matrix that is initialized again.

10.7
Chapter Summary

We began this chapter with the definition of the problem. The problem that we are trying to solve is to design a filter that will be able to estimate the desired signal and to minimize the error between the desired signal and the estimate of the desired signal. This was the same problem that we tried to solve in Chapter 9 where we studied the Weiner filter. The difference between the Weiner filter and the adaptive filter is the method that we use to solve the problem. Adaptive filters are used in many different ways. The application differs in how you define the required signal and what error you are trying to minimize.

The first method that we studied is the method of steepest descent. In this method, we first defined an error cost function. The error cost function defined a bowl-shaped surface. The method of steepest descent determines the gradient on the bowl surface and then moves in the direction opposite to the steepest gradient. This gets us to the bottom of the bowl where the error is the smallest possible.

To compute the gradient we saw that we need to solve the Weiner–Hopf equations. To solve these equations we need to compute the autocorrelation matrix and the cross-correlation vector. The idea of the adaptive filter is to not directly compute the autocorrelation matrix and the cross-correlation vector but to estimate the required quantities by using instantaneous quantities. As we saw in Eq. (10.14), the estimate that we use is the product of the error and the input vector.

The method of steepest descent uses the step size parameter to control the convergence rate of the algorithm. Making the step size too large could possibly make the algorithm unstable. To make sure that the algorithm is stable, we limit the step size to be less than two times the reciprocal of the largest eigenvalue as shown in Eq. (10.28). As a further modification to the selection of the step size, we saw that we could choose the step size as $\mu < 2/M\gamma_{xx}(0)$, where $\gamma_{xx}(0)$ is the energy in the signal.

The time it takes for the algorithm to converge is given in Eq. (10.30). This time is dependent on the smallest eigenvalue in the autocorrelation matrix. So the stability of the algorithm is governed by the largest eigenvalue while the time it takes to reach the stable solution is governed by the smallest eigenvalue. This tells us that when the ratio of the eigenvalues is very large, then it could take a very long time for convergence. However, this method is very simple and robust and can follow various types of signals with a minimum amount of computation.

There are many other variations to the stochastic gradient algorithm. Each of these algorithms attempts to adjust either the speed of convergence or the noise that is present in the error signal. Each method either computes the gradient in a different way or adjusts the step size in a different way. One method used to adjust the step size is to use the energy that is present in the signal.

Two examples and the MATLAB script were given to demonstrate the steepest descent algorithm.

The second method of designing the adaptive filters is the recursive least squares algorithm. In this method, we use weighted error to adjust the filter coefficients. With weighting the error vector, the Weiner–Hopf equations that we solve to reach the optimum value have to be modified slightly. This modification is shown in Eq. (10.38), where the modified autocorrelation matrix and the modified cross-correlation vectors are used. This method requires more computations than the steepest descent algorithm but the gain that we get is that this algorithm converges faster than the LMS algorithm. The convergence rate is faster due to the fact that there is a separate adjustment parameter for each mode in the autocorrelation matrix.

To begin this algorithm we assume that the modified autocorrelation matrix has some unique value as shown in Eq. (10.37). This matrix is inverted and a new estimate of the filter coefficients is determined. Using this estimate of the optimum filter and a new input, a new error is computed. Since we have a new input, a new inverse of the new modified autocorrelation matrix has to be computed. The computing of the inverse of the modified autocorrelation matrix could become problematic. This problem is avoided by the use of the matrix inversion lemma. The inversion of the modified autocorrelation matrix is performed recursively as shown in Eq. (10.43). Performing the matrix inversion recursively relieves us of a large burden in computation.

In this algorithm we use the forgetting factor to emphasize the more recent error while we try to forget what the error was a long time ago. The forgetting factor is always positive and it is always less than 1. Choosing a forgetting factor close to 1 implies that we are going to forget slowly while a forgetting factor close to zero implies that we forget very quickly.

We demonstrated the algorithm using the same two examples that we used to demonstrate the LMS algorithm.

Further Reading

Introduction to Stochastic Processes

1 Wiener, N. (1949) *Exploration, Interpolation and Smoothing of Stationary Time Series*, MIT Press/John Wiley & Sons, Inc., Cambridge, MA/New York.
2 Goren, C.H. (1969) *Go with the Odds*, Macmillan, New York.
3 Papoulis, A. (1984) *Probability, Random Variables and Stochastic Processes*, 2nd edn, McGraw-Hill, New York.
4 Blake, I.F. (1979) *An Introduction to Applied Probability*, John Wiley & Sons, Inc., New York.
5 Cooper, G.R. and McGillem, C.D. (1986) *Probabilistic Methods of Signal and System Analysis*, 2nd edn, Holt, Rinehart and Winston, New York.
6 Peevbles, P.Z., Jr. (1993) *Probability, Random Variables and Random Signal Principles*, 3rd edn, McGraw-Hill, New York.
7 Walpole, R.E. and Myers, R.H. (1985) *Probability and Statistics for Engineers and Scientists*, 3rd edn, Macmillan Publishing Company, New York.
8 Blackman, R.B. (1965) *Linear Data-Smoothing and Prediction in Theory and Practice*, Addison-Wesley, Reading, MA.
9 Douglas, C.M. and Runger, G.C. (1999) *Applied Statistics and Probability for Engineers*, 2nd edn, John Wiley & Sons, Inc., New York.

Discrete Time Signals and Systems

1 Ahmed, N. and Natarajan, T. (1983) *Discrete Time Signals and Systems*, Reston Publishing, Reston, VA.
2 Gabel, R.A. and Roberts, R.A. (1980) *Signals and Linear Systems*, 2nd edn, John Wiley & Sons, Inc., New York.
3 Reid, J.G. (1983) *Linear System Fundamentals*, McGraw-Hill, New York.
4 Seely, S. and Poularikas, A.D. (1991) *Signals and Systems*, 2nd edn, PWS-Kent, Boston, MA.
5 Stanley, W.D., Dougherty, G.R., and Dougherty, R. (1984) *Digital Signal Processing*, 2nd edn, Reston Publishing, Reston, VA.
6 Neff, H.P., Jr. (1984) *Continuous and Discrete Linear Systems*, Harper & Row, New York.
7 Jong, M.T. (1982) *Methods of Discrete Signal and System Analysis*, McGraw-Hill, New York.
8 Chen, C. (1979) *One-Dimensional Digital Signal Processing*, Marcel Dekker, New York.

Filter Design

1 Terrell, T.J. (1980) *Introduction to Digital Filters*, The Macmillan Press Ltd., London.
2 Zeimer, R.E., Tranter, W.H., and Fannin, D.R. (1983) *Signals and Systems: Continuous and Discrete*, Macmillan, New York.

Digital Filters: Theory, Application and Design of Modern Filters, First Edition. Rajiv J. Kapadia.
© 2012 Wiley-VCH Verlag GmbH & Co. KGaA. Published 2012 by Wiley-VCH Verlag GmbH & Co. KGaA.

3 Stanley, W.D., Dougherty, G.R., and Dougherty, R. (1984) *Digital Signal Processing*, 2nd edn, Reston Publishing, Reston, VA.

4 Neff, H.P., Jr. (1984) *Continuous and Discrete Linear Systems*, Harper & Row, New York.

5 Jackon, L.B. (1996) *Digital Filters and Signal Processing*, 3rd edn, Kluwer Academic Publishers, Boston, MA.

6 Parks, T.W. and Burns, C.S. (1987) *Digital Filter Design*, John Wiley & Sons, Inc., New York.

Z-Transforms

1 Seely, S. and Poularikas, A.D. (1988) *Elements of Signals and Systems*, PWS-Kent, Boston, MA.

2 Kuo, B.C. (1995) *Automatic Control Systems*, 7th edn, Prentice-Hall, Englewood Cliffs, NJ.

3 Jury, E.I. (1964) *Theory and Application of the Z-Transform Method*, John Wiley & Sons, Inc., New York.

4 Papoulis, A. (1962) *The Fourier Integral and Its Applications*, McGraw-Hill, New York.

5 Morrison, N. (1994) *Introduction to Fourier Analysis*, John Wiley & Sons, Inc., New York.

6 Proakis, J.G. and Manolakis, D.G. (1988) *Introduction to Digital Signal Processing*, Macmillan, New York.

7 Vich, R. (1987) *Z-Transform Theory and Applications*, D. Reidel Publishing, Dordrecht, The Netherlands.

8 Oppenheim, A.V. and Schafer, R.W. (1999) *Discrete Time Signal Processing*, 2nd edn, Prentice-Hall, Upper Saddle River, NJ.

Computing the DFT

1 Cooley, J.W. and Tukey, J.W. (1965) An algorithm for the machine calculation of complex Fourier series. *Mathematics of Computation*, **19**, 297–301.

2 Cooley, J.W. (1992) How the FFT gained acceptance. *IEEE Signal Processing Magazine*, **9** (1), 10–13.

3 Gold, B. and Rader, C.M. (1969) *Digital Processing of Signals*, McGraw-Hill, New York.

4 Antiniou, A. (1993) *Digital Filters: Analysis Design and Applications*, 2nd edn, McGraw-Hill, New York.

5 Mitra, S.K. (1998) *Digital Signal Processing: A Computer-Based Approach*, McGraw-Hill, New York.

6 Rabiner, L.R. and Gold, B. (1975) *Theory and Application of Digital Signal Processing*, Prentice-Hall, Englewood Cliffs, NJ.

7 Hamming, R.W. (1998) *Digital Filters*, 3rd edn, Dover Publications, Mineola, NY.

8 Bellanger, M. (1989) Chapter 1, in *Digital Processing of Signals*, 2nd edn, John Wiley & Sons, Inc., New York.

9 Brigham, E.O. (1974) *The Fast Fourier Transform*, Prentice-Hall, Englewood Cliffs, NJ.

Multirate Signal Processing

1 Strang, G. and Nguyen, T. (1997) *Wavelets and Filter Banks*, Wellesley-Cambridge Press, Wellesley, MA.

2 McClellan, J.H., Schafer, R.W., and Yoder, M.A. (1998) *DSP First: A Multimedia Approach*, Prentice Hall, Upper Saddle River, NJ.

3 Tretter, S.A. (1976) *Introduction to Discrete Time Signal Processing*, John Wiley & Sons, Inc., New York.

4 Hamming, R.W. (1998) *Digital Filters*, 3rd edn, Dover Publications, Mineola, NY.

5 Clark, G.A., Mitra, S.K., and Parker, S.R. (1981) Block implementation of adaptive digital filters. *IEEE Transactions on Circuits and Systems*, **28**, 584–592.

6 Mitra, S.K. (1998) *Digital Signal Processing: A Computer-Based Approach*, McGraw-Hill, New York.

MATLAB

1 Linz, P. and Wang, R.L.C. (2003) *Exploring Numerical Methods: An Introduction to Scientific Computing Using MATLAB*, Jones and Bartlett Publishers, Boston, MA.

2 Etter, D.M. (1993) *Engineering Problem Solving with MATLAB*, Prentice-Hall, Englewood Cliffs, NJ.

3 Proakis, J.G., Rader, C.M., Ling, F., Nikias, C.L., Moonen, M., and Proudler, I.K. (2002) Chapter 2, in *Algorithms for Statistical Signal*

Processing, Prentice-Hall, Upper Saddle River, NJ.

4 Roberts, M.J. (2004) *Signals and Systems: Analysis Using Transform Methods and MATLAB*, McGraw-Hill, New York.

5 Poularikas, A.D. and Ramadan, Z.M. (2006) *Adaptive Filtering Primer with MATLAB*, CRC Press, Boca Raton, FL.

Adaptive Filters

1 Clary, J. (1984) Robust algorithms in adaptive control. Ph.D. thesis, Stanford University.

2 Bryson, A.E. (1978) Kalman filter divergence and aircraft motion estimators. *AIAA Journal of Guidance and Control*, **1** (1), 71–79.

3 Bryson, A.E. and Ho, Y.C. (1975) *Applied Optimal Control*, Halsted Press, Washington, DC.

4 Balakrishnan, A.V. (1984) *Kalman Filtering Theory*, Optimization Software Inc., New York

5 Haykin, S. (1986) *Adaptive Filter Theory*, Prentice-Hall, Englewood Cliffs, NJ.

6 Haykin, S.S. (1991) *Adaptive Filter Theory*, 2nd edn, Prentice-Hall, Englewood Cliffs, NJ.

7 Wedrow, B. *et al.* (1975) Adaptive noise canceling: principles and applications. *Proceedings of the IEEE*, **63** (12), 1692–1716.

8 Sorenson, H.W. (1970) Least squares estimation: from Gauss to Kalman. *IEEE Spectrum*, **7**, 63–68.

Weiner Filters

1 Bode, H.W. and Shannon, C.E. (1950) A simplified derivation of linear least squares smoothing and prediction theory. *Proceedings of the IRE*, **38**, 417–425.

2 Haykin, S. (1986) *Adaptive Filter Theory*, Prentice-Hall, Englewood Cliffs, NJ.

3 Widrow, B. and Stearns, S.D. (1985) *Adaptive Signal Processing*, Prentice-Hall, Englewood Cliffs, NJ.

4 Levinson, N. (1947) The Weiner RMS error criterion in filter design and prediction. *Journal of Mathematical Physics*, **25**, 261–278.

5 Satorius, E.H. and Pack, J.D. (1981) Application of least squares lattice algorithm for adaptive equalization. *IEEE Transactions on Communications*, **COM-29**, 136–142.

6 Parlett, B.N. (1980) *The Symmetric Eigenvalue Problem*, Prentice-Hall, Englewood Cliffs, NJ.

7 Wilkinson, J.H. (1965) *The Algebraic Eigenvalue Problem*, Clarendon Press, Oxford.

8 Lawson, C.L. and Hanson, R.J. (1974) *Solving Least Squares Problem*, Prentice-Hall, Englewood Cliffs, NJ.

Discrete Time Systems in the Frequency Domain

1 Ludeman, L. (1986) *Fundamentals of Digital Signal Processing*, John Wiley & Sons, Inc., New York.

2 Cunningham, E.P. (1992) *Digital Filtering: An Introduction*, Houghton Mifflin Co., Boston, MA.

3 Seely, S. and Poularikas, A.D. (1991) *Signals and Systems*, 2nd edn, PWS-Kent, Boston, MA.

4 Antoniou, A. (1993) *Digital Filters: Analysis Design and Applications*, 2nd edn, McGraw-Hill, New York.

5 Cappellini, V., Constantinides, A.G., and Emiliani, P. (1978) *Digital Filters and Their Applications*, Academic Press, London.

6 Szidarovszky, F. and Bahill, A.T. (1992) *Linear Systems Theory*, CRC Press, Boca Raton, FL.

Appendix A: Mathematical Identities

1) Complex numbers:

$$j = \sqrt{-1}$$

$$z = x + iy = Ae^{j\theta}$$

$$z^* = x - iy = Ae^{-j\theta}$$

$$z + z^* = 2\,\mathrm{Re}\,(z) = 2x$$

$$z - z^* = 2\,\mathrm{Im}\,(z) = j2y$$

$$z\,z^* = |z|^2 = x^2 + y^2$$

$$A = \sqrt{x^2 + y^2}$$

$$\theta = \tan^{-1}\left(\frac{y}{x}\right)$$

$$x = A\cos(\theta)$$

$$y = A\sin(\theta)$$

2) Euler's identity:

$$e^{\pm j\theta} = \cos(\theta) \pm \sin(\theta)$$

$$\cos(\theta) = \frac{e^{j\theta} + e^{-j\theta}}{2}$$

$$\sin(\theta) = \frac{e^{j\theta} - e^{-j\theta}}{j2}$$

Digital Filters: Theory, Application and Design of Modern Filters, First Edition. Rajiv J. Kapadia.
© 2012 Wiley-VCH Verlag GmbH & Co. KGaA. Published 2012 by Wiley-VCH Verlag GmbH & Co. KGaA.

3) Trigonometric identities:

$$\cos^2(\theta) + \sin^2(\theta) = 1$$

$$\cos(\theta \pm \phi) = \cos(\theta)\cos(\phi) \mp \sin(\theta)\sin(\phi)$$

$$\sin(\theta \pm \phi) = \cos(\theta)\sin(\phi) \pm \sin(\theta)\cos(\phi)$$

$$\sin(2\theta) = 2\cos(\theta)\sin(\theta)$$

$$\cos(2\theta) = \cos^2(\theta) - \sin^2(\theta)$$

$$\cos^2(\theta) = \frac{1 + \cos(2\theta)}{2}$$

$$\sin^2(\theta) = \frac{1 - \cos(2\theta)}{2}$$

$$\sin(\theta)\cos(\phi) = \frac{\sin(\theta + \phi) + \sin(\theta - \phi)}{2}$$

$$\cos(\theta)\cos(\phi) = \frac{\cos(\theta + \phi) + \cos(\theta - \phi)}{2}$$

$$\sin(\theta)\sin(\phi) = \frac{\cos(\theta - \phi) - \cos(\theta + \phi)}{2}$$

Appendix B: Transform Tables

B.1
Fourier Series

$$x_a(t+T) = x_a(t)$$

$$x_a(t) = \sum_{k=\infty}^{\infty} c_k \, e^{(j2\pi kt/T)}$$

$$c_k = \frac{1}{T} \int_T x_a(t) e^{(-j2\pi kt/T)} dt$$

$$x_a(t) = \frac{a_0}{2} + \sum_{k=\infty}^{\infty} a_k \cos\left(\frac{2\pi kt}{T}\right) + \sum_{k=\infty}^{\infty} b_k \sin\left(\frac{2\pi kt}{T}\right)$$

$$a_k = \frac{2}{T} \int_T x_a(t) \cos\left(\frac{-j2\pi kt}{T}\right) dt = 2\mathrm{Re}\{c_k\}$$

$$b_k = \frac{2}{T} \int_T x_a(t) \sin\left(\frac{j2\pi kt}{T}\right) dt = -2\,\mathrm{Im}\{c_k\}$$

$$x_a(t) = \frac{a_0}{2} + \sum_{k=\infty}^{\infty} a_k \cos\left(\frac{2\pi kt}{T} + \theta_k\right)$$

$$a_k = \sqrt{a_k^2 + b_k^2} = 2|c_k|$$

$$\theta_k = \tan^{-1}\left(\frac{-b_k}{a_k}\right) = \tan^{-1}\left(\frac{\mathrm{Im}\{c_k\}}{\mathrm{Re}\{c_k\}}\right)$$

Digital Filters: Theory, Application and Design of Modern Filters, First Edition. Rajiv J. Kapadia.
© 2012 Wiley-VCH Verlag GmbH & Co. KGaA. Published 2012 by Wiley-VCH Verlag GmbH & Co. KGaA.

B.2
Fourier Transform

The transform: $X_a(f) = \int_{-\infty}^{\infty} x_a(t)e^{-j2\pi ft}dt$

The inverse transform: $x_a(t) = \int_{-\infty}^{\infty} X_a(f)e^{j2\pi ft}df$

Fourier transform pairs:

x(t)	X(f)	Identifying name
$e^{-at}u(t)$	$\dfrac{1}{a+j2\pi f}$	Causal exponential
$e^{j2\pi f_0 t}$	$\delta(f-f_0)$	Shift in frequency
$te^{-at}u(t)$	$\dfrac{1}{(a+j2\pi f)^2}$	Multiplication by linear ramp
$\cos(2\pi f_0 t)$	$\dfrac{\delta(f+f_0)+\delta(f-f_0)}{2}$	Cosine
$\sin(2\pi f_0 t)$	$j\dfrac{\delta(f+f_0)-\delta(f-f_0)}{2}$	Sin
$e^{-at}\sin(2\pi f_0 t)u(t)$	$\dfrac{2\pi f_0}{(a+j2\pi f)^2+(2\pi f_0)^2}$	Damped Sin
$e^{-at}\cos(2\pi f_0 t)u(t)$	$\dfrac{a+j2\pi f}{(a+j2\pi f)^2+(2\pi f_0)^2}$	Damped sin
$\delta(t)$	1	Unit impulse
1	$\delta(f)$	Constant

Fourier transform properties:

x(t)	X(f)	Property
Real	$X^*(f)=X(-f)$	Symmetry
$ax_1(t)\pm\beta x_2(t)$	$\alpha X_1(f)\pm\beta X_2(f)$	Linearity
$x(at)$	$\dfrac{1}{\lvert a\rvert}X\left(\dfrac{f}{a}\right)$	Scaling in time
$x(-t)$	$X(-f)$	Reflection in time
$x(t-t_0)$	$e^{-j2\pi ft_0}X(f)$	Shift in time

(*Continued*)

$x(t)$	$X(f)$	Property
$e^{j2\pi f_0 t}x(t)$	$X(f-f_0)$	Shift in frequency
$\dfrac{d^k x(t)}{dt^k}$	$(j2\pi f)^k X(f)$	Differentiation in time
$(t)^k x(t)$	$\left(\dfrac{1}{2\pi}\right)^k \dfrac{d^k X(f)}{df^k}$	Differentiation in frequency
$\displaystyle\int_{-\infty}^{\infty} x_1(\tau)x_2(t-\tau)d\tau$	$X_1(f)X_2(f)$	Convolution in time
$x_1(t)x_2(t)$	$\displaystyle\int_{-\infty}^{\infty} X_1(\alpha)x_2(f-\alpha)d\alpha$	Convolution in frequency
$\displaystyle\int_{-\infty}^{\infty} x_1(\tau)x_2^*(t+\tau)d\tau$	$X_1(f)X_2^*(f)$	Cross correlation
$\displaystyle\int_{-\infty}^{\infty} x_1(t)x_2^*(t)dt$	$\displaystyle\int_{-\infty}^{\infty} X_1(f)X_2^*(f)df$	Parseval's relation

B.3
Laplace Transform

The transform: $X_a(s) = \displaystyle\int_0^{\infty} x_a(t)e^{-st}dt$

The inverse transform: $x_a(t) = \dfrac{1}{j2\pi}\displaystyle\int_{c-j\infty}^{c+j\infty} X_a(s)e^{st}ds$

Fourier transform pairs:

$x(t)$	$X(s)$	Identifying name
$\delta(t)$	1	Unit impulse
$u(t)$	$\dfrac{1}{s}$	Unit step
$e^{-at}u(t)$	$\dfrac{1}{s+a}$	Causal exponential
$t^m u(t)$	$\dfrac{m!}{s^{(m+1)}}$	Polynomial
$e^{-at}t^m u(t)$	$\dfrac{m!}{(s+c)^{(m+1)}}$	Damped polynomial
$\sin(2\pi f_0 t)u(t)$	$\dfrac{2\pi f_0}{s^2 + (2\pi f_0)^2}$	Sine

(Continued)

$x(t)$	$X(s)$	Identifying name
$\cos(2\pi f_0 t)u(t)$	$\dfrac{s}{s^2 + (2\pi f_0)^2}$	Cosine
$e^{-at}\sin(2\pi f_0 t)u(t)$	$\dfrac{2\pi f_0}{(s+c)^2 + (2\pi f_0)^2}$	Damped sine
$e^{-at}\cos(2\pi f_0 t)u(t)$	$\dfrac{(s+c)}{(s+c)^2 + (2\pi f_0)^2}$	Damped cosine

Laplace transform properties:

$x(t)$	$X(s)$	Property
$x^*(t)$	$X^*(s^*)$	Complex conjugate
$ax_1(t) \pm \beta x_2(t)$	$aX_1(s) \pm \beta X_2(s)$	Linearity
$x(at)$	$\dfrac{1}{a}X\left(\dfrac{s}{a}\right)$	Scaling in time
$x(t-t_0)u(t-t_0)$	$e^{-st_0}X(s)$	Shift in time
$tx(t)$	$-\dfrac{d(X(s))}{ds}$	Multiplication by linear ramp
$\dfrac{d(x(t))}{dt}$	$sX(s)-x(0^+)$	Differentiation in time
$\displaystyle\int_0^t x(\tau)d\tau$	$\dfrac{X(s)}{s}$	Integration in time
$\displaystyle\int_0^t x(\tau)y(t-\tau)d\tau$	$X(s)Y(s)$	Convolution

B.4
Z-Transform

The transform: $X_a(z) = \displaystyle\sum_{k=0}^{\infty} x(k)z^{-k} \qquad R_- < |z| < R_+$

The inverse transform: $x_a(k) = \dfrac{1}{j2\pi}\displaystyle\int_c X_a(z)z^{k-1}dz \qquad |k| = 0, 1, \cdots$

Z-transform pairs:

$x(t)$	$X(z)$	Identifying name
$\delta(k)$	1	Unit impulse
$u(k)$	$\dfrac{z}{z-1}$	Unit step
$ku(k)$	$\dfrac{z}{(z-1)^2}$	Unit ramp
$\alpha^k u(k)$	$\dfrac{z}{z-\alpha}$	Exponential
$\sin(\alpha k)u(k)$	$\dfrac{z\sin(\alpha)}{z^2-2z\cos(\alpha)+1}$	Sine
$\cos(\alpha k)u(k)$	$\dfrac{z(z-\cos(\alpha))}{z^2-2z\cos(\alpha)+1}$	Cosine
$\beta^k\sin(\alpha k)u(k)$	$\dfrac{\beta z\sin(\alpha)}{z^2-2\beta z\cos(\alpha)+\beta^2}$	Damped sine
$\beta^k\cos(\alpha k)u(k)$	$\dfrac{z(z-\beta\sin(\alpha))}{z^2-2\beta z\cos(\alpha)+\beta^2}$	Damped cosine

Z-transform properties:

$x(t)$	$X(z)$	Property
$x^*(k)$	$X^*(z^*)$	Complex conjugate
$\alpha x_1(k)\pm\beta x_2(k)$	$\alpha X_1(z)\pm\beta X_2(z)$	Linearity
$x(-k)$	$X(z^{-1})$	Reversal in time
$x(k-r)$	$z^{-r}X(z)$	Shift in time
$kx(k)$	$-z\dfrac{d(X(z))}{dz}$	Multiplication by linear ramp
$\alpha^k x(k)$	$X\left(\dfrac{z}{\alpha}\right)$	Scaling in frequency
$\displaystyle\sum_{r=-\infty}^{\infty}y(r)x(k-r)$	$X(z)Y(z)$	Convolution

B.5
Discrete Fourier Transform

The transform: $X_a(k) = \sum_{n=0}^{N} x_a(n)e^{-(j2\pi nk/N)}$ $0 \le k < N$

The inverse transform: $x_a(n) = \dfrac{1}{N}\sum_{k=0}^{N} X_a(k)e^{j2\pi nk/N}$ $0 \le n < N$

Discrete Fourier transform properties:

$x(n)$	$X(k)$		Property
$\alpha x_1(\mathbf{n}) \pm \beta x_2(n)$	$\alpha X_1(k) \pm \beta X_2(k)$		Linearity
$x(-n)$	$X^*(k)$	Time reversal	$x(n)$ is periodic and real
$x(n-r)$	$e^{-j2\pi rk/N}X(k)$		Circular shift
$x(k)y(k)$	$X(k)Y(k)$		Circular convolution

Appendix C: Introduction to MATLAB

C.1
Introduction

MATLAB is a high-level programming language that is ideally geared toward evaluating mathematical complex numerical algorithms. I have tried to write all the MATLAB scripts with the assumption that the student reading this book is not familiar with this programming language. Here, I provide a quick overview of the programming language and at the same time I try to point out some programming tricks that help things along.

The language consists of functions that are either built in or they can be written by the user and executed. Each MATLAB script known as M-files consists of sequences of commands the interpreter will execute; as we have seen in the book, a complete new M-file can be written and executed with a few MATLAB commands.

MATLAB works with three different types of windows. The first window is the Command window, identified by the heading Command. In the Command window, you can write the commands that are executed immediately. The second window is the Editor window, identified by the name of the M-file or it is titled Untitled if this is a new file that has not been saved yet. You can execute the M-files either from the Editor window or from the Command window. The third window is the Figure window, identified by the title Figure No 1. Any graphs that you draw are drawn in the Figure window.

C.2
Numbers and Data Representation

MATLAB represents numbers in the conventional way in which we write numbers. The limits for the largest and the smallest number that can be represented in MATLAB are the smallest number being $\pm 10^{-308}$ and the largest number being ± 10308. Representing numbers in MATLAB is done as follows. All the numbers above are equal and can be entered in MATLAB shown as follows:

$$135246.0 \qquad 135.246E-3 \qquad 1.35246e2$$

Digital Filters: Theory, Application and Design of Modern Filters, First Edition. Rajiv J. Kapadia.
© 2012 Wiley-VCH Verlag GmbH & Co. KGaA. Published 2012 by Wiley-VCH Verlag GmbH & Co. KGaA.

All data in MATLAB are represented as a rectangular matrix. The matrix does not require dimensioning as MATLAB will dimension the matrix on the fly. While trying the dimension, the matrix on the fly does slow the speed of operation, and assigning adequate memory space for the matrix first does help. The elements of the matrix can be real or complex or mixed. For example, a vector can be represented in MATLAB as

$$z = [3.5 \quad 4+3^*j \quad -5-3^*j \quad \cos(28) \quad \exp(-2)]$$

In the given example, z is a row vector of length 5. The elements of Z range from $z(1)$ to z (5). Notice we have represented values of the vector as a real constant in element $z(1)$, as complex constants in elements $z(2)$ and $z(3)$, and as functions that MATLAB will evaluate before assigning the value to elements $z(4)$ and $z(5)$. To enter a number as a complex value, we use the symbol * followed by the letter j; if you choose to, you may use the letter I also. In the above example, we have separated the various elements of the vector by spaces; instead of spaces, we can separate the individual elements by commas also. If we wanted a row vector, then we would separate the elements by semicolons. Thus, we get the method to use to enter a multirow matrix. This is done as follows:

$$Y = [1, 2, 3; 4, 5, 6; 7, 8, 9] = \begin{bmatrix} 1 & 2 & 3 \\ 4 & 5 & 6 \\ 7 & 8 & 9 \end{bmatrix}$$

In the example above, we have defined a 3×3 matrix. Matrix operations can now be performed on this matrix; for example, the transpose operator "will give us the transpose of the matrix so, for example, the command $W = Y$" will give us a matrix:

$$W = \begin{bmatrix} 1 & 4 & 7 \\ 2 & 5 & 8 \\ 3 & 6 & 9 \end{bmatrix}$$

Or, an arithmetic operation on matrices can be performed as

$$S = Y + W = \begin{bmatrix} 2 & 6 & 10 \\ 6 & 10 & 14 \\ 10 & 14 & 18 \end{bmatrix}$$

$$D = Y - W = \begin{bmatrix} 0 & -2 & -4 \\ 2 & 0 & -2 \\ 4 & 2 & 0 \end{bmatrix}$$

Notice that all the operations are performed as you would expect them to be executed. Multiplication and divisions of the matrices can be performed in two different ways. For example, we may want to perform matrix multiplication according to the rules of matrix multiplications or we may want to perform element-by-element multiplication. The difference between the two is shown below. Matrix MM is obtained by performing

Matrix Multiplication, while matrix EM is obtained by using element-by-element multiplication. Element-by-element multiplication means you take element (i,j) from each of the two matrices and multiply the elements, the result of multiplication is the (i,j) element of the resulting matrix. The MM and the EM matrices are computed as follows:

$$MM = Y^*W = \begin{bmatrix} 14 & 32 & 50 \\ 32 & 77 & 122 \\ 50 & 122 & 194 \end{bmatrix}$$

$$EM = Y.^*W = \begin{bmatrix} 1 & 8 & 21 \\ 8 & 25 & 48 \\ 21 & 48 & 81 \end{bmatrix}$$

Notice the difference in the results. The matrix MM is the result of matrix multiplication, while the matrix EM is the result of element-by-element multiplication. The difference is communicated to MATLAB by adding the "dot" before the multiplication command. The "dot" tells the MATLAB interpreter that the command that follows is to be performed on the element-by-element basis.

Matrix division is a little more complex. There are four different possibilities depending on the size and the shape of the two matrices involved in the division process. Say that X is a square matrix and Y is a row matrix with the same number of elements as the number of rows of X, then matrix LEFT division $X\backslash Y$ is equivalent to $inv(X)^*Y$, this is how you would solve a system of simultaneous equations. On the other hand, if Y is a column vector with the same number of elements as the number of columns in X, then matrix RIGHT division Y/X is equivalent to $Y^*inv(X)$. The Right division operation can also be carried out if one of the elements is a scalar.

If X and Y are both square matrices, then LEFT division $X\backslash Y$ is equivalent to $inv(X)^*Y$, while RIGHT division X/Y is equivalent to $X^*inv(Y)$. In division also, to perform element-by-element operation, you precede the division operator by the "dot" operator.

C.3
Control Flow

Just as all other programming languages, MATLAB also has control flow commands. These are `if`, `else`, `elseif`, `end`, `for`, and `while`. The control flow commands allow us to selectively execute a group of statements when the defined logical statement evaluates to "True" value. The command `for` or `while` are used to execute a loop several times, the command `if`, `else`, and `elseif` are used to select one of the many possible commands to execute. These commands are demonstrated as follows:

a) **The FOR command**: We want to get a vector of Fibonacci sequence for the first 10 terms. The sequence begins with the first two elements equal to 1, and the remaining sequences are generated by adding the previous two sequences. The MATLAB script would be as follows:

```
x(1) = 1;
x(2) = 1;
for n = 3:10
x(n) = x(n-1)+x(n-2)
end
```

b) **The WHILE Command:** We want to get the same vector of Fibonacci sequence for the first 10 terms as we did with the for sequence. The sequence begins with the first two elements equal to 1, and the remaining sequences are generated by adding the previous two sequences. The MATLAB script would be as follows:

```
x(1) = 1;
x(2) = 1;
n=3;
while (n < 11)
x(n) = x(n-1)+x(n-2)
   n = n + 1
end
```

c) **The IF, ELSE, and ELSEIF Command:** When you have one of, say, the three different sequences that you have to execute depending on the value of a variable, the IF command is particularly useful. This command is executed as follows:

```
if (x < x1)
```

Execute the first group of commands when variable X is less than constant $x1$.

```
elseif ((x>x1) & (x<x2))
```

Execute the second group of commands when variable X is more than constant $x1$ and it is less than constant $x2$.

```
else (x > x2)
```

Execute the last group of commands when variable X is more than constant $x2$.

```
End
```

C.4
Special Operators and Predefined Variables

MATLAB uses several special characters and operators to function as special commands. We have already seen that the variables i or j can be used to indicate complex quantities. Remember that these same letters can be used to represent a variable also. Just so as to be sure of the use of the variables, it is a good idea to use only one of the variables to indicate an imaginary quantity and not use that same variable to represent a normal variable.

Another variable that is predefined is the word pi used to represent the mathematical quantity π that is equal to 3.14 159...; other word that is predefined as a variable is eps that is defined as a value of 2^{-52}. This allows us to define a very small value or to take care of rounding errors in many computations. The variable ans stores the most recent result.

The square brackets [ ] are used to enter matrices and vectors in the program. As shown earlier, the elements of the matrix are separated either by a space or by a comma. The different rows of a matrix are separated by semicolons; so, a semicolon also indicates an end of a row.

The colon (:) symbol is used to generate regular vectors. For example, $y = a{:}b$ will generate a vector y with its first value being a and every subsequent value is incremented by 1 till the final value b is reached; so, the vector will be $y = [a\ a + 1\ a + 2 \ldots b]$. This implies $a > b$. When a third number is present on the right-hand side of the example like $y = a{:}k{:}b$, the increment from a to b is by an increment value of k instead of 1; so, the vector will be $y = [a\ a + k\ a + 2k \ldots b]$.

C.5
Drawing Plots in MATLAB

MATLAB has the ability to draw either 2D or 3D graphs to display the result of computations. Some examples of drawing 2D plots are given below. The plots can be drawn with linear axis or logarithmic axis for either one axis or both the axes. A title can be added to the plot along with titles for the x- and the y-axis.

The commands to draw a plot that we have used in this book mostly are "Plot" or "Stem." The Plot command will draw a curve from point to point. You specify the specific points as a combination of (x,y) pairs of numbers as two vectors. The values of the x-axis are specified as the first vector and then the values of y-axis are specified as the second vector. The Stem command will draw a stair step graph with a stem drawn at the location of the x-axis value, with the stem height being specified in the vector of the Y-values.

It is also possible to draw several different graphs on the same graph paper. The easiest way to do this is to use the command "*SUBPLOT.*" The command of subplot divides the graph paper in a grid and each graph is drawn in one block of the grid. To specify the grid and which block of the grid the graph is drawn, we use three numbers. The first two numbers define the grid, the first number defines the number of rows in the grid and the second number defines the number of columns in the grid. Once the grid is determined, the blocks on the grid are numbered starting from 1 and going across rows one by one. So, for example subplot (347) defines a grid with three rows with four columns in each row and the graph that follows will be in the seventh block that for this example will be the third column and second row.

C.6
Some Special Commands Used in this Book

In this book, we have used MATLAB commands and functions to demonstrate various concepts. Here, we discuss the commands used in this book. The discussion here is not complete, but it is sufficient for the student to use the command. For complete details of how the command is used and all its variations, the student is referred to the help documentation in the MATLAB program.

a) **Randn**: Generates a matrix of random numbers that have a Gaussian distribution.

$$\text{Syntax} : r = \text{randn} (n, m)$$

Description: This command returns an *n*-by-*m* matrix containing pseudorandom values drawn from the Gaussian distribution. The numbers generated have a zero mean and a variance of 1.

b) **Zeros**: Generates a matrix of zero values. This is very useful to allocate memory space for vectors so that MATLAB can run faster.

$$\text{Syntax} : r = \text{zeros} (n, m)$$

Description: This command returns an *n*-by-*m* matrix containing all zero values.

c) **Ones**: Generates a matrix with all the elements initialized to 1. This is very useful to allocate memory space for vectors so that MATLAB can run faster and at the same time assign a constant value to all the elements of the matrix. All the elements will be assigned the same constant value.

$$\text{Syntax} : r = \text{ones}(n, m)$$

Description: This command returns an *n*-by-*m* matrix with all the elements of the matrix being assigned the value of 1.

d) **Filter**: Defines a one-dimensional digital filter. This filter can have the FIR or the IIR form.

$$\text{Syntax} : Y = \text{filter} (B, A, X)$$

Description: *Y* is the output from a digital filter defined by the vectors *B* and *A* that will filter the data *X*. *B* represents the polynomial of the numerator and *A* represents the polynomial of the denominator. Coefficient $a(1)$ is required to be 1. If it is not, then the other coefficients of the filter will be adjusted accordingly.

e) **Conv:** This command defines Convolution and polynomial multiplication.

   Syntax : $Y = \text{conv}(B, A)$

   Description: If A and B are vectors of polynomial coefficients, convolving them is equivalent to multiplying the two polynomials.

f) **CConv:** Modulo N circular convolution

   Syntax : $Y = \text{cconv}(A, B, N)$

   Description: Performs circular convolution of vectors A and B and returns the result vector of length N. If the length N equals the length of A plus the length of B minus 1, the circular convolution is equivalent to the linear convolution.

g) **FFT:** Computes the discrete Fourier Transform (DFT) of the vector.

   Syntax : $Y = \text{fft}(X, N)$

   Description: Computes the N point discrete Fourier transform of the vector X. If the parameter N is omitted, then the DFT is the same length as the vector X. If N is greater than the length of vector X, then X is zero padded to be of length N.

h) **IFFT:** Computes the inverse DFT of the vector.

   Syntax : $Y = \text{ifft}(X, N)$

   Description: Computes the N point inverse discrete Fourier transform of the vector X. If the parameter N is omitted, then the inverse DFT is the same length as the vector X. If N is greater than the length of vector X, then X is zero padded to be of length N.

i) **Length:** Computes the length of the vector.

   Syntax : $Y = \text{length}(X)$

   Description: Computes the length of the vector X. If X is a matrix, then this will give you the length of the longest dimension of the matrix.

j) **Residue:** Computes the residue for partial fraction expansion.

   Syntax : $[R\ P\ K] = \text{residue}(B, A)$

   Description: Determines the residues, the poles, and the direct term of a partial fraction expansion of a ratio of the two polynomials. The polynomial B is the numerator polynomial and A is the denominator polynomial. R is the vector of residues at the location of the poles specified in the vector of P and K is the direct term.

k) **Freqs:** Computes the frequency response of a filter defined by the polynomials B and A at the frequency locations specified in the vector W.

Syntax : $Y = \text{freqs}(B, A, W)$

Description: Computes the complex frequency response for the filter defined by the vectors B and A at frequency locations W. If the vector W is omitted, then the frequency response is computed at 200 locations.

l) **Freqz**: Computes the digital filter frequency response of a filter defined by the polynomials B and A at N equally spaced points around the upper half of the unit circle.

Syntax : $Y = \text{freqz}(B, A, N)$

Description: Computes the discrete complex frequency response for the filter defined by the vectors B and A. The frequency response is computed at N equally spaced locations around the upper half of the unit circle. If the parameter N is omitted, 512 equally spaced points are computed.

m) **CZT**: Computes the Chirp Z-transform of the sequence X.

Syntax : $Y = \text{czt}(X, M, W, A)$

Description: Computes the M element Z-transform of the vector X. M is the length of the Z-transform, W is a complex angular spacing between the points of the Z-transform, and A is the location that defines first point of the transform. The contour is defined by the equation $z = A^* W^{-(0:(M-1))}$. The parameters M, W, and A are optional, and their default values are $M = \text{length}(X)$, $W = e^{-j2\pi/M}$ and $A = 1$.

Index

Digital Filters: Theory, Application and Design of Modern Filters, First Edition. Rajiv J. Kapadia.
© 2012 Wiley-VCH Verlag GmbH & Co. KGaA. Published 2012 by Wiley-VCH Verlag GmbH & Co. KGaA.